Springer Geology

Springer Geology Field Guides

FieldGuides

Series Editor
Soumyajit Mukherjee, Earth Sciences, IIT Bombay, Mumbai, Maharashtra, India

Springer Geology Field Guides is a book series that provides the details of both well-known and little known transects to discover the beauty and knowledge of Geology, worldwide. Springer Geology Field Guides aims to bring geology field trips to professionals, students, and amateurs to find the most interesting geology worldwide. This series includes carefully crafted guidebooks that help generations of geologists explore the terrain with minimum or no guidance. In this series, the audience will also find field methodologies and case studies as examples.

This book series will welcome both authored and edited field guides of all geology disciplines, including structural geology, tectonics, sedimentology, stratigraphy, paleontology, economic geology, among others. Photo-atlases are also welcome.

More information about this subseries at https://link.springer.com/bookseries/16656

Paul D. Ryan
Editor

A Field Guide to the Geology of Western Ireland

The Birth and Death of an Ocean

Editor
Paul D. Ryan
School of Natural Sciences, Earth
and Ocean Sciences
National University of Ireland
Galway, Ireland

ISSN 2197-9545 ISSN 2197-9553 (electronic)
Springer Geology
ISSN 2730-7344 ISSN 2730-7352 (electronic)
Springer Geology Field Guides
ISBN 978-3-030-97478-7 ISBN 978-3-030-97479-4 (eBook)
https://doi.org/10.1007/978-3-030-97479-4

© The Editor(s) (if applicable) and The Author(s), under exclusive license to Springer Nature Switzerland AG 2022
This work is subject to copyright. All rights are solely and exclusively licensed by the Publisher, whether the whole or part of the material is concerned, specifically the rights of translation, reprinting, reuse of illustrations, recitation, broadcasting, reproduction on microfilms or in any other physical way, and transmission or information storage and retrieval, electronic adaptation, computer software, or by similar or dissimilar methodology now known or hereafter developed.
The use of general descriptive names, registered names, trademarks, service marks, etc. in this publication does not imply, even in the absence of a specific statement, that such names are exempt from the relevant protective laws and regulations and therefore free for general use.
The publisher, the authors and the editors are safe to assume that the advice and information in this book are believed to be true and accurate at the date of publication. Neither the publisher nor the authors or the editors give a warranty, expressed or implied, with respect to the material contained herein or for any errors or omissions that may have been made. The publisher remains neutral with regard to jurisdictional claims in published maps and institutional affiliations.

This Springer imprint is published by the registered company Springer Nature Switzerland AG
The registered company address is: Gewerbestrasse 11, 6330 Cham, Switzerland

This Guide is dedicated to the geologists who carried out detailed field studies in western Ireland, their families who supported them and to the people of the West for their hospitality and understanding.

Foreword

The western mountains and seaboard of the Counties Mayo and Galway afford a unique and spectacular insight into the geology of a zone of early Ordovician collision of an oceanic island arc with the rifted continental margin of Laurentia, followed by a subduction polarity reversal, which established a continental margin arc, then a late Silurian continent-continent collision. Major deformations occurred during arc collision and obduction to form the Grampian Orogeny, within the arc during the late Ordovician Mayoian shortening, and sinistral-oblique continental collision during the late Silurian Erian Orogeny. Neither the Grampian nor the Erian Orogenies generated significant mountains. The Grampian Orogeny involved thrust stacks of rifted margin sediments with the three aluminosilicates beneath an obducting arc and fore-arc ophiolite, which flexurally loaded the whole margin to below sea level. The Erian Orogeny was generated by the terminal sinistral-oblique closure of the Iapetus Ocean with a transition from transpression, through strike-slip, to transtension; there was little crustal thickening and very low-grade metamorphism. Mayoian shortening generated broad long wavelength, high amplitude structures. The region was covered by a blanket of lower Carboniferous strata and uplifted and dissected by faulting and erosion during the Cenozoic. This tectonic evolution is richly illustrated by a wide variety of lithologies and structures that are well-exposed in an accessible and magnificent landscape.

Until the middle to late 1960s, it would have been impossible to write this explanation of the evolution of western Ireland in actualistic terms. We were able to recognise that the rocks of western Ireland are a small part of the Appalachian-Caledonian Orogen that stretches some twelve thousand kilometres from Florida to Spitsbergen and that was continuous until the Atlantic Ocean opened. Marshall Kay made the bold suggestion that the Appalachians were developed at the early Palaeozoic margin of Laurentia, but it was not until 1965, when Tuzo Wilson outlined the new theory of plate tectonics, that we were able to reconstruct the Orogen in actualistic terms, a fundamental breakthrough simultaneous with similar analyses of other Phanerozoic orogens.

The earliest geological mapping of western Ireland was done by Officers of the Valuation Department of Ireland, under its Director, Sir Richard Griffiths, during the

1830s. The most notable Field Officer was Patrick Ganly who produced in the early 1840s the first known six inches to the mile geological maps, which were discovered by Jean Archer in 1980 in the Museum of the Geology Department of the National University of Ireland, Galway. Sir Roderick Murchison visited Ganly in the field in his quest for the base of the Silurian System. Under the first Director (Sir Henry James), of the Irish Geological Survey, then a part of the British Geological Survey, established in 1845 by Act of Parliament under its first Director, Sir Henry De la Beche, undertook an ambitious programme of geological mapping of Ireland at the six inches to the mile scale (1:10,560) and produced a comprehensive series of geological maps at the one inch to the mile scale (1:63,360). These maps, hand-coloured in rich watercolour and gold faults watercolour by Mrs. E. Williams in London, are among the most useful and beautiful ever made. The maps, accompanied by Memoirs describing in great detail the geology, of Co. Galway and Co. Mayo, published in the second half of the nineteenth century, were the starting point for all subsequent geological investigations of western Ireland. The finest work was by James Kilroe, George Kinahan, Hugh Leonard, Joseph Nolan, and Richard Symes, who not only made accurate maps but also discovered, described, and interpreted many original geological features such as graded and current bedding used as way-up criteria. However, these great geologists were occasionally puzzled by an outcrop, as in 1878 when George Kinahan described a complicated fault gouge along an unconformity criss-crossed by a maze of small faults as "grits, sandstones, slates, schist, conglomerates, breccias, and serpentine, all more or less deformed and metamorphosed promiscuously together". That is what I call "covering the bases".

From the meticulous mapping of Charles Gardiner and Sidney Reynolds in the early 1900s to the present day, at least forty academic geologists and their doctoral students, to my knowledge, have worked in west Mayo and west Galway, some before and some after the discovery of plate tectonics. This is not counting the host of undergraduates who did their undergraduate honours mapping projects in western Ireland, many of which are first-class, and the search for economic mineral deposits onshore and petroleum offshore. Although the basic geological mapping and derivative observations in western Ireland have been very good and steadily improving for a hundred and ninety years, the advent of plate tectonics allowed us to think of many new questions and find new things in the rocks. Even today, there are many features and parts of the western Ireland Caledonides that we do not fully understand although these are mostly on a sub-plate tectonic scale. As that great geologist Robert Shackleton, who mapped the Twelve Bens of Connemara, once said "an area or outcrop is never completely and fully mapped and understood". The authors of this book have spent a collective 500 or more field seasons in western Ireland; we are still going there and finding things out.

John F. Dewey
University College
High Street, Oxford, UK

Preface

The West of Ireland is one of the few regions in the world where detailed geological mapping has been carried out for nearly 200 years. Much of the knowledge we now have is based upon the work of Ph.D. students and individual researchers in the latter half of the last century. The contributors to this book wish to preserve this knowledge, while they still can, for the benefit of future researchers and students of geology. We present this as an example of how detailed geological field studies, coupled with modern analytical methods, can deepen our understanding of fundamental geological processes.

Galway, Ireland   Paul D. Ryan

Acknowledgements

The contributors would like to thank their colleagues, listed below, whose careful and constructive reviews represent a significant contribution to this work.

Ian Alsop, University of Aberdeen
Joaquina Alvarez-Marron, Jaume Almera, Consell Superior d'Investigacions Científiques
Alan Boyle, University of Liverpool
Mark Cooper, Geological Survey of Northern Ireland
Simon Harley, University of Edinburgh
David Harper, University of Durham
Peter Haughton, University College Dublin
Robert Holdsworth, University of Durham
Laurent Jolivet, Sorbonne Université
Paul Kapp, University of Arizona
Graham Leslie, British Geological Survey
Brian McConnell, Geological Survey Ireland
Pat Meere, University College Cork
Brendan Murphy, St. Francis Xavier University
Terry Pavlis, The University of Texas at El Paso
Steven Whitmeyer, James Madison University
Robert Wintsch, Indiana University
Nigel Woodcock, University of Cambridge

Contents

Introduction .... Paul D. Ryan	1
The Basement and Dalradian Rocks of the North Mayo Inlier .... J. Stephen Daly, David M. Chew, Michael J. Flowerdew, Julian F. Menuge, Rose Fitzgerald, Claire A. McAteer, and Ray Scanlon	9
The Slishwood Division and Its Relationship with the Dalradian Rocks of the Ox Mountains .... J. Stephen Daly, Michael J. Flowerdew, David M. Chew, and Ian S. Sanders	73
The Central Ox Mountains .... Paul D. Ryan, Ken McCaffrey, David M. Chew, John R. Graham, and Barry Long	107
The Ordovician Arc Roots of Connemara .... Bruce W. D. Yardley and Robert A. Cliff	131
The Ordovician Fore-Arc and Arc Complex .... Paul D. Ryan, John F. Dewey, and John R. Graham	179
The South Connemara Group .... Paul D. Ryan and John F. Dewey	229
The Silurian of North Galway and South Mayo .... John R. Graham, John F. Dewey, and Paul D. Ryan	245
The Late Silurian to Upper Devonian Galway Granite Complex (GGC) .... Martin Feely, William McCarthy, Alessandra Costanzo, Bernard E. Leake, and Bruce W. D. Yardley	303
Devonian and Carboniferous .... John R. Graham	363

The Geology of Western Ireland: A Record of the 'Birth' and 'Death' of the Iapetus Ocean 405
Paul D. Ryan, David M. Chew, Robert A. Cliff, Alessandra Costanzo,
J. Stephen Daly, John F. Dewey, Martin Feely, Rose Fitzgerald,
Michael J. Flowerdew, John R. Graham, Bernard E. Leake,
Barry Long, Claire A. McAteer, Ken McCaffrey, William McCarthy,
Julian F. Menuge, Ian S. Sanders, Ray Scanlon, and Bruce W. D. Yardley

Data Sources for use in Pre- and Post-trip Exercises 423
Paul D. Ryan

Contributors

David M. Chew Department of Geology, Trinity College Dublin, Dublin 2, Ireland

Robert A. Cliff School of Earth and Environment, University of Leeds, Leeds, UK

Alessandra Costanzo School of Natural Sciences, Earth and Ocean Sciences, National University of Ireland Galway, Galway, Ireland

J. Stephen Daly UCD School of Earth Sciences, University College Dublin, Belfield, Dublin 4, Ireland

John F. Dewey University College, Oxford, UK

Martin Feely School of Natural Sciences, Earth and Ocean Sciences, National University of Ireland Galway, Galway, Ireland

Rose Fitzgerald UCD School of Earth Science, Science Centre West, University College Dublin, Belfield, Dublin, Ireland

Michael J. Flowerdew CASP, West Building, Madingley Rise, Cambridge, UK

John R. Graham Department of Geology, Trinity College Dublin, Dublin, Ireland

Bernard E. Leake School of Earth and Ocean Sciences, Main Building, Cardiff University, Cardiff, Wales, UK

Barry Long Geological Survey Ireland, Dublin, Ireland

Claire A. McAteer UCD School of Earth Sciences, University College Dublin, Belfield, Dublin 4, Ireland

Ken McCaffrey Department of Earth Sciences, Durham University, Durham, UK

William McCarthy School of Earth and Environmental Science, Bute Building, St. Andrews, UK

Julian F. Menuge UCD School of Earth Sciences, University College Dublin, Belfield, Dublin 4, Ireland

Paul D. Ryan School of Natural Sciences, Earth and Ocean Sciences, National University of Ireland Galway, Galway, Ireland

Ian S. Sanders Department of Geology, Trinity College Dublin, Dublin 2, Ireland

Ray Scanlon Geological Survey Ireland, Dublin, Ireland

Bruce W. D. Yardley School of Earth and Environment, University of Leeds, Leeds, UK

Introduction

Paul D. Ryan

1 Organisation

This guide presents the field geology of western Ireland, extending from Donegal Bay in the north to Galway Bay in the south. The Proterozoic to early Palaeozoic rocks of this region, of approximately 140 km east–west and 160 km north–south (Fig. 1), preserve a record of the evolution of the Iapetus Ocean adjacent to Laurentia (North America and Greenland) from before its 'birth' during Neoproterozoic rifting to after its 'death' in late Silurian continent–continent collision. Our current understanding of this region is primarily based upon detailed geological mapping at a scale of 1:10,560 (6 inches to the mile), much of it carried out in the latter half of the 20th Century. The resultant detailed knowledge of field relationships, coupled with easy accessibility and generally good exposure on the recently glaciated rift shoulder of the north-east Atlantic have made this a classic area for the field study of ancient orogens.

The enormous variety of lithologies and processes available for study include: fluviatile to deep-sea sediments; layered ultramafic intrusions to reverse-zoned granite batholiths; zeolite to eclogite facies metamorphic assemblages; continental rifting; subduction processes; island arc evolution; arc-continent collision; Andean margin development; and continent–continent collision and subsequent cratonization.

Each chapter follows a similar structure, with an introduction to the geology of the region followed by one or more excursions and a bibliography. Exercises, which may form the basis for student workbooks or simply promote discussion amongst participants, are supplied at appropriate places in each chapter. Chapters are organized, as far as is practical, in order of the geological age of the various

P. D. Ryan (✉)
School of Natural Sciences, Earth and Ocean Sciences, National University of Ireland Galway, University Road, Galway, Ireland
e-mail: paul.ryan@nuigalway.ie

© The Author(s), under exclusive license to Springer Nature Switzerland AG 2022
P. D. Ryan (ed.), *A Field Guide to the Geology of Western Ireland*, Springer Geology Field Guides https://doi.org/10.1007/978-3-030-97479-4_1

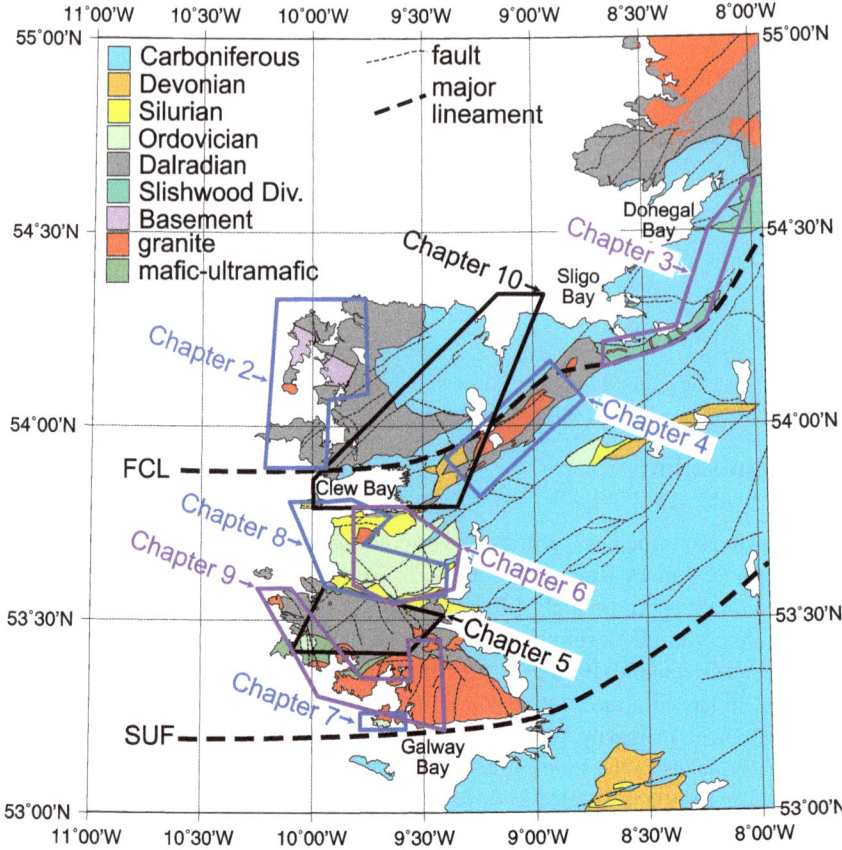

Fig. 1 Outline geological map of the west of Ireland based upon the Geological Survey Ireland 1:1,000,000 geological map of Ireland (Geological Survey Ireland 2015) with minor modifications showing the regions covered in each chapter. Simplifications are as follows. The Dalradian includes the Clew Bay and Deer Park Complexes. Basement refers to the Annagh Gneiss Complex. Granite includes all granitoid rocks. Mafic–ultramafic includes igneous rock complexes of that composition. Slishwood Div. = Slishwood Division. FCL = Fair Head–Clew Bay Line. SUF = Southern Uplands Fault

rock units. These generally decrease towards the south. This introduction will firstly review, very briefly, the geology to be examined and its likely tectonic setting. The reader is referred to the bibliography in the relevant chapters. The second section will offer advice on practical matters relevant to the planning and the safe conduct of geological field trip work in Ireland. The third section will outline the considerable resources offered by Geological Survey Ireland to support geological field work. Finally, there is a list of Irish organisations that support tourism.

2 Brief Review of the Geology of Western Ireland

The region is divided by two crustal-scale structures, the Fair Head–Clew Bay line (FCL) which runs through Clew Bay and the Southern Uplands Fault (SUF) which runs through Galway Bay (Fig. 1). Chapters "The Basement and Dalradian Rocks of the North Mayo Inlier" and "The Slishwood Division and Its Relationship with the Dalradian Rocks of the Ox Mountains" deal with rocks north of the FCL, attributed to Laurentia, which comprise Proterozoic basement gneisses overlain unconformably by the Dalradian Supergroup (herein after referred to as Dalradian). The Dalradian is interpreted as a Neoproterozoic rift-related sequence deposited during the rift and early drift phases of the opening of the Iapetus Ocean. These rocks were subject to the short-lived mid-Ordovician Grampian Orogeny which is ascribed to the obduction of an intra-Iapetus fore-arc and arc. This arc developed during the Cambrian above a Laurentia-facing subduction zone within Iapetus. The FCL, although subsequently highly modified, locally preserves features attributed to the suture to this collision. Chapter "The Central Ox Mountains" deals with Dalradian rocks that straddle this major lineament. Chapter "The Ordovician Arc Roots of Connemara" discusses the geology of the Dalradian of Connemara, which preserves gabbroic masses that advected heat during the Grampian event. Connemara lies south of the FCL and this position has been attributed either to strike-slip disruption and, or to it occupying a thrust-bound tectonic window within the overriding arc. The remnants of this arc and a remarkably preserved fore-arc basin, the South Mayo Trough, are the subject of chapter "The Ordovician Fore-Arc and Arc Complex". The Grampian arc-continent collision event was followed by subduction flip, which developed a magmatic arc along the Laurentian margin in late Ordovician times. Evidence for the associated trench which lay along the line of the Southern Uplands Fault is presented in chapter "The South Connemara Group". Silurian volcanism, sedimentation and strike-slip modification of earlier Grampian structures along this active continental margin are discussed in chapter "The Silurian of North Galway and South Mayo". The Iapetus Ocean in Ireland was finally closed by the docking of the Avalonian microcontinent during the latest Silurian/earliest Devonian. This transpressive event was not associated with major nappe formation or metamorphism. The lack of major post-Grampian deformation and the preservation of the obducted arc complexes make the West of Ireland unique within the Caledonides for understanding the Grampian Orogeny. The Devonian Galway Granite Batholith, described in chapter "The Late Silurian to Upper Devonian Galway Granite Complex (GGC)", was emplaced in a thermal event, associated perhaps with northward subduction of the remaining Iapetus plate and subsequent slab break-off. A smaller Devonian batholith, the Ox Mountains Igneous Complex discussed in chapter "The Central Ox Mountains", preserves unique evidence for sheet-like emplacement during regional transpression along the FCL. Devonian sedimentation coeval with this magmatic activity is described in chapter "Devonian and Carboniferous" as are the Carboniferous clastic rocks and limestones that marked gradual post-orogenic cratonization.

3 Practical Matters

3.1 Access to Private Lands

The situation regarding access to lands during geological fieldwork in Ireland is summarised by Boyle et al. (2009). In essence, there is no "right to roam" legislation. Permission is generally needed for access onto private lands. There is, however, a public right of recreation along the foreshore, but access may require crossing private land. It is always better to ask when uncertain. This Guide has, where possible, used easily accessible outcrops along the roadside or the shore or adjacent to public rights of way.

3.2 Road Network

Ireland has an extensive road network and none of the outcrops visited require vehicles with off-road capability. Access to many localities is provided by local (minor) roads, designated 'L'. These can be single track with passing places and have weight limited bridges or sharp bends, which may make them unsuitable for larger vehicles without the driver having local knowledge. Such roads are divided into primary (numbered L1000–4999), secondary (L5000–9999) or tertiary (L10000+). All localities are accessible by car or minibus.

3.3 Location

Localities and relevant parking places are located using latitude and longitude given to 6 decimal places suitable for use with GPS systems. A Google Earth kmz file is available with pins placed at all such locations. Satellite navigation systems selecting the shortest route may not choose the best route in terms of road quality. For example, the shortest route may choose tertiary local roads where a longer route would use primary local roads. Ordnance Survey Ireland Discovery Series 1:50,000 maps, available for purchase online (Table 1) are a valuable aid for route planning.

3.4 Climate

The west of Ireland has a temperate oceanic climate. Mean temperatures range from 5 to 7 °C in the winter to 14 to 16 °C in the summer, despite its northerly latitude.

Weather is humid and changeable, with periods of high rainfall and, or strong winds. Forecasts, warnings and information on seasonal and spatial trends are

Table 1 A list of useful organisations that either support tourism in Ireland or supply useful services with their websites

Organisation	Service	Web site
Discover Ireland	Tourism	https://www.discoverireland.ie/about-discover-ireland
Geological Survey Ireland	Geological data	https://www.gsi.ie/en-ie/Pages/default.aspx
Health Service Executive	Health	https://www.hse.ie/eng/
Irish Sailing		https://www.sailing.ie/Tides/Galway
Leave No Trace Ireland	Countryside Code	https://www.leavenotraceireland.org/about/why-leave-no-trace/
MET éireann	Weather	https://www.met.ie/forecasts/national-forecast
MET éireann	Sea condition	https://www.met.ie/forecasts/marine-inland-lakes/waves
Ordnance Survey Ireland	Geographic Maps	https://store.osi.ie/index.php/paper-products/discovery-and-discoverer.html
Road Safety Authority	Road safety	https://rsa.ie/en/RSA/Pedestrians-and-Cyclists/Pedestrian-safety/
Wild Atlantic Way	Tourism	https://www.discoverireland.ie/Wild-Atlantic-Way

provided by Met Éireann, the Irish Meteorological Service (Table 1). Daylight hours vary significantly with season. There are nearly 17 h daylight during the summer solstice and less than 8 h during the winter solstice. Field work is limited by available light during the 4 winter months (November–February).

3.5 Safety Considerations

Despite the mild climate, rain coupled with wind-chill can be a hazard. Warm waterproof clothing should be always available. Hillsides are generally covered in waterlogged grasses, heathers or mosses and appropriate footwear is essential. Ireland has a large tidal range with spring tides being more than 5 m and neaps around 3 m. Tide tables (Table 1) should be consulted before any shore traverse, which are best conducted on an ebb tide. Rogue waves constitute a rare but considerable hazard (O'Brien et al. 2013) especially on exposed shores during or after Atlantic storms. Rocks on the foreshore can be slippery and cliffs should be avoided due to the danger of rock falls. The Road Safety Authority (Table 1) provides safety advice for pedestrians on Irish roads. The Health Service Executive gives advice on the avoidance of Lyme disease caused by bites from a tick infected with *Borrelia*. About 200 cases are reported annually, mostly amongst those whose leisure or work takes them onto heathland, light woodland, or grassy areas. Mobile phone coverage is generally good, although signal quality may vary. It is advisable to always carry a mobile or cell phone. Emergency services can be contacted by dialing 112 or 999.

3.6 Country Code

Leave No Trace Ireland provides advice on how to behave when visiting the outdoors (Table 1). The seven basic principles are: Plan Ahead and Prepare; Be Considerate of Others; Respect Farm Animals and Wildlife; Travel and Camp on Durable Ground; Leave What You Find; Dispose of Waste Properly; and Minimise the Effects of Fire. In addition, there is in general no need to deface outcrops by hammering, especially at sites covered by the GSI geoheritage programme (see Sect. 4.3). Samples for further study can easily be collected from loose material.

3.7 Tourism

The west of Ireland is a major tourist destination termed 'The Wild Atlantic Way' in an area of natural beauty rich in cultural heritage. Visitors interested in augmenting their geological studies, or passing a rainy day, are advised to consult the Discover Ireland website (Table 1).

4 Geological Survey Ireland Products

4.1 GSI Maps and Reports

Bedrock geology and Quaternary sediment and geomorphology mapping can be viewed and queried online on the GSI Map Viewer and downloaded for free at https://www.gsi.ie/en-ie/data-and-maps/Pages/default.aspx.

Printed bedrock geology 1:100,000 scale map sheets 6 (North Mayo), 7 (Sligo-Leitrim), 10 (Connemara), 11 (South Mayo) and 14 (Galway Bay) cover the area of this field guide; accompanying booklets describe each rock unit and their geological history.

4.2 GSI Tellus Data

The GSI Tellus project is a national survey collecting airborne geophysical data and surface geochemistry data over Ireland. Magnetic, electromagnetic and radiometric data are available for the whole area covered by this guide. At the time of writing, stream sediment and soil geochemistry are available for part of the area. The data can be viewed and downloaded at https://www.gsi.ie/en-ie/data-and-maps/Pages/default.aspx. These data can be used to supplement fieldwork with information on contacts and lineaments, physical and chemical contrasts between rock units, and buried geology.

4.3 Geoheritage and Site Protection

The Geological Survey Ireland geoheritage programme identifies important geological and geomorphological sites throughout the country, with the most scientifically significant sites selected by a panel of theme experts and 'audited' on a county basis to establish location, boundary, access and any specific vulnerability issues. These 'County Geological Sites' are now routinely included in County Development Plans and the planning process, to ensure the recognition and appropriate protection of geological heritage. County Geological Sites for Sligo, Mayo and Galway, including sites covered in this guide book, can be viewed online under the Geological Heritage tab on the GSI Map Viewer (https://dcenr.maps.arcgis.com/apps/MapSeries/index.html?appid=a30af518e87a4c0ab2fbde2aaac3c228) at https://www.gsi.ie/en-ie/data-and-maps/Pages/default.aspx. In addition to the usual good field work practices, particular care should be taken to avoid hammering and preserve these sites.

4.4 GSI Field Guides

The GSI field guidebook 'Geology of South Mayo' (Graham 2021) provides a wider selection of localities in the Ordovician and Silurian rocks of South Mayo and adjacent parts of northern Co. Galway than this book. It is, therefore, a complementary guide for those users who wish to focus in more detail on the early Palaeozoic geology of the region. The guide is available from local bookshops, visitor attractions and the GSI website. In a similar manner, care has been taken to ensure that the localities visited in the GSI field guide to the Galway Granite (Feely et al. 2006) complement those presented in this work. This guide is available and can be ordered from the GSI website (Table 1).

5 Useful Organisations

See Table 1.

References

Boyle AP, Ryan PD and Stokes A (2009) External drivers for changing fieldwork practices and provision in the UK and Ireland. In: Whitmeyer SJ, Mogk DW, Pyle EJ (eds) Field geology education: historical perspectives and modern approaches. Geol Soc Am Spec Pap: 461. https://doi.org/10.1130/2009.2461(23)

Feely M, Leake B, Baxter S, Hunt J, Mohr P (2006) A geological guide to the granites of the Galway Batholith, Connemara, western Ireland. Geological survey Ireland, field guide Series. Dublin, Ireland, 70pp. ISBN 1-899702-56-3

Graham JR (2021) Geology of South Mayo: a field guide. Geological survey Ireland, field guide Series Dublin, Ireland, 158pp. ISBN 978-1-899702-68-7

Geological Survey Ireland (2015) Bedrock geology 1:1,000,000 Ireland. https://secure.dccae.gov.ie/arcgis/rest/services/Bedrock/BedrockGeology1Million/MapServer

O'Brien L, Dudley JM, Dias F (2013) Extreme wave events in Ireland: 14 680 BP-2012. Nat Hazards EaRth Syst Sci 13:625–648

The Basement and Dalradian Rocks of the North Mayo Inlier

J. Stephen Daly, David M. Chew, Michael J. Flowerdew, Julian F. Menuge, Rose Fitzgerald, Claire A. McAteer, and Ray Scanlon

Abstract Basement rocks of the North Mayo inlier are unique in Ireland in displaying a series of Palaeoproterozoic to Mesoproterozoic orthogneisses (Annagh Gneiss Complex) affected by the polyphase Grenville Orogeny. Together with their cover of Dalradian rocks, they represent the Laurentian margin of the remnant Rodinia supercontinent, whose protracted break-up started in the Tonian (1000–720 Ma), and ultimately led to the opening of the Iapetus Ocean in the Ediacaran (635–541 Ma). The magmatic manifestation of the opening of the Iapetus Ocean is seen in North Mayo as a $c.$ 600 $m.y.$ old basaltic dyke swarm represented as dolerite intrusions that cut both the Annagh Gneiss Complex basement and its Dalradian cover. It is inferred that these dykes fed volcanic deposits as the Dalradian sediments accumulated on the extending continental margin. Hyperextension of the Laurentian margin during Iapetus opening is manifest by the presence in the Dalradian rocks on Achill of megablocks of serpentinite sourced from the sub-continental lithospheric mantle, as well as exposures of the basement itself. Indirect evidence that the Annagh Gneiss Complex is both the source of abundant Dalradian detritus, as well as its basement, is now strongly supported by the recent discovery of a probable original unconformity between them. Closure of the Iapetus Ocean resulted in the collision of the Annagh Gneiss Complex basement and its Dalradian cover with intra-oceanic arc terranes to the south. Consequent shortening and tectonic burial led to the development of Alpine-scale fold nappes and both Barrovian and contemporaneous blueschist-facies metamorphism—another unique feature of this sector of the Grampian orogen. The basement rocks also experienced Barrovian metamorphism as well as extensive deformation during the Grampian orogeny ($c.$ 470–460 Ma) with

J. S. Daly (✉) · J. F. Menuge · R. Fitzgerald · C. A. McAteer
UCD School of Earth Sciences, University College Dublin, Belfield, Dublin 4, Ireland
e-mail: stephen.daly@ucd.ie

D. M. Chew
Department of Geology, Trinity College Dublin, Dublin 2, Ireland

M. J. Flowerdew
CASP, West Building, Madingley Rise, Madingley Road, Cambridge CB3 0UD, UK

R. Scanlon
Geological Survey Ireland, Beggars Bush, Haddington Road, Dublin D04 K7X4, Ireland

the Iapetus-related dyke swarm acting as a structural marker allowing Grampian and Grenvillian events to be separated. Late Grampian tectonism resulted in modification (downbending) of the Grampian nappe pile close to the Laurentian margin, which was also the locus of late Caledonian (Devonian) strike slip motion.

1 Introduction

The North Mayo Inlier (Fig. 1) is made up of the Palaeo- to Mesoproterozoic orthogneisses of the Annagh Gneiss Complex (Kennedy and Menuge 1992; Menuge and Daly 1994) together with its cover of Neoproterozoic metasediments of the Dalradian Supergroup (McAteer et al. 2010).

The Dalradian Supergroup is generally thought to have been deposited on the Laurentian margin of the Rodinia supercontinent (Cawood et al. 2003, 2007a, b) as a sequence of clastic sediments, glacial sediments, limestones and basic volcanics. An original depositional contact has been inferred in North Mayo (Chew et al. 2003; McAteer et al. 2010) between the Dalradian rocks and the structurally underlying Annagh Gneiss Complex. Evidence for a preserved unconformity (Fig. 14) on the west coast of the Mullet Peninsula is presented here for the first time. Exposures of the Annagh Gneiss Complex are controlled by two major south-east-plunging anticlines on the Mullet peninsula and the adjacent 'mainland' (Figs. 1, 2).

Close to the exposed Annagh Gneiss Complex, the Dalradian rocks were deformed and metamorphosed under Barrovian amphibolite-facies conditions (Yardley 1980) during the Grampian Orogeny. At deeper levels, the Annagh Gneiss Complex rocks were partially reworked during the Grampian (Daly and Flowerdew 2005; Henrichs et al. 2018) but they experienced a somewhat higher metamorphic grade with widespread migmatization earlier during the Grenville Orogeny (van Breemen et al. 1978; Daly 1996). Farther south in southernmost Achill and on the island of Achillbeg (Fig. 1), the Dalradian rocks reached blueschist-facies conditions during the Grampian Orogeny (Yardley et al. 1987; Chew et al. 2003).

2 Annagh Gneiss Complex

The Annagh Gneiss Complex is well exposed along the west coast of the Mullet peninsula, but exposures are poor inland especially in the east on the mainland between Doolough and Tristia (Fig. 3).

Early investigations established that the Annagh Gneiss Complex comprises intermediate to acid and partially migmatitic orthogneisses. In many areas the gneisses exhibit a strong northwest-southeast to west northwest-trending gneissose banding defined by alignment of biotite and alternating 1–10 cm thick layers rich and poor in mafic minerals. In addition there is a generally east southeasterly plunging lineation defined by mineral aggregates and occasionally by amphibole alone. The overall

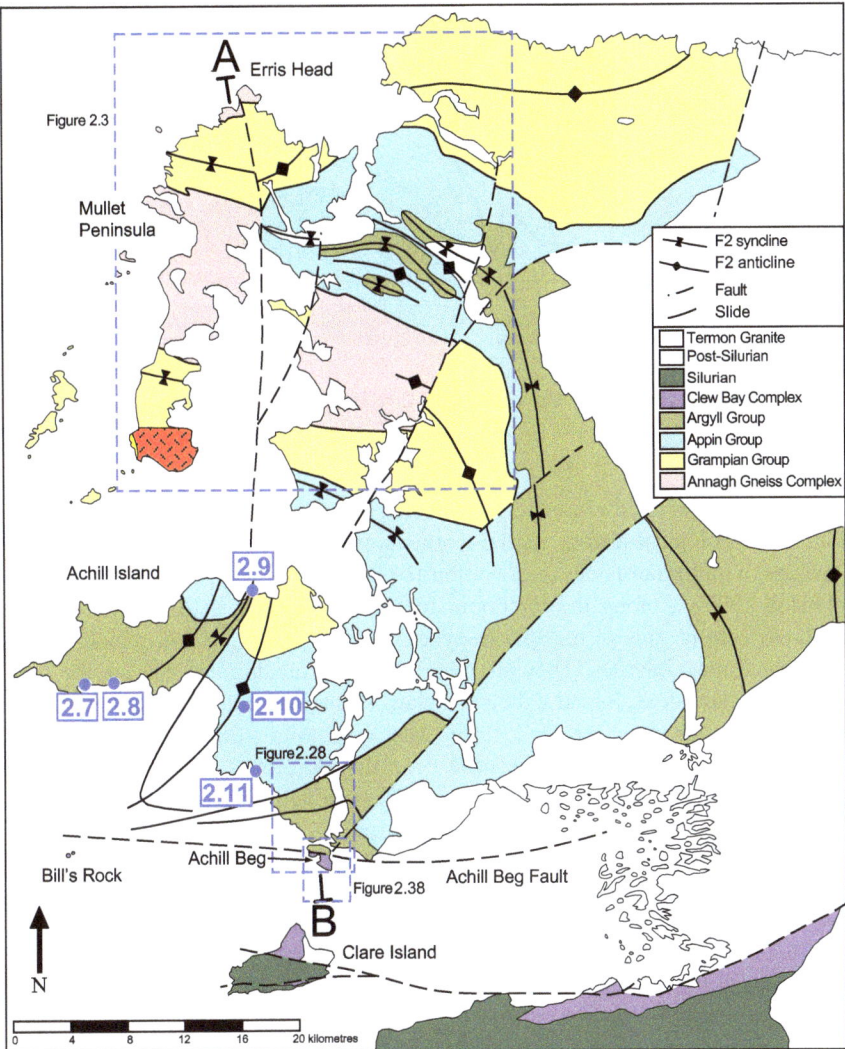

Fig. 1 Sketch geological map of north Mayo geology (after Chew 2003) showing the line of section A−B in Fig. 2, the locations of Figs. 3, 28 and 38 as well as selected localities on Achill Island

composition is granodioritic to granitic (Winchester and Max 1984). Early amphibolitized basic bodies are generally concordant with the main gneissose foliation and represent dykes or sills. Several generations of granitoid (and pegmatite) sheets up to a few metres wide range from syn- to post-tectonic with respect to the main foliation. Generally the strong gneissose fabric tends to obscure the relationships between the various lithologies because they all tend to be aligned parallel to it.

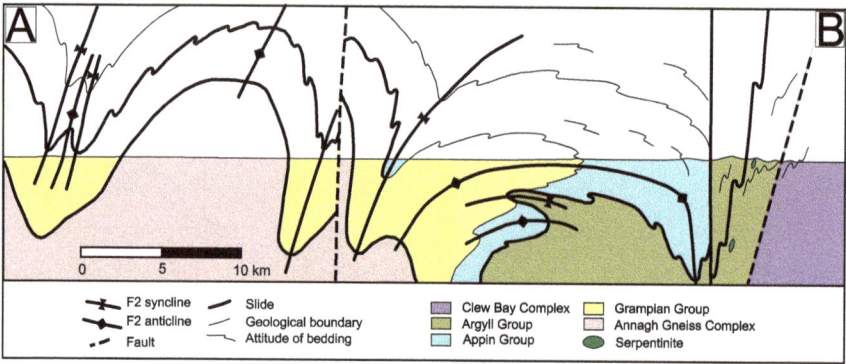

Fig. 2 North–south geological cross section from Erris Head to Achillbeg along the line of section shown in Fig. 1 (after Chew 2003)

However, in several places, e.g., at Annagh Head, Cross Point and Doolough (Fig. 3), a geological history can be worked out on the basis of cross-cutting relationships of minor intrusions (e.g., Sutton 1972; Phillips 1981; Max and Long 1985; Johnston 1995). Among the important features to be seen at Annagh Head is a late cross-cutting suite of metabasite dykes (Fig. 4). These are now unmigmatized but foliated metadolerites. They are correlated with metabasites that cut the overlying Dalradian rocks. Because their correlatives share all deformational events that affect the Dalradian (e.g. Loc. 2.2), the late metabasite suite cutting the Annagh Gneiss Complex is considered to have been first deformed during the Grampian Orogeny. The metabasites may thus be used to distinguish between pre-Grampian and Grampian events.

The pre-Grampian relationships are particularly clear at Doolough, where late Grenvillian granite pegmatites cross-cut folded Doolough gneisses, themselves of early Grenvillian age (Fig. 5, see below). By correlating the youngest ductile structures between Doolough and Annagh Head, the effects of the Grenville Orogeny can be divided into three deformational events, D_G1–D_G3 (Daly 1996; Daly and Flowerdew 2005).

The pre-Grampian deformational and minor intrusive chronology has been calibrated using U–Pb zircon dates (Daly 1996 and unpublished). This chronology (Table 1), together with geological mapping, and geochemical and isotopic data from outcrop and borehole samples, results in a threefold subdivision of the Annagh Gneiss Complex (Fig. 3) into c. 1.75 Ga Mullet gneisses, c. 1.27 Ga Cross Point gneisses and c. 1.18 Ga Doolough gneisses (Daly 1996; Fitzgerald et al. 1996). Unfortunately none of the geological boundaries between the three sets of gneisses is exposed, but it is thought likely that the original contacts between the Cross Point and Mullet gneisses were intrusive. Most of the Annagh Gneiss Complex has Sm–Nd model ages (t_{DM}, DePaolo 1981) in the c. 1.8–2.0 Ga range (Menuge and Daly 1990; Fitzgerald et al. 1996). The Mullet gneisses and Cross Point gneisses are just distinguishable in terms of Sm–Nd model ages with t_{DM} values of 1.9–2.0 Ga and

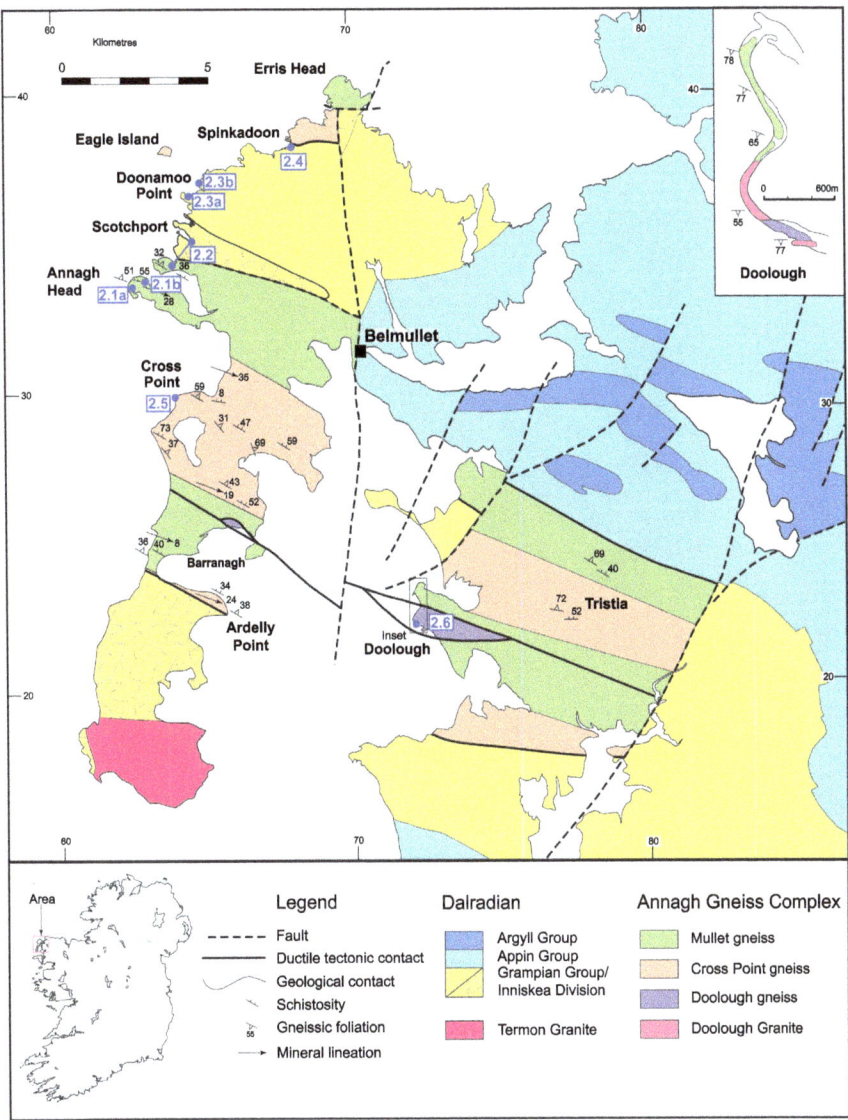

Fig. 3 Geological sketch map of the Annagh Gneiss Complex and Dalradian of North Mayo show excursion localities on the Mullet Peninsula and at Doolough (after Daly 2009)

1.7–1.8 Ga respectively (Fitzgerald et al. 1996). Correlation of the Annagh Gneiss Complex with the Lewisian of north-west Scotland is precluded, given that the latter consists principally of Archaean crust. Early studies using Sm–Nd data suggested that the Annagh Gneiss Complex developed in two distinct episodes with crustal

Fig. 4 Photograph of the Mullet gneisses on a vertical cliff face at Annagh Head (Loc. 2.1a) looking east (from Daly 2009). The vertical dimension of the photograph is about 20 m. The main foliation (**a**) in the *c.* 1750 Ma gneiss (Table 1) wraps around boudins of early amphibolite (**b**). The gneiss and amphibolite were migmatized (**c**) at 995 ± 6 Ma. A deformed granite pegmatite (**d**) cuts the foliation and migmatitic leucosomes in the gneisses and has a U–Pb zircon age of 980 + 38/ − 12 Ma. Grampian strain is inferred to be low at this locality because the metadolerite (**e**), which cuts the *c.* 980 Ma pegmatite, is only weakly deformed. However, just north and south of the area in the photograph, the metadolerite, together with the other rocks, is strongly deformed in a north–north-westwards trending Grampian shear zone (Loc. 2b). U–Pb ages from Daly (1996)

extraction ages of *c.* 1.30–1.40 Ga and *c.* 1.71–1.99 Ga, termed 'Doolough' and 'Mullet' gneisses, respectively (Menuge and Daly 1990, 1994).

Further Sm–Nd model age dating (Fitzgerald et al. 1996) using samples from a wider area has shown that the Doolough gneiss (Fig. 3) is much less extensive than depicted by Menuge and Daly (1990) and is actually restricted to small areas around Doolough and Barranagh (Fig. 3).

In the following sections, the main components of the Annagh Gneiss Complex and the events affecting them are discussed in chronological order, oldest first.

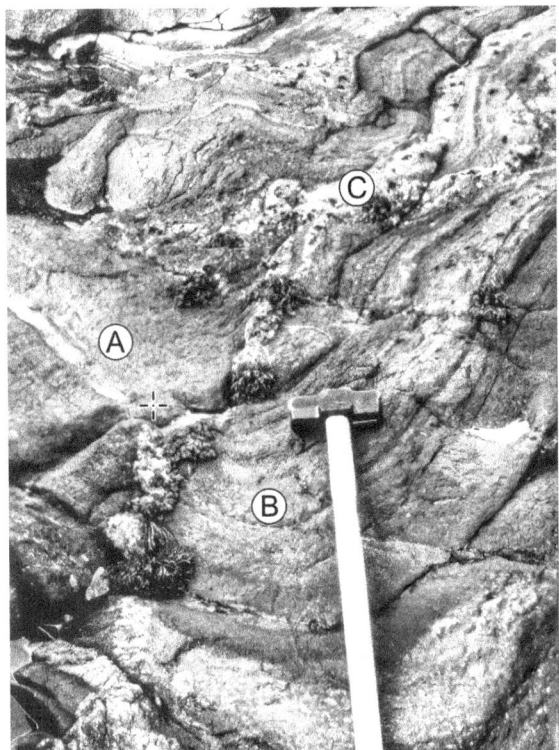

Fig. 5 Photograph of deformed Doolough gneiss at Doolough (Loc. 2.6) from Daly (2009). Here, the Doolough gneiss (**a**) has a U–Pb zircon age of 1177 ± 4 Ma (Table 1). An early deformation event (D_G1) produced a strong foliation and produced or accentuated the compositional banding. The banding was subsequently folded into F_G2 folds with strongly curved hinges, here with a vertical plunge and steep north–south axial plane (**b**). Both events predate 1015 ± 4 Ma, the U–Pb zircon age obtained from the Doolough Granite (Daly 1996). The F_G2 fold is cut by an undeformed pegmatite (**c**) with a U–Pb zircon age of 995 ± 6 Ma. U–Pb ages from Daly (1996)

2.1 Mullet Gneiss

The Mullet gneiss occurs in two northwest-southeast trending bands extending from the central part of the Mullet peninsula and apparently continuing onto the mainland north and south of Tristia (Fig. 3). A northern basement area at Erris Head is undated but has been assigned to the Mullet gneiss on the basis of Nd isotopic data. Here, the Mullet gneiss forms the sparse host rock to an extensive injection complex of intrusive (presumably Grenvillian) granitic pegmatites. The Mullet gneiss is well displayed on the Mullet peninsula near Annagh Head (Figs. 3, 4, Loc. 2.1). Formation of the Mullet gneiss protoliths at Annagh Head (Fig. 4) has been dated at 1753 ± 3 Ma (Daly 1996, Table 1). These rocks have juvenile initial ε_{Nd} values (Fitzgerald et al. 1996) similar to those obtained from the Rhinns Complex of Islay and Inishtrahull (Marcantonio et al. 1988; Daly et al. 1991).

The main lithology is a migmatized quartzofeldspathic orthogneiss (i.e. originally igneous) locally with more micaceous basic bands. Compositionally the Mullet gneiss comprises metaluminous I-type quartz monzodiorites, granodiorites and granites with granodiorite predominating (Daly 2009). In a few places, e.g., at Annagh Head and north of Doolough, the Mullet gneiss contains K-feldspar megacrysts, which probably originated as phenocrysts. Although alkali-calcic, the trace element

Table 1 Pre-Caledonian history of the Annagh gneiss complex

U–Pb zircon age (Ma)	Dated sample	Magmatic, metamorphic and structural events
1753 ± 3	Mullet gneiss at Annagh Head	Crust formation from juvenile magma with mantle-like Nd (evidently part of a major crust-forming event seen also at Islay, Inishtrahull and Rockall)
1271 ± 6 1287 + 38/-35	Cross Point gneiss at tristia Cross Point gneiss at Cross Point	Anorogenic magmatism
1177 ± 4	Doolough gneiss at Doolough	Addition to the crust of juvenile magma Early Grenville deformation. Formation of foliation (D_G1) in the Doolough gneiss and later folding (D_G2). Possible tectonic juxtaposition of Doolough gneiss with the older Cross Point gneiss and Mullet gneiss
1015 ± 4	Doolough Granite at Doolough	Peralkaline granite intrusion
995 ± 6 999 + 29/-9 993 ± 6 984 ± 6 980 + 38/-12	Leucosome at Annagh Head Pegmatite at Tristia Pegmatite at Doolough Pegmatite at Cross Point Pegmatite at Annagh Head	Migmatisation and injection of discordant pegmatites
958 + 17/-19	Metamorphic zircon in Cross Point gneiss	Late Grenville deformation (D_G3). Discordant pegmatites foliated Further metamorphism
603 ± 11	Mam sill, NW Donegal (assuming a correlation with the NW Mayo metadolerites)	Regional crustal extension. Deposition of Dalradian sediments

signature of the Mullet gneiss is compatible with a volcanic arc (subduction-related) setting. The Mullet gneiss contains boudins of coarse amphibolite, now generally concordant but originally emplaced as dykes because they are locally cross-cutting. The age of these dykes is unknown but they are succeeded by several episodes of Grenvillian deformation (see below). Generally the gneissose fabric is composite resulting from combined Grenvillian and later (probably Grampian) strain (cf. Johnston 1995), but as may be seen at Annagh Head (Loc. 2.1a) it is predominantly a Grenvillian feature (see below). Both before and after the emplacement of the amphibolites, the Mullet gneiss was migmatized. The later migmatitic leucosomes are sometimes pinkish and often pegmatitic. Some discrete pegmatite sheets (of Grenvillian age, see below) appear to post-date the migmatization but the ages of the two events are currently not distinguishable within age uncertainty.

2.2 Cross Point Gneiss

The main belt of Cross Point gneiss extends in a broad northwest-southeast trending belt from Cross Point on the Mullet Peninsula to Tristia on the mainland lying between bands of Mullet gneiss (Fig. 3). Smaller outcrops also occur south of Erris Head and along the southern margin of the Annagh Gneiss Complex in tectonic contact with the Dalradian at Ardelly Point (Fig. 3). The Cross Point gneisses have been dated at two localities (Daly 1996). At Tristia (Fig. 3) a near concordant age of 1271 ± 6 Ma was obtained while at Cross Point (Fig. 3, Loc. 2.5) severe disturbance during Grenvillian metamorphism resulted in an upper discordia intercept age of $1287 + 36/-35$ Ma. These ages are indistinguishable within error. However, the more precise *c.* 1270 Ma age is taken as dating the Cross Point gneiss igneous protoliths.

The Cross Point gneiss is syenitic to granitic in composition and has A-type within-plate geochemical characteristics (Fitzgerald et al. 1996). It is commonly banded on various scales depending in part on the state of strain and in part on the development of migmatitic leucosomes. Near Cross Point the felsic gneiss is interbanded on a decimetric to centimetric scale with foliated amphibolites. At least one larger layered mafic body occurs (Fig. 20) that is presumably related to these, but its field relations are ambiguous. The fact that it is migmatized can be used to distinguish it from the post-Grenvillian metadolerites. Occasionally the felsic gneiss contains mafic clots or tabular mafic enclaves suggestive of mixing between felsic and mafic magmas. In places, e.g. at Tristia (Fig. 3), the Cross Point gneiss is relatively homogenous, resembling a foliated granitoid intrusion. This rock is shown on Griffith's (1839) map with an ornament similar to that employed for the widespread late Caledonian granites, which it superficially resembles. Away from Tristia, the Cross Point gneiss is more strongly foliated or lineated and although it is much younger, its deformational and intrusive history is similar to that seen in the Mullet gneisses.

Geochronology and geochemical modelling (Fitzgerald et al. 1996) together with limited exposures of the contacts suggest that the Cross Point gneiss may be intrusive into, and is likely derived from the Mullet gneiss, probably under anorogenic conditions. This is consistent with its younger age, similar initial ε_{Nd} (Fig. 6) and ε_{Hf} (Fig. 7) values compared with the Mullet gneiss, and its A-type chemistry. At 1270 Ma, the Cross Point gneiss although strongly overlapping, actually has slightly more positive average ε_{Nd} values (Fig. 6) than the Mullet gneiss, suggesting an additional mantle component in its genesis. A possible intrusive relationship between the Cross Point and Mullet gneisses is consistent with the map pattern even though the contacts between the two units are not exposed. Strain (probably of Grampian age) intensifies in both units astride the inferred position of the contact north of Cross Point so that the original relationships may be impossible to decipher.

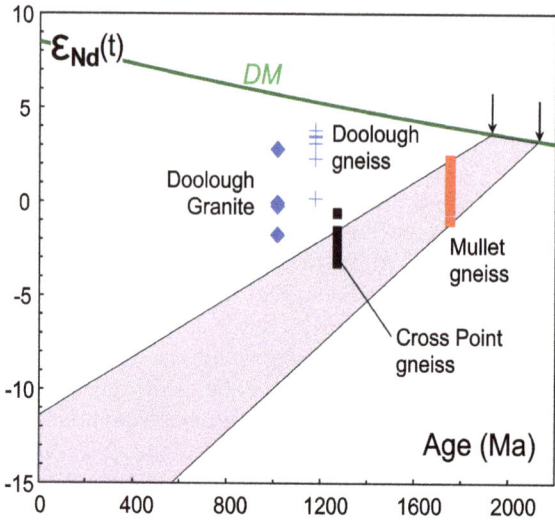

Fig. 6 Sm–Nd isotopic evolution diagram (from Daly 2009) for selected rocks from the Annagh Gneiss Complex (Menuge and Daly 1990 and unpublished). Calculated ε_{Nd} values for the Mullet, Cross Point and Doolough gneisses and the Doolough Granite are plotted at their time of formation (Table 1) together with the range of Nd evolution lines (magenta) for the c. 1750 Ma Mullet gneiss. At c. 1270 Ma, ε_{Nd} values of most of the Cross Point samples overlap the range of the Mullet gneisses while some are more positive. Hence the Cross Point gneiss could have formed by melting the Mullet gneisses with the addition of some basic mantle-derived material, consistent with field evidence such as mafic enclaves in felsic gneiss and interbanding of felsic and mafic lithologies. In contrast, at c. 1177 Ma, the Doolough gneisses have positive ε_{Nd} values indicating that they represent juvenile crust derived mainly from the mantle but possibly contaminated by continental crust

2.3 Doolough Gneiss and Doolough Granite

The Doolough gneiss (Figs. 3, 5, Loc. 2.6) is made up mainly of sub-alkaline granite and trondhjemite with subordinate amounts of amphibolite. Mafic and felsic varieties of Doolough gneiss are inter-banded on a decimetric scale. They are highly deformed and the small outcrops at Doolough (Loc. 2.6) and Barranagh (Fig. 3) may coincide with a major Grenvillian shear zone along which the Doolough gneisses were juxtaposed with the rest of the Annagh Gneiss Complex. If this is true the Doolough gneiss outcrops might only represent a small fraction of their original extent (Menuge et al. 1995). Formation of the Doolough gneiss protolith has been dated at 1177 ± 4 Ma (Daly 1996, Table 1). Both a subduction-related and an anorogenic origin are compatible with their geochemistry and with a juvenile origin for these rocks, as suggested by their positive initial ε_{Nd} (Fig. 7, Menuge and Daly 1990, 1994) and ε_{Hf} values (Fig. 8). The Doolough gneiss makes up only a small area of the Annagh Gneiss Complex outcrop and unless originally intruded into Mullet gneiss, was possibly juxtaposed with the rest of the Annagh Gneiss Complex during early Grenvillian deformation (D_G1) when it acquired its high strain fabric.

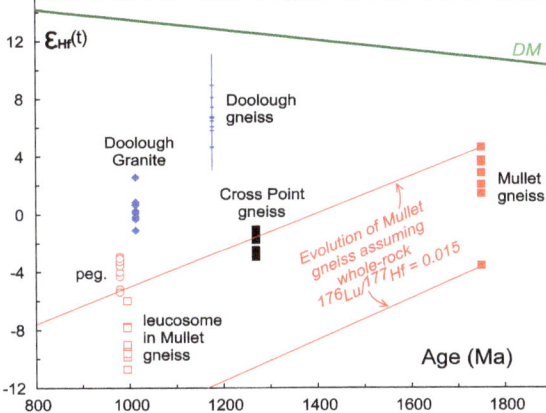

Fig. 7 Lu–Hf evolution diagram (from Daly 2009) for the Annagh Gneiss Complex. Laser-ablation analyses of dated zircons in *c.* 995 Ma migmatitic leucosomes within the Mullet gneiss have an identical range of Hf isotopic compositions to those of their host at 995 Ma (evolving between the red lines, assuming a typical crustal Lu/Hf ratio), suggesting that the leucosomes derived their Hf by in situ melting of the Mullet gneiss. In view of their strongly overlapping compositions at *c.* 1270 Ma, the Cross Point gneiss probably also formed by melting the *c.* 1750 ma Mullet gneiss, in broad agreement with inferences made from the whole-rock Nd isotope data (Fig. 6). In contrast, zircons from the *c.* 1177 Ma Doolough gneiss (blue error crosses) have more radiogenic Hf, requiring a juvenile (mantle) source. Pegmatites (peg.) cutting the Mullet gneiss at *c.* 980 Ma also require an additional input of radiogenic Hf, which could have come from tectonically juxtaposed Doolough gneiss. DM = depleted mantle

Fig. 8 Grenvillian pegmatite (parallel to hammer) at Cross Point (Loc. 2.5), dated at 984 ± 6 Ma, cuts 1270 Ma Cross Point gneiss and the foliation (D_G1 or D_G2) affecting them. The pegmatite is later deformed by D_G3 (foliation parallel to hammer), which also affects the gneisses (e.g. foliation parallel to hammer but discordant to layering). Metamorphic zircons aligned parallel to the D_G3 fabric have been dated at *c.* 958 Ma (Table 1). Mafic and felsic banding (visible in bottom right and top left of photograph) within the Cross Point gneiss is in part original, reflecting the mixed nature of these rocks—partly granitoid melts from the Mullet gneiss and partly a mantle-derived basic component (see also Figs. 6 and 7). Hammer head is 17.5 cm long

Migmatization is sporadic in the Doolough gneiss but at least two generations of granitoid intrusives may be distinguished—the peralkaline Doolough Granite (Winchester and Max 1987a; Aftalion and Max 1987), which postdates two deformational events in the Doolough gneisses and a later suite of granite pegmatites (see below).

The Doolough Granite appears to stitch a tectonic contact between the Doolough gneisses to the south and the Mullet gneisses to the north (Fig. 3). The gneiss is cut by the Doolough Granite and also occurs within it as partially digested xenoliths (Fig. 23). Sm–Nd model ages from the Doolough Granite vary from 1450 to 1758 Ma and tend to be older towards the northern contact, consistent with contamination of the Doolough Granite by the different wall rocks on the northern and southern margins. Magmatic zircons from the Doolough Granite have yielded a U–Pb age of 1015 ± 4 Ma (Daly 1996). A second population of zircons, with a poorly defined age of $c.$ 1220 Ma has likely been inherited from the $c.1177$ Ma old Doolough gneiss. This is supported by whole-rock Sm–Nd and by zircon Lu–Hf data (Figs. 6, 7). The $c.$ 1015 Ma old magmatic zircons display resorbed cores (as yet undated), but their Hf isotopic composition (Fig. 7) is consistent with a source stored in grains equivalent to those of the Doolough gneiss, which are envisaged to have dissolved and regrown during crystallisation of the Doolough Granite.

Aftalion and Max (1987) dated large unabraded zircon fractions from the Doolough Granite, which yielded a discordia intercept age of 1093 ± 8 Ma, which probably reflects a mixture of 1015 Ma and older zircons, affected by recent lead loss. By air-abrading the zircons (Krogh 1982) it was possible to remove the effects of lead loss (Daly 1996). Moreover by analysing small fractions and single grains, as well as in situ methods (Daly unpublished) it has been possible to deconvolute the mixture, thus revealing the correct age.

2.4 Grenvillian Deformation and Metamorphism

Evidence for three Grenvillian deformational events, denoted D_G1, D_G2 and D_G3 etc., can be deduced in the field at many localities and is supported by U–Pb dating of the intrusive bodies that separate them (Daly 1996; Table 1).

At Doolough (Figs. 3, 5, Loc. 2.6) all three Grenvillian deformation events may be seen in the field. F_G2 folding of a D_G1 high strain foliation in the Doolough gneisses must predate the 1015_{Ma} Doolough Granite which contains xenoliths of foliated Doolough gneiss, displaying F_G2 folds. Also at Doolough an undeformed pegmatite cutting folded, foliated Doolough gneiss (Fig. 5) was dated at 993 ± 6 Ma. This result confirms a minimum age for D_G2. Locally these pegmatites are deformed either by D_G3 or perhaps by Grampian movements. The generally weak deformation seen in the Doolough granite and in the pegmatites at Doolough suggests that D_G3 was less important here than elsewhere in the complex.

At Cross Point (Fig. 3, Loc. 2.5) the Grenvillian deformation history is also quite well constrained. Since the Cross Point gneisses are $c.$ 1270 Ma old, all deformation must be Grenvillian or younger. Two phases of Grenvillian deformation can be recognised and although it is not possible to distinguish between D_G1 and D_G2 at Cross Point, the later deformation there is probably the D_G3 event. A pegmatite cutting the Cross Point gneisses and an early (D_G1 or D_G2) foliation affecting them has yielded a zircon age of 984 ± 6 Ma. The D_G3 foliation affects this pegmatite (Fig. 8) and also transects the layering and the early foliation in the gneisses at a high angle. This locality is unusual in that the foliation throughout the complex is generally parallel to compositional banding. One group of zircons from the $c.$ 1270 Ma old Cross Point gneiss, interpreted to be metamorphic, are aligned parallel to the D_G3 foliation (Fig. 8) and yield an age of 958 + 17/-19 Ma (Table 1). This age is (just) younger than that obtained from the cross-cutting pegmatite. These metamorphic zircons grew during development of the D_G3 foliation possibly as the result of Zr release from amphibole. Post-tectonic titanite from the same locality yields a concordant U–Pb age of $c.$ 943 Ma (Daly and Flowerdew 2005).

At Tristia (Fig. 3) a post-D_G1 or -D_G2 pegmatite cutting foliated 1270 Ma old Cross Point gneiss yielded a U–Pb zircon age of 999 + 29/-9 Ma. The pegmatite is itself deformed by D_G3, which is thus constrained to be younger than $c.$ 999 Ma in keeping with the results from Cross Point and Doolough (Table 1). On Annagh Head (Fig. 4, Loc. 2.1a), zircons from a migmatitic leucosome cutting the 1753 Ma Mullet gneiss have yielded an age of 995 ± 6 Ma. At the same locality a deformed pegmatite cutting the main foliation in the Mullet gneiss has a U–Pb zircon age of 980 + 38/-12 Ma. This pegmatite and the foliation affecting it are cut by a weakly deformed metadolerite. If it is assumed that the metadolerite deformation is Grampian, then some pre-Grampian, presumably Grenvillian, deformation must have occurred at Annagh Head after $c.$ 980 Ma. This is presumably the DG_3 deformation, which is also present at Cross Point.

In several places on the Mullet Peninsula, kinematic indicators such as oblique extensional crenulation and asymmetric feldspar porphyroclasts indicate sinistral shear in a west-north-westwards direction before the emplacement of pegmatitic granite sheets dated between 993 and 984 Ma (Table 1).

Metamorphic titanite from several locations within the Annagh Gneiss Complex probably grew at $c.$ 980 Ma (Daly and Flowerdew 2005). The U–Pb data are slightly discordant with some lead loss apparently occurring during the Caledonian. However, these data suggest that prolonged Caledonian heating is unlikely to have exceeded $c.$ 600 °C, the closure temperature for Pb diffusion in titanite (Heaman and Parrish 1991). In contrast, apatite (U–Pb closure temperature of $c.$ 450 °C) in the Mullet gneiss at Annagh Head yields Grampian ages (Henrichs et al. 2018), and thus was likely reset during Grampian orogenesis.

Grenvillian events may be summarised as follows: D_G1 produced the main foliation in the Doolough gneisses, which were folded by F_G2 folds (Figs. 5, 22 and 23). These events occurred after 1177 Ma and before 1015 Ma. Very high strains affected the Doolough gneisses and may also have been responsible for its juxtaposition with the rest of the Annagh Gneiss Complex during DG_1 or DG_2. Further tectonothermal activity resulted in migmatization and the intrusion of discordant pegmatites (at Annagh Head, Tristia and Cross Point) in the interval 995–980 $_{Ma}$. DG_3 probably took place at c. 960 Ma. This event produced a foliation in the pegmatites and locally developed a foliation in the gneisses (e.g. at Cross Point). Late titanite grew at or before c. 940 Ma at Cross Point (Daly and Flowerdew 2005).

The evolution of the Annagh Gneiss Complex may be summarised as follows. Much of the Annagh Gneiss Complex originated as juvenile Palaeoproterozoic crust represented by the 1753 ± 3 Ma old calc-alkaline Mullet gneisses. Late Mesoproterozoic Cross Point gneisses with A-type geochemistry and Palaeoproterozoic t_{DM} ages were emplaced as anorogenic granitoids at 1271 ± 6 Ma, probably by melting of the pre-existing Palaeoproterozoic Mullet gneisses with the addition of a mantle-derived mafic component. The Doolough gneisses comprise a small volume of juvenile granitoids and associated basic rocks, which formed at 1177 ± 4 Ma. The Doolough gneisses occur as a tectonic sliver, possibly incorporated along a Grenvillian high strain zone. They may be a further manifestation of anorogenic magmatism or they may represent a much larger volume of juvenile subduction-related material. Grenvillian deformation occurred in two stages, between 1177–1015 Ma and from 995–960 Ma. These events were separated by the intrusion of the Doolough peralkaline granite at 1015 ± 4 Ma and by migmatization and pegmatite emplacement between 995 and 980 Ma. Post- Grenvillian metadolerites help to distinguish Caledonian strain from earlier deformation.

Despite isotopic evidence for the possible presence of Archaean crust at depth (Flowerdew et al. 2009; Kirkland et al. 2013), it is likely that much of north-western Ireland is underlain by juvenile Palaeoproterozoic crust formed c. 1.75–1.8 Ga ago. This Palaeoproterozoic crust may be correlated westwards with rocks of similar age and geochemistry on the Rockall Plateau (Daly et al. 1995), in the Ketilidian terrain of south Greenland, and with large areas of central North America. The Proterozoic of NW Ireland and SW Scotland may have formed part of a continuous belt along the accreting continental margin of Laurentia in the Palaeoproterozoic (Muir et al. 1992; Park 1995). Anorogenic magmatism, similar in age to the c. 1.27 Ga protoliths of the Cross Point gneiss, is widely recognised in Laurentia and Baltica (Cadman et al. 1993; Romer 1996; Gower 1996; Gower and Krogh 2002). The D_G1 and D_G2 deformational events affecting the Annagh Gneiss Complex correlate broadly with the Grenvillian orogeny in Laurentia and with the Sveconorwegian in Baltica. This early deformation may correspond to the c. 1050 Ma Ottawan Orogeny (cf. McLelland et al. 1996; Rivers et al. 2012) in the western Grenville Province, and may also correlate with events in NW Scotland and Baltica. Later (D_G3) deformation possibly corresponds to the Rigolet phase of the Grenville Orogeny in Canada (Rivers et al. 2012; Rivers 2015) although deformation affecting the Annagh Gneiss Complex may have continued until c. 960 Ma as in the Sveconorwegian Province of SW Sweden (cf. Page et al. 1996).

3 Dalradian Supergroup

In Ireland the mainly Neoproterozoic Dalradian Supergroup occurs in three major inliers in Donegal, in north Mayo (Fig. 1) and in Connemara. Smaller inliers occur in northeast Antrim, in Tyrone and in the Ox Mountains (Fig. 1). The Dalradian Supergroup is a thick sequence of deformed and metamorphosed clastic sediments, limestones and basic volcanics with a distinctive stratigraphy (Harris and Pitcher 1975; Harris et al. 1994), divided into four major groups—Grampian, Appin, Argyll and Southern Highland (Fig. 9). Although a complete succession is not preserved, the North Mayo inlier is unique in Ireland because all four stratigraphic groups are present (Fig. 1). Although the disposition of Dalradian stratigraphic units is generally

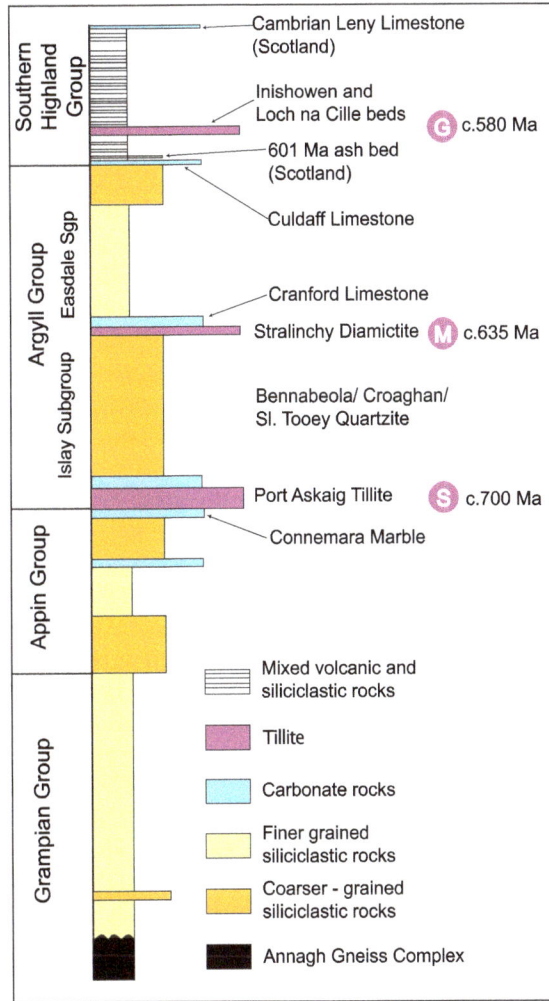

Fig. 9 Schematic stratigraphy of the Dalradian Supergroup (after McCay et al. 2006) highlighting the three main episodes of glaciation recorded by the Dalradian sequence. S = Sturtian; M = Marinoan; G = Gaskiers

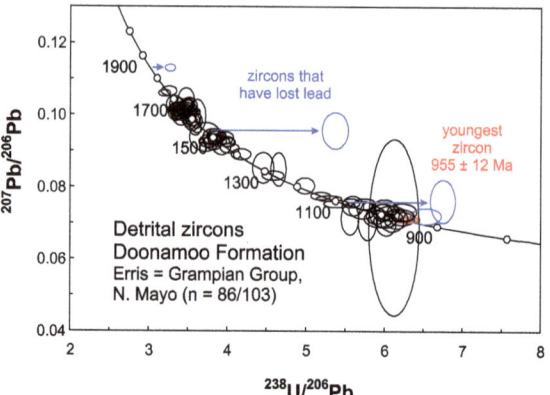

Fig. 10 Tera-Wasserburg Concordia diagram for detrital zircons (McAteer et al. 2010) from the Doonamoo Formation at Spinkadoon (Loc. 2.4a)

accepted at group level (Fig. 1), the effects of Grampian folding and the paucity of distinctive Dalradian marker horizons such as the Loch Tay Limestone or the Tayvallich Lavas leaves room for stratigraphic revision and diverse opinions. In this account, the Dalradian stratigraphic nomenclature follows that of Long et al. (1992) as depicted on the corresponding Geological Survey of Ireland map (Max et al. 1992). One important difference is that, following Kennedy and Menuge (1992), the Inishkea Division is now considered to be an integral part of the Grampian Group (known as the Erris Group in Mayo). The Inishkea Division has previously been referred to as "Scotchport schists" and was for some time considered to represent tectonically reconstituted Annagh Gneiss Complex (Sutton and Max 1969, see below). The Erris Group comprises mainly psammitic, locally quartz-rich, rocks that occur spatially close to and generally younging away from the Annagh Gneiss Complex. Correlation of the Erris Group with the Grampian Group is supported by geochemical evidence (Winchester et al. 1988).

One distinctive Dalradian stratigraphic marker is a widespread, although poorly exposed diamictite of inferred glacial origin, known as the Briska Boulder Bed Formation (Max et al. 1992) in the north of the inlier and as the Doogort Boulder Bed on Achill (Loc. 2.9). This unit is correlated with the Port Askaig Tillite Formation of the Argyll Group in SW Scotland and is assumed to represent the Sturtian glaciation (Fig. 9). Local successions have been constructed based on its presence (Long et al. 1992). Assigning units higher in the local successions to the upper part of the Argyll Group or to the Southern Highland Group is more problematic. Metagreywacke sequences with graphitic pelites, cherts, pillow lavas and limestones occur on south Achillbeg Island, Clare Island and in parts of the Ox Mountains. These were previously included in the Southern Highland Group of the Dalradian but are more likely part of the Clew Bay Complex (see below) and are generally correlated with the Highland Border Complex of Scotland.

Dalradian deposition began in an ensialic rift basin (Harris et al. 1978), which started to form in the Tonian on the continental margin of Laurentia, with the development of extreme extension and eventual development of oceanic crust in the Ediacaran

(Chew 2001; Chew and van Staal 2014). The initial rifting is thought to be expressed in the development of the Grampian Group, the oldest exposed rocks of the succession which outcrop in north-west Mayo. These comprise a repetitive sequence of quartzofeldspathic meta-arkoses. Later as rifting diminished (Soper and England 1995) a more lithologically diverse association of shallow water, probably marine sediments was deposited in an upwards-shallowing sequence in response to thermal subsidence, forming the Appin Group. The succeeding Argyll Group is a response to a renewed phase of rifting and was deposited initially in confined basins, and later in more laterally continuous sequences as the basins deepened. Eventually, volcanics and turbidites filled the basin near the top of the sequence, which may prograde onto oceanic crust (Anderton 1985; Soper 1994). Intrusive metadolerites which are locally numerous in the north of the North Mayo inlier but rarer farther south are all affected by the D2 fabric. They are considered to represent a single suite of pre-D1 intrusions and are probably related to volcanicity in the upper parts of the succession. As discussed above these metadolerites are probably equivalent to the late metadolerite suite cutting the Annagh Gneiss Complex and are interpreted as relating to the opening of the Iapetus Ocean.

3.1 Age of the Dalradian Supergroup

The depositional age of the Dalradian succession is difficult to establish because of the dearth of datable fossils, none of which are known from Ireland. From a field geological perspective only very broad constraints may be placed on the timing of Dalradian deposition. A minimum age is provided by the unconformable relationship with Upper Llandovery (Telychian) sandstones and probable Upper Llandovery volcanics (Leake and Tanner 1994) in east Connemara. Thus Dalradian deposition and the early development of the Caledonian orogeny must be older than c. 433 Ma. Indirect evidence for a pre-Llandeilo age is suggested by the presence of presumed Dalradian clasts with metamorphic muscovite Rb–Sr ages of 462 Ma and 471 Ma (Graham et al. 1991) in the late Llanvirn Derryveeney Formation (Clift et al. 2009) in South Mayo. This is supported by an abundance of Ar–Ar and Rb–Sr metamorphic mineral ages from various Dalradian rocks (Dempster 1985; Flowerdew et al. 2000; Chew et al. 2003, 2004).

Field evidence placing an older limit on Dalradian sedimentation is now available in the form of the newly recognised unconformity (Fig. 14). On this basis a post-Grenvillian, i.e. post c. 960 Ma (Table 1) age for the deposition of the Grampian (= Erris) Group is indicated. In Scotland, Smith et al. (1999) have argued that the stratigraphically lowest Glenshirra Subgroup of the Grampian Group was likely deposited unconformably on previously deformed basement metasediments of the Glen Banchor succession. Somewhat younger Grampian Group metasediments of the Ben Alder succession and the Pityoulish Semipelite Formation were also interpreted by Smith et al. (1999) to have been deposited unconformably on similar but geographically distinct basement metasediments of the Dava Succession, from which

Highton et al. (1999) obtained a U–Pb zircon age of 840 ± 11 Ma, interpreted as dating migmatization. If the proposed unconformity is correct, then deposition of the Grampian Group must have taken place after 840 Ma. A maximum age of c. 955 Ma for the Erris (= Grampian) Group in North Mayo is indicated by the age of the youngest detrital zircon (Fig. 9; McAteer et al. 2010) at Spinkadoon (Loc. 2.4a) in the Doonamoo Formation.

Geochronological investigations in Scotland have been critical in constraining Dalradian deposition to be mainly of Neoproterozoic age. The Tayvallich Volcanics of Scotland, near the top of the Argyll Group, have yielded a zircon U–Pb age of 595 ± 4 Ma (Halliday et al. 1989) interpreted to date their eruption. In addition the Ben Vuirich Granite, which cuts Appin and Argyll Group metasediments in Perthshire, has yielded a high precision conventional U–Pb age of 590 ± 2 Ma (Rogers et al. 1989) and a SHRIMP ion microprobe U–Pb age of 597 ± 11 Ma (Pidgeon and Compston 1992). When the Ben Vuirich granite yielded this unexpectedly old age, misinterpretation of the field relations as indicating a post-D2 age lead to the notion of a Neoproterozoic orogeny affecting the Dalradian, with the formation of major crustal scale nappe structures. It also led to the suggestion that Dalradian deposition may have taken place on the Gondwana continent rather than in Laurentia (Bluck and Dempster 1991). Re-investigation of the field relations showed that the Ben Vuirich granite predates major nappe formation (Tanner and Leslie 1994) and post-dates an early D1 foliation in the enclosing metasediments. Tanner (1996) suggested that the foliation cut by the Ben Vuirich Granite may be mimetic after bedding and hence has no orogenic significance.

A Sm–Nd mineral isochron of 625 ± 47 Ma (Kirwan et al. 1989) for the Mam sill, a metadolerite cutting Appin Group sediments in Donegal, confirms the Precambrian age for at least the Appin Group in Ireland. This age has been confirmed (at 603 ± 11 Ma) by ion microprobe U–Pb dating of zircon from a gabbro pegmatite from the Mam sill (Daly unpublished). Kirwan et al. (1989) regarded the Mam sill as pre-deformational, intruded into sediments of the lower part of the Appin Group in Donegal, which they interpreted to have been semi-consolidated at the time of intrusion. Geochemical investigations of the Donegal metadolerites (Kirwan written communications 1988, 1996) are consistent with a rift origin for the magmatism. The c. 600 Ma age for the Mam sill is similar to other more precise estimates for magmatism related to the rifting of Iapetus (i.e. 615 Ma for the Long Range dykes in Labrador, Kamo et al. 1989 and c. 595 Ma for the Tayvallich volcanics, Halliday et al. 1989).

Carbon isotope excursions and unradiogenic $^{87}Sr/^{86}Sr$ values in carbonate formations above and below the Port Askaig Tillite Formation (which defines the base of the Argyll Group in Scotland) have been used to correlate it with the global Sturtian glaciation (McCay et al. 2006), which is believed to have had a synchronous global onset at c. 717 Ma (MacLennan et al. 2018). Rooney et al. (2011) obtained a Re-Os whole-rock isochron age of 659.6 ± 9.6 Ma for the Ballachulish slate in Scotland (Appin Group). This provides a maximum age for the overlying Port Askaig Tillite Formation and implies it may be correlative with the c. 650 Ma end-Sturtian glaciations of Australia (Rooney et al. 2011). A maximum depositional age of 687 ± 12 Ma was obtained by Chew et al. (2019) for the three youngest U–Pb detrital zircon ages in a psammite from the Port Askaig Tillite Formation in Donegal.

3.2 Structure and Metamorphism in the N Mayo Inlier

The major Grampian folds affecting the North Mayo Dalradian are D2 structures with associated penetrative S to L-S fabrics. There is a primary facing divergence of the F2 folds either side of the Annagh Gneiss Complex basement core. To the north of this 'root zone' shallowly inclined F2 folds face north with axial planes trending WNW-ESE; to the south, the axial planes of the F2 folds are progressively rotated through nearly 180°, resulting in recumbent F2 folds that face south (Chew 2003). Approaching the Achill Beg Fault, the south-facing F2 antiform is rotated into a downward-facing orientation, analogous to the Highland Border Downbend in Scotland. The F2 fold nappes deform an earlier foliation and locally refold F1 folds. There are no major reversals of D2 facing direction to indicate the presence of large-scale F1 fold closures but local reversals do occur showing the presence of F1 folds, the largest of which have wavelengths of a few hundred metres. Early deformation has led to the formation of high strain zones (traditionally known as tectonic slides) that excise significant parts of the stratigraphic succession, potentially originating as extensional syn-depositional faults (Soper and Anderton 1984). The main ductile strain fabric in these slide zones is deformed by the D2 structures.

There is a broad consensus that the major structure in the Dalradian rocks of Achill Island (Fig. 2) is a major, recumbent, southward-facing F2 fold, which is progressively "downbent" towards the south (Sanderson et al. 1980; Harris 1993a, b, 1995; Johnston 1995; Johnston and Phillips 1995). This "downbending" results in predominantly vertical rocks adjacent to the northern margin of Clew Bay, including south Achill Island and Achillbeg (Fig. 2). The structural studies in South Achill listed above have in general concentrated on the kinematics and timing of formation of this steep zone. Although there is a general consensus that it is produced by E-W dextral shearing, for which there is abundant evidence in South Achill, the timing relative to the main D2 nappe-forming event remains contentious. Later crenulation cleavages affect the S2 foliation in many parts of the inlier, particularly in the more pelitic rocks in the south. Johnston (1995), Johnston and Phillips (1995) and Harris (1995, 1993a, b) all recognise an east–west trending L2 stretching lineation. Johnston (1995) argued that D2 structures represent northwestward-vergent thrusts over a crustal-scale ramp zone. Harris (1995) favoured a westward direction of transport for the recumbent D2 nappes, contemporaneous with E-W dextral transpression in the steep zone adjacent to the Achillbeg Fault. Alternatively, Sanderson et al. (1980) and Chew (2003) consider the strike-swing and generation of downward-facing folds to be the result of later modification by an E-W dextral shear zone along the north margin of Clew Bay. Structures related to E-W dextral shear which cut the main S2 nappe fabric have grown fabric-forming muscovite, and isotopic dating of these D3 shear-related structures are distinctly younger than the main S2 fabric (Chew et al. 2004).

The southern portion of the inlier, adjacent to the Achillbeg fault is characterized by the development of early (MP1) blueschist-facies assemblages (Chew et al. 2003; Gray and Yardley 1979). Textural evidence suggests that the blueschist-facies metamorphism developed contemporaneously with Barrovian metamorphism in the Dalradian farther north in central Achill. The latter yielded PT estimates of 525 ± 45 °C, 6.5 ± 1.5 kbar, while the blueschist-facies rocks of southern Achill yielded lower temperatures and higher pressures (460 ± 45 °C, 10.5 ± 1.5 kbar) (Chew et al. 2003). The grade of Barrovian metamorphism increases northwards. Near Erris Head (Fig. 3) Yardley et al. (1987) reported PT conditions of 620 ± 20 °C and 8 ± 2 kbar in Dalradian metapelites, while Flowerdew (2000) calculated similar PT values of 9–11 kbar at 650 °C in a metadolerite dyke on Eagle Island (Fig. 3) demonstrating the effects of the Grampian metamorphism on the Annagh Gneiss Complex basement.

Thermobarometry also showed that South Achill experienced blueschist-facies metamorphism as a relatively coherent crustal block, and is not composed of a series of metamorphic terranes exhumed from different crustal levels. Combined with evidence (Chew 2001) that South Achill represents distal Dalradian sedimentation on the Laurentian margin, it appears likely that the blueschist-facies assemblages were developed on the leading edge of the subducting Laurentian plate during early Grampian orogenesis, and were overridden from the south by a supra-subduction ophiolite (the Deerpark Complex; Sect. 2.5) and associated intra-oceanic arc. During D2 times, the blueschist-facies assemblages of South Achill were progressively replaced by greenschist-facies minerals.

4 The Clew Bay Complex

The Clew Bay Complex comprises a series of low-grade turbiditic metasediments, and higher grade metamorphosed basic and ultrabasic rocks. The turbiditic portions of the Clew Bay Complex include the Ballytoohy and Portruckagh Formations on Clare Island, the South Achillbeg Formation on the island of Achillbeg and the Killadangan Formation on the south coast of Clew Bay. The Clew Bay Complex as a whole shares many similarities with the Highland Border Complex of Scotland. The two components of the Clew Bay Complex, the low-grade deep marine metasediments and the higher grade metamorphosed basic and ultrabasic rocks, are always found in tectonic contact with each other, and the contact between the Clew Bay Complex and the Dalradian is marked by the Achillbeg Fault (Fig. 1; Fig. 38, loc. 2.24). Hence the original relationship between these units is somewhat problematic. However, the ophiolitic affinity of the high grade metabasites and serpentinites (Ryan et al. 1983) means that the Clew Bay Complex figures prominently in tectonic models of the Grampian orogeny (e.g. Dewey and Mange 1999), with the low-grade turbidites interpreted as an accretionary complex and the high grade metabasites and serpentinites representing a supra-subduction ophiolitic mélange.

Chew (2003) concluded that the Dalradian and the Clew Bay Complex on the island of Achillbeg (Fig. 1) share the same structural history across the Achillbeg Fault. A D1 high strain event is common to both units, and is associated with the development of tectonic slides. The D2 event is responsible for the formation of crustal-scale nappes, and in both units, beds are consistently downward facing on the S2 foliation. The above structural studies all suggest contemporaneous deformation of the Dalradian and the Clew Bay Complex during a mid-Ordovician Grampian orogenic event, particularly as the S2 fabric on South Achillbeg has been dated at c. 460 Ma (Chew et al. 2003).

An alternative interpretation of the structure for parts of the Clew Bay Complex is given by Max (1989) and Williams et al. (1994) who interpreted the whole of the Ballytoohy Formation on Clare Island as a mélange. The large, inland exposures of greywacke were regarded as megablocks (up to 500 m long) enclosed within the mélange matrix. The mélange matrix has yielded trilete miospores interpreted to be of Silurian (Wenlock) age (Williams et al. 1994, 1996). These microfossil discoveries are contentious as they preclude the involvement of the Clew Bay Complex in the Ordovician Grampian orogeny.

Pelitic lithologies in the Clew Bay Complex of South Achillbeg typically contain phengitic muscovite + chlorite + albite + quartz. Estimating peak metamorphic temperature in low greenschist facies assemblages such as these is challenging due to the difficulty in ensuring that the metamorphic phases present are in chemical equilibrium. Thermobarometric estimates based on chlorite and phengite pairs (Chew et al. 2004) yield P–T conditions c. 325–400 °C, 10 kbar. These P–T estimates support models (e.g. Dewey and Mange 1999) that interpret the low grade-turbidites as an accretionary complex, which would be expected to have experienced high-pressure, low-temperature metamorphic conditions.

5 Excursions

5.1 Description of Localities in the Belmullet Area

Localities on the Mullet Peninsula and on the mainland at Doolough (Fig. 3) are covered by the Ordnance Survey of Ireland Discovery Series map sheet 22 (1:50,000). Much of the area is designated as a Gaeltacht (Irish-speaking) and many road signs are in the Irish language only. Convenient access to some localities is over private land (as indicated where appropriate below). In the authors' experience, permission to cross private land will be granted if politely requested. Great care should be taken on coastal exposures particularly at Annagh Head (Loc. 2.1), Termoncarragh (Loc. 2.2) and Spinkadoon (Loc. 2.4) where there are dangerous cliffs. Some outcrops are at least partly tide dependent. Localities 2.1–2.6, conveniently grouped together on the Mullet Peninsula, describe features in the Annagh Gneiss Complex and their relationship to the Dalradian cover rocks. They can be visited in any order but it is recommended to visit localities 2.1 (Annagh Head) and 2.4 (Cross Point) on the same day. Locality 2.6 is located on the mainland and if the focus is on the Annagh Gneiss Complex, it is probably best appreciated after first visiting Annagh Head and Cross Point.

5.2 Locality 2.1: Annagh Head (54.24168°, −10.107999°)

Access and other information Parking is available at the end of the stone-metalled road (54.241563 °, −10.102281°). Cross through the style and walk uphill westwards towards the summit of Annagh Head, noting the spectacular storm beach to the north (Fig. 11) and the lighthouse on Eagle Island. The Inishkea islands and beyond them, the Duvillaun islands, can be seen to the southwest as well as the cliffs on Achill to the south.

This locality provides clear evidence for the separation of Grenvillian from Grampian events and is accordingly divided into two parts.

Fig. 11 Spectacular storm beach at Annagh Head (Loc. 2.1a) discussed by Dewey and Ryan (2017). Loc. 2.1b is visible to the left of the storm beach in the top left corner

5.2.1 Locality 2.1a: Mullet Gneiss, Granitic Pegmatites and Late Metadolerites—Mainly Pre-Grampian Events. (54.241582°, −10.107890°)

This is the type locality of the c. 1.75 Ga old Mullet gneiss (Menuge and Daly 1990; Daly 1996), the oldest rock identified in mainland Ireland. The main lithology is a migmatized quartzofeldspathic gneiss, probably an orthogneiss (i.e. originally igneous) with occasionally more micaceous darker bands. Occasionally deformed K-feldspar megacrysts are present, interpreted as originally phenocrysts or antecrysts. The Mullet gneiss (A, Fig. 4) yields c. 1.9 Ga Sm–Nd model ages (Fig. 6) and a U–Pb zircon discordia IDTIMS age of 1753 ± 3 Ma (Table 1, Daly 1996). The Mullet gneiss encloses boudins of coarse amphibolite (B, Fig. 4), now mainly concordant but possibly originally emplaced as dykes. All of these lithologies are cut (migmatized) by pinkish, often pegmatitic, leucosomes (C, Fig. 4) dated at 995 ± 6 Ma. Following the intense deformation responsible for the dominant northwest-southeast fabric, discrete sheets of granite and pegmatite were emplaced at c. 980 Ma (Daly 1996). After further deformation (Fig. 12), the complex was intruded by a suite of unmigmatized but variably foliated metadolerites (D, Fig. 4).

The dominant structure here is a strong WNW-ESE trending gneissose fabric, accompanied be an ESE-plunging lineation. At first these fabrics appear to obscure relationships between the various lithologies. However, owing to the excellent exposure at this locality, in many places it is possible to work out a detailed deformation and minor intrusion chronology (e.g. Sutton 1972; Phillips 1981; Max and Long

Fig. 12 Foliated granitic pegmatite cutting Mullet gneiss, mafic bands and granitic leucosomes at Annagh Head (Loc. 2.1a). The deformation affecting the pegmatite is likely of Grenvillian rather than Grampian age because when followed c. 20 m to the left of the photograph, the pegmatite and its foliation are cut by a pre-Grampian metadolerite dyke, that locally displays chilled margins, despite itself being affected by a later Grampian fabric. The camera faces north and the person in the top right of the photograph provides a scale

1985; Daly 1996, 2009). Hence Annagh Head was chosen for U–Pb dating (Daly 1996). Importantly, the Grampian strain here is low. Hence the discordant metadolerite dykes, which generally display a subtle foliation and lineation, can be used as a structural marker to distinguish Grenvillian from Grampian strain effects. They record Grampian Ar–Ar hornblende ages (Daly and Flowerdew 2005).

Exercises

(1) It is relatively straightforward to distinguish the early migmatized from the (usually) discordant unmigmatized late metabasites? Carefully assess the contacts of the migmatized variety. Are any of them discordant? What features do they cut?

(2) What is the relative age of the granitic pegmatites and the migmatitic leucosomes?

(3) Close to as well as away from a discordant metadolerite, measure and compare the fabrics in them with the fabrics in the Mullet gneiss. Why are they so similar?

5.2.2 Locality 2.1b: Highly Strained Mullet Gneiss, Granitic Pegmatites and Late Metadolerites—Strong Grampian Reworking. (54.242857°, −10.105362°)

Access and other information To reach Locality 2.1b, walk back towards the parking place and skirt around the storm beach (Fig. 11) along its eastern edge. At about 54.242437°, −10.103318°, turn towards the west-northwest and cross the storm beach where the boulders are generally well-compacted.

Starting on the wave-cut platform, and traversing generally north–north-eastwards (across the strike and in the direction of dip) several variably deformed unmigmatized metadolerites are highly strained together with the enclosing gneisses in a Grampian greenschist-facies sinistral shear zone (Fig. 13).

Exercises

(1) Search for kinematic indicators in the deformed Mullet gneiss close to the highly strained metadolerite. What was the original protolith of the K-feldspar porphyroclasts?

(2) Compare the orientations of the Grampian foliation and lineation with those of the earlier Grenvillian foliation and lineations at Loc. 2.1a.

Fig. 13 Photograph of a sinistrally-sheared K-feldspar porphyroclast, taken looking north-northeast (Loc. 2.1b). The plane of the photograph is parallel to the lineation within a Grampian shear zone at Loc. 2.1b affecting Mullet gneiss and post-Grenvillian metadolerites

5.3 Locality 2.2: Termoncarragh (54.246331°, −10.074065°) to Scotchport (54.255119°, −10.074170°).

Access and other information This locality involves a coastal traverse from Termoncarragh to Scotchport. It should only be attempted in good weather and in good light—the cliffs are steep and unstable in places and there are unfenced collapsed blowholes and caves (including the remarkable *Poll na Sean-tine*). Although parking is easier at Scotchport, it is recommended to start at Termoncarragh where there is limited parking where the road comes closest to the coast at the head of Portnafrankagh (54.246331°, −10.074065°). To reach the start of the traverse, walk north-westwards along the north coast of Portnafrankagh to reach exposures of generally highly deformed Mullet gneiss (similar to locality 2.1b). Head northwards across the strike and down dip, following the indented coastline. If possible, before the start, park one or more vehicles at Scotchport, which the drivers can use to retrieve any vehicles parked at Portnafrankagh.

This is a traverse from Mullet gneiss to Dalradian metasediments (Scotchport schist—"Inishkea Division") which demonstrates an inaccessible but visible unconformity between the Dalradian Scotchport schist and the underlying Mullet gneiss of the Annagh Gneiss Complex. The traverse begins in generally highly strained Mullet gneiss at the north end of the first bay north of Portnafrankagh at 54.249489°, −10.082889°. From here follow the coast north-northeastwards (in the direction of dip) comparing the lithologies with those seen at Annagh Head (Loc. 2.1) and seeking out examples of lower strain where cross-cutting relations are preserved. This section of coast provides an opportunity to examine rocks whose origin was controversial in the past, where the highly strained orthogneisses were at one time interpreted as phyllonites or tectonic schists, derived by severe tectonic re-working of the Annagh Gneiss Complex (Sutton and Max 1969).

However, the contact between highly strained Mullet gneiss in the south and the Dalradian semi-pelitic schists (Scotchport schist) to the north is quite sharp. The contact between the two is mainly a north-dipping reverse fault (Fig. 14a), in whose hanging wall a contact is visible between Scotchport schist and Mullet gneiss

Fig. 14 a Photograph at Loc. 2.2 taken from 54.252436°, −10.084382° at low tide in exceptionally calm sea conditions. The area outlined in white, shown in Fig. 14b includes the inaccessible exposure of the contact between the Dalradian (Scotchport schist, Inishkea Division) and the underlying Mullet gneiss of the Annagh Complex. The unconformity is exposed in the hanging wall of the main fault plane (in shade), which strikes WNW-ENE and dips steeply north-northeastwards **b** Photograph taken from 54.252680°, −10.085290° (Loc. 2.2) at low tide in exceptionally calm sea conditions of the area outlined in white in Fig. 14a. The cliff face here is c. 25 m high. Bedded Dalradian metasediments of the Scotchport schist (Inishkea Division) are interpreted to lie unconformably on migmatized Mullet gneiss displaying pinkish leucosomes, boudinised metabasite bodies and discordant granitic pegmatites. The preservation of these features stands in marked contrast to the mylonitised slice of Mullet gneiss that is interleaved with the Scotchport schist at Locality 2.2b **c** Photograph taken from 54.252436°, −10.084382° (Loc. 2.2) of the area outlined in black in Fig. 14a

(Fig. 14b), which the first author (Daly, in preparation) interprets as an unconformity. The unconformity has so far defied attempts to reach it because even in exceptionally calm summer weather, there is a substantial sea swell.

Exercises

(1) At cliff top level, the contact is steep and sharp between the Scotchport schist to the north and the Mullet gneiss to the south. It is marked on Sheet 6 (Max et al. 1992) as a north-northeast-dipping fault. Is it a reverse or normal fault?

(2) What features preserved in the rocks to the south support the interpretation that they are highly strained orthogneisses (*in extremis*, phyllonites, or tectonic schists) as opposed to metasediments? Such was the debate 50–60 years ago (e.g. Sutton and Max 1969).

Between here and Scotchport, and especially at Scotchport itself (54.255119°, −10.074170°), the compositional layering, the presence of cryptic sedimentary structures suggestive of cross-bedding, and the lack of typical relict gneissic features in these rocks all suggest that the reworked gneiss hypothesis for the Inishkea Division is erroneous. The Scotchport schist was included within the Dalradian and assigned to the Erris (=Grampian) Group by Kennedy and Menuge (1992).

In the low cliffs along the northern shore of the bay, a sliver of mylonitised granitoid, probably a tectonic slice of Mullet gneiss, is enclosed in typical Scotchport schist.

Farther to the northwest, a separate "sliver of gneiss" is associated with a band of porphyroblastic amphibole-epidote-chlorite rock ("garbenshiefer") with rare fuchsite. North of this, concordant sheets of metadolerite have a strong fabric parallel to that of the enclosing Scotchport schist similar to the relationship at Loc. 2.1b.

Exercises

(1) Bedded Dalradian metasediments of the Scotchport schist (Inishkea Division) rarely display clear evidence of way-up. However fresh exposures on the cliff face (Fig. 14c) reveal some beds with strongly asymmetrical contacts, which are sharper on their structurally lower surfaces, suggesting right-way-up graded bedding. This implies that at this locality, the Scotchport schist youngs away from its inferred depositional contact with the underlying Mullet gneiss. Do you agree?

(2) At Scotchport Bay, does the intrusive metadolerite body exhibit the same deformational history as the enclosing Inishkea metasediments?

(3) Can you suggest an alternative origin? To test the "gneiss hypothesis", search for preserved features such as migmatitic banding, pods of pegmatite and metadolerite. Metadolerite could be represented by thin garnetiferous micaceous bands

5.4 Locality 2.3: Doonamoo: (54.26921°, −10.0761°) and Aughernagalliagh (54.27496°, −10.06978°)

Access and other information Park off road in the tarmac car park at 54.246331°, −10.074065°. This locality demonstrates metadolerite dykes (and Cenozoic dolerites) cutting Erris Group (Doonamoo Formation) psammite.

5.4.1 Locality 2.3a: Doonamoo Dykes (54.26921°, −10.0761°).

Access and other information Walk southwards to the clearly signposted Dún na mBó sculpture, which was made by Travis Price and is one of many contributions to the Tír Sáile Sculpture Trail (https://www.northmayoarttrail.com/en/home-english) around North Mayo. Dún na mBó is arranged around a natural blowhole. Doonamoo is also the location of one of many promontory forts described by Westropp (1912), which extended from Doonamoo Point to the sculpture site. From here, cross the fence via the gate and head towards 54.26921°, −10.0761° descending from the cliff top parallel to or along the gully marking an eroded Cenozoic dolerite dyke (Fig. 15).

Locality 2.3 (a and b) is one of several places throughout the North Mayo Dalradian (Max 1970; Max et al. 1992) displaying field evidence for the intrusion of basic dykes (now metadolerites) before any tectonic deformation had affected the Dalradian host rocks. These dykes are not directly isotopically dated but they are considered to be the magmatic products of extension that heralded the birth of the Iapetus ocean. Correlation with a sill cutting Appin Group Dalradian rocks in Donegal suggests an age of *c.* 590 Ma (Daly, Whitehouse, Menuge and Kirwan unpublished).

At Loc. 2.3a, the metadolerite dyke is subparallel to the bedding in the Doonamoo Formation psammite. In addition to the metadolerites there are several NW–SE trending dolerite dykes at this locality. These are characterised by chilled margins, igneous textures, including flow banding defined by concentrations of plagioclase and lack of foliation and occasional development of micro-amygdales and conspicuous spheroidal weathering. They are presumed to be of Cenozoic age and related to the Porth na Sceitheach sub-swarm of the Killala dyke swarm (Mohr 1987), dated at *c.* 58 Ma (Thompson 1985).

Fig. 15 Cenozoic dyke at Loc. 2.3a (54.26921, −10.0761) cutting the contact between a pre-Grampian metadolerite (below) and psammite (above) of the Doonamoo Formation (Dalradian). Hammer is 40 cm long

Exercises

(1) Both the metadolerite and psammite are foliated. Compare their orientations and comment.
(2) Examine the dolerite contacts searching for brecciated country rocks. Are the breccia fragments cemented? What material forms the breccia matrix?
(3) Describe and account for the xenoliths. Do these resemble the country rocks or are they exotic? Compare their shape with that of the breccia fragments along the contacts.

5.4.2 Locality 2.3b: Aughernagalliagh Dyke (54.27496°, −10.06978°)

Access and other information From the car park, walk northwards until the road turns east-west, passing the concrete bollards (designed to protect the road from coastal erosion). Close to the CCTV mast (presumably placed to monitor coastal erosion rather than detecting geologists), cross the fence and the stream and continue uphill to cross the second fence via the gate. Walk westwards until outcrop is reached, then head northwestwards parallel to the strike. Turn north-northeastwards at 54.272375°, −10.070844° and skirt the coast line along the cliff top and past the deep gully towards the dyke locality (Fig. 16) at 54.27496°, −10.06978°

Loc. 2.3b is located at the upper (hanging wall) contact of a NW–SE trending metadolerite dyke that forms the cliff edge close to 54.27496°, −10.06978°. The dyke is strongly discordant to the bedding in the Doonamoo Formation psammite

Fig. 16 A youthful Michael Flowerdew (then a PhD student at UCD) seated on a pre-Grampian metadolerite dyke at Loc. 2.3b. The dyke cuts the bedding in the adjacent Doonamoo Formation psammite but shares its deformational fabric and hence is interpreted as a syn-depositional intrusion related to the opening of the Iapetus Ocean. Camera faces east southeast

(Fig. 16). The relationship between the dyke and the foliation affecting it and its host rock is critical.

The authors correlate these metadolerites with those that cut the Annagh Gneiss Complex and which can be seen to act as clear structural markers at Locs 2.1 and 2.5.

Exercise
Carefully examine the foliation affecting the metadolerite and the psammite (Fig. 16). Is there any evidence of an earlier fabric affecting the psammite? Do you agree that it passes into the surrounding psammites and that this is the first fabric-forming deformational event affecting them?

5.5 *Locality 2.4: Grampian Group Metasediments (54.287520°, −10.026507°) and Contact with Cross Point Gneiss (54.20858°, −10.08410°)*

Access and other information Drive slowly downslope on the gravel-surfaced road beyond the end of the tarmac, open and close the gate, and park (at 54.283000°, −10.025195°) just before the track crosses the bridge over a small stream. Be sure to close the gate securely and ensure that no sheep escape while it is open. To the northwest, metadolerite sheets cutting pink Cross Point gneisses of the northern basement inlier can be seen in the distance along the coast. The gneisses here have not been dated - they were assigned to the Cross Point gneiss based on Sm–Nd data.

5.5.1 Locality 2.4a: Spinkadoon (54.287520°, –10.026507°)

Along the foreshore to the north of the mouth of the stream, well-exposed Grampian Group psammites contain occasional pebbly beds (Fig. 17). These have been investigated by McAteer et al. (2010), who dated several of the clasts.

Exercises

(1) Identify the clast compositions and comment on their provenance
(2) In several places, where the strain is lower, the way-up can be determined (using grading, cross-lamination and hummocky cross-stratification). From the way up and observing the main flat-lying foliation (Fig. 18), work out the facing direction and infer the nature of the major structure.
(3) Some pelite beds contain dark porphyroblasts with light-coloured rims. What are these and what reaction do they suggest?

Fig. 17 Pebbly semi-pelite bed within the Doonamoo Formation at Loc. 2.4a. Note that the base of the pebbly bed truncates the heavy mineral banding in the underlying psammite. Penknife is 11 cm long

Fig. 18 Although staurolite-kyanite grade metamorphism has been reported from here (Yardley 1980), the psammites are non-diagnostic and the pelites appear to be retrogressed chlorite-biotite-muscovite schists. Importantly the subhorizontal orientation of the foliation, readily observed in the pelitic schists at Loc. 2.4a, is key to understanding the overall structure. At first this foliation may be difficult to see in the psammites. Hammer head is 17.5 cm long

5.5.2 Loc. 2.4b: Spinkadoon (54.287520°, −10.026507°)

Access and other information The cliffs are particularly dangerous at this locality and farther north, within the gneisses, there are unfenced blowholes! <u>Only</u> attempt this locality in good weather. From the parking place, walk uphill along the road and cross the fence via the gate and examine the complicated contact with the gneisses in the next small headland, starting in Cross Point gneiss at (54.28775°, −10.02711°).

Here, slivers of gneiss are interleaved with psammite in a c. 100 m wide mylonitic shear zone (Fig. 19a) with well-developed upright isoclinal folds which die out away from the contact. This shear zone also affects one of the late metadolerites (Fig. 19b). Outside the shear zone to the north many of the features evident in the gneisses at Cross Point (Loc. 2.5) may be recognised, although there is less amphibolite.

5.6 Locality 2.5: Cross Point (54.20858°, −10.08410°)

Access and other information Park on the east side of the road by the graveyard wall at Cross Point (southwards from 54.208834°, −10.082660°). Starting where the road meets the shore, walk westwards onto good outcrop and then walk clockwise around the graveyard along the wave-cut platform in the general north-eastward direction of dip of the layering. This locality is partially covered at high tide but the essentials are not tide-dependent. The far northwestern exposures (54.210381°,

Fig. 19 a Photograph (taken at 54.287456°, −10.026337° looking west) of part of a c. 100 m wide late Grampian (? D3) shear zone (at Loc. 2.4b) affecting Doonamoo Formation (Dalradian) psammites and pre-Grampian metadolerite dykes (to the left) and Cross Point gneiss (to the right on the higher ground, e.g., at 54.28775°, −10.02711°). Field of view is c. 60 m wide **b** Late Grampian (presumably F3) asymmetrical (S) isoclinal folds of pre-Grampian metadolerite dykes at Loc. 2.4b that also affect the enclosing Dalradian psammites of the Doonamoo Formation (Erris = Grampian Group) within a c. 100 m wide shear zone at the contact with the Cross Point gneiss (to the left of the photograph, taken at 54.20747°, −10.02682° and looking to the east-southeast). Hammer is 40 cm long

Fig. 20 Layered basic body at Cross Point (54.209413°, −10.083290°, Loc. 2.5). The reef in the distance to the west, accessible only at low tide, displays particularly clear field relationships between the Cross Point gneiss, discordant granite pegmatites (dated here at 984 ± 6 Ma, Daly 1996, 2009 and a Grenvillian foliation dated at c. 960 Ma (Table 1) and is illustrated in Fig. 8. Hammer is 40 cm long

−10.085256°), which are only accessible at low tide, should only be visited under calm sea conditions and should never be visited unaccompanied.

Felsic bands (granitoid and syenitic) from this locality yield Sm–Nd model ages from 1804 to 1878 Ma, typical of the *c.* 1270 Ma old Cross Point gneiss. These are interleaved on a scale of metres to centimetres with mafic bands. The mafic component also occurs as enclaves within the felsic bands. All of these lithologies display migmatitic leucosomes on various scales. Several intrafolial isoclinal folds deform the layering and are cut by granite pegmatites and later by unmigmatized metadolerites.

At low tide, on the farthest point north-westwards (54.210381°, −10.085256°; Fig. 3), the structural relationships between the Cross Point gneiss, discordant granitic pegmatites and Grenvillian deformation fabrics can be examined. These have yielded U–Pb zircon (Table 1) and titanite data (Daly and Flowerdew 2005), which provide constraints on the chronology of Grenvillian events. Cross Point is unusual because the late Grenvillian (D_G3) foliation here is sometimes highly oblique to the banding in the gneisses (Fig. 8).

A large migmatized early metabasite body (54.209413°, −10.083290°) is present close to the northern end of the graveyard. This rock is petrographically distinct in exhibiting magmatic layering (Fig. 20). It is readily distinguished from later (?) cross-cutting metadolerites at this locality but its contacts are not exposed and hence its field relationships are unclear. The trondhjemitic gneiss, from which Aftalion and Max (1987) obtained imprecise U–Pb data, was collected from this body.

Several post-Grenvillian metabasite dykes are present. With luck, depending on the distribution of beach boulders and seaweed banks, their contact relationships are sometimes clearly displayed (Fig. 21).

Exercises

(1) Compare the lithologies and their relative abundances at this locality with those at Loc. 2.1a, noting the similarities and differences.

(2) Carefully examine the relationships between the mafic and felsic bands and the mafic enclaves. What do these relationships suggest? Are they consistent with the Sm–Nd (Fig. 6) and zircon Lu–Hf (Fig. 7) data?

(3) A strong lineation lies within the plane of the layering and generally plunges shallowly east-southeastwards. Does it affect all of the lithologies. The authors interpret its age as Grenvillian based on its relationship with the late metadolerites. Do you agree?

Fig. 21 Discordant post-Grenvillian metadolerite dyke cutting *c.* 1270 Ma Cross Point gneiss and Grenvillian foliation (Loc. 2.5), itself affected by a Grampian foliation. Hornblende defining this foliation in a similar metadolerite at Cross Point yielded a 474.5 ± 4.4 Ma ^{40}Ar-^{39}Ar plateau age (Daly and Flowerdew 2005). Hammer is 40 cm long

Fig. 22 Grenvillian F_G2 folds affecting Doolough gneiss at Loc. 2.6a (54.141800°, -9.957290°). Hammer head is 7.5 cm long

5.7 Locality 2.6: Doolough Gneiss and Peralkaline Granite (54.141467°, -9.9524546°)

Access and other information Parking is available off road where the road to Doolough passes closest to the sea at (54.141467°, -9.9524546°). Alas, history shows

Fig. 23 Ghosted remnants of $F_G 2$ folds (e.g., to the right and below the lens cap) in a semi-digested xenolith of Doolough gneiss within the Doolough Granite at 54.143453°, -9.9601078°. Camera lens is 5.5 cm in diameter

that this is an ephemeral carpark. A quick glance along the beach reveals the remnants of previous concrete restraining walls. It is clear that a revetment would be a more appropriate coastal defence here.

5.7.1 Loc. 2.6a: Doolough Gneiss (54.142309°, -9.956758°)

Access and other information From the carpark walk westwards as far as the fork in the road and walk across the boulder-strewn wave-cut platform to Locality 2.6a, which is best visited at low tide.

Examine wave-cut platform and beach exposures of *c.* 1180 *Ma* Doolough gneiss and the *c.* 1015 Ma intrusive Doolough peralkaline granite (Table 1). Mafic and felsic varieties of Doolough gneiss are inter-banded on a decimetric scale. High strain has transposed this layering into parallelism prior to folding into tight to isoclinal folds with very variable plunge directions (Figs. 5, 22). The mafic bands are readily distinguished from the later metadolerites that cut the Grenvillian deformation. Several generations of ductile structures and minor intrusives can be distinguished.

Returning towards the beach fresh exposures of migmatitic gneiss are sometimes accessible depending on recent storm movement of boulders and beach shingle. Many of the leucosomes are garnet-bearing or otherwise display stictolithic (flecky) textures, suggesting that the principle mechanism of formation was dehydration melting.

> **Exercise**
> Describe the folds (Figs. 5 and 22) and comment on their relative structural age and genesis. Are these refolded folds or sheath folds?

5.7.2 Locality 2.6b: Doolough (Peralkaline) Granite (54.143460°, -9.960351°)

Access and other information This locality can be accessed by walking along the high water mark parallel to the coast.

Moving northwestwards along the coast, exposures of the Doolough gneiss are sometimes covered in storm debris. However even with good exposure, the gneissic banding seems gradually to disappear. Eventually a fine-grained pink to orange-weathering homogeneous magnetite-bearing granite is encountered whose contacts are very difficult to place precisely. After several traverses, it becomes apparent that some exposures of the gneiss are actually ghosted remnants of metasomatized xenoliths of Doolough gneiss. Importantly several of these take the form of steeply plunging folds that deform remnant centimetric to decimetric gneissose banding (Fig. 23). Moreover the composition of the xenoliths overrepresents the mafic component of the Doolough gneiss, owing to its resistance to alteration.

Aftalion and Max (1987) obtained a U–Pb zircon age of 1093 ± 7 Ma from the weakly deformed Doolough granite at this locality. This result differs markedly from the preferred age of 1015 Ma (Daly 1996, and see above) and demonstrates the pitfalls inherent in U–Pb analysis of large (multi-grain) zircon fractions.

Sm–Nd model ages from the Doolough Granite vary from 1450 to 1758 Ma (Fig. 6) and tend to be older towards the northern contact. North of the Doolough Granite, highly strained gneisses are encountered. Based on their chemistry, Sm–Nd model ages (c. 2.0 Ga) and the occasional occurrence of K-feldspar megacrysts similar to those at Annagh Head, these rocks have been assigned to the Mullet gneiss. It is suggested that the Doolough Granite stitches a tectonic contact between the Mullet and Doolough gneisses. Contamination by both can explain the range of Sm–Nd and Lu–Hf data from the Doolough Granite (Figs. 6, 7).

> **Exercise**
> Attempt to locate the contact and search for semi-digested xenoliths. From their structure, what do you conclude about the relative age of the Doolough Granite and Doolough gneiss.

5.8 Description of Localities on Achill Island

The Ordnance Survey 1:50,000 Discovery Series Map 30 covers all of Achill Island. Achill is designated as a Gaeltacht (Irish-speaking) area and some road signage is only in Irish. Access is generally not an issue but care should be taken on the Atlantic cliff exposures. Some outcrops are tide dependent.

The island of Achill and the small islet of Achillbeg offer a superb structural cross section through the Grampian orogenic belt, from south of the basement gneisses of the Mullet peninsula (Figs. 1, 2, 28, 38) through to the contact with the accretionary complex rocks of the Clew Bay Complex. In addition, the south Achill stratigraphy (Fig. 28) records the transition from rifting to hyperextension of the Laurentian margin during the opening of the Iapetus Ocean in the late Neoproterozoic. These hyperextended Laurentian margin rocks are also the only portion of the Dalradian Supergroup that has experienced high-pressure, low-temperature (blueschist-faces) metamorphism which likely developed on the leading edge of the subducting Laurentian plate during early Grampian collision. While the Atlantic coastal exposure is excellent, inland exposure is sparse, and there remain difficulties in correlating the stratigraphy from west Achill to south Achill. Localities 2.7 to 2.9 are in West Achill, 2.10–2.17 in South Achill and 2.18–2.24 on the small islet of Achillbeg (Fig. 1).

5.9 Locality 2.7: Keem Bay (53.96957°, −10.19019°)

Access and other information Low tide is important for this locality, but with a bit of scrambling over coastal exposures and boulders the key outcrops can be accessed two hours either side of low tide. Keem Bay is very picturesque and the lowermost car park at the beach (53.96715°, −10.19527°) is often full in summer. Temporary toilets were available at the lower car park in July 2021. There is a permanent toilet at the upper car park (53.96950°, −10.19406°). Walk down to the beach and head for the secluded coves at the northeast of the beach (53.96957°, −10.19019°). This locality examines pebble provenance and the Grampian structural sequence.

A deformed conglomerate unit (the Keem Conglomerate of the North Achill Dalradian Succession) at localities 2.7 and 2.8 offers a glimpse of the local basement at the time of its deposition. Sedimentary structures in the Keem Conglomerate Formation imply that this succession is inverted. The clast content of this conglomerate unit is very diverse. The majority are vein quartz and quartzite, but there are abundant K-feldspar clasts and black magnetite-quartz clasts. Less common clasts are coarse quartz porphyry and rare sedimentary and metamorphic clasts. The source is similar in age (*c.* 1.8 Ga) to exposed areas of Palaeoproterozoic basement in Scotland and NW Ireland but is so different petrologically and chemically as to be ruled out as a plausible source area (Scanlon 1998). The black magnetite rich pebbles have chemical and petrological characteristics similar to a Banded Iron Formation. The beach sands, particularly when reworked by streams, contain abundant heavy

Fig. 24 F2 folds in pelitic schists at Keem Bay (Loc. 2.7; 53.96850°, −10.19216°) with axial planes dipping shallowly to the NE. Photo taken looking east-southeast, rucksack is 45 cm high

mineral bands which are derived from the Keem conglomerate and adjacent schists. The prolate pebble shape fabric and F2 fold axes (Fig. 24) both plunge *c.* 20° towards 080. Bedding typically strikes NW–SE and dips *c.* 35° towards the NE. The axial planes of F2 folds and the associated S2 crenulation cleavage strike NW–SE and dip *c.* 20° towards the NE.

Exercise
Determine way-up in the rocks (e.g. grading, channel geometries) and facing directions on the main (S1 and S2) fabrics

5.10 Locality 2.8: Rusheen Point (53.96573°, −10.15524°)

Access and other information Park at layby on north side of the road (53.96915°, −10.15662°). A larger layby suitable for a coach is available 200 m to the west (53.96949°, −10.15981°). Although not as scenic as Keem Bay, this outcrop is better for larger groups and is easier to access as it is not tide dependent. Walk down the slope from the car parking spot, passing the '59 Eire' sign (53.96725°, −10.15622°). This is a World War II lookout post, of which 82 were constructed primarily along Ireland's west coast from Cork to Donegal. These large makeshift markers were made from stone and painted white with the words ÉIRE and the identifying number of the local lookout post, and informed overflying military aircraft of the neutrality of the land below. This locality examines pebble provenance and the Grampian structural sequence.

This locality is one of the best places in Achill Island for establishing the structural deformation sequence, as the various deformation sequences (D1, D2 and D3) are orientated at relatively high angles to each other. The prolate pebble shape fabric (Fig. 25a) plunges c. 10° towards 080. Bedding (and the bedding-parallel S1 schistosity) in the eastern part of the outcrop strikes E-W and dips c. 55° N. The effects of D2 deformation are relatively weak, and the S2 crenulation cleavage strikes E-W and dips c. 35° S. Late, E-W trending F3 folds (Fig. 25b) with a wavelength of 5 to 10 cm and vertical to sub-vertical axial planes are also visible and are associated with a locally developed S3 crenulation cleavage.

Exercise

Determine way-up in the rocks (e.g. grading, channel geometries) and facing directions on the main (S1 and S2) fabrics, and reconstruct the deformation chronology at this outcrop.

Fig. 25 A prominent prolate pebble shape fabric in E-W striking beds of the Keem Conglomerate at Rusheen Point (Loc. 2.8; 53.96725°, −10.1562°). Notebook is 16 cm long. **b** E-W striking, sub-vertical F3 folds and associated S3 crenulation cleavage in metasandstone beds of the Keem Conglomerate at Rusheen Point (Loc. 2.8). Photo taken looking east; pen is 13 cm long

5.11 Locality 2.9: Doogort, Boulder Bed Formation (54.01434°, −10.01769°)

Access and other information park either above the beach (54.01280°, −10.01909°) or use the main beach car park. The outcrops are accessible about 90 min either side of low tide by scrambling along the loose boulders on the coastline. These boulders are extremely slippery when wet.

This locality, which examines Cryogenian glacial deposits, is very important for stratigraphic correlation purposes within the Dalradian Supergroup. The cobbles of granite embedded in a grey (meta-)siltstone matrix (diamictite) of the Doogort Boulder Bed (Fig. 26) have been interpreted as an ice-rafted glacial deposit, equivalent to the Port Askaig Tillite Formation that can be traced throughout the Dalradian Supergroup of Scotland and Ireland (Fig. 9). Unfortunately in North Achill, the Doogort Boulder Bed is bounded by two faults/shear zones and thus does not significantly aid regional correlation of the Achill Dalradian stratigraphy. The succession is inverted, with bedding striking N-S and dipping *c.* 15° E, and hence the outcrops on the foreshore between the boulders (a series of pelites, quartzites and a pale orange dolomite bed (54.01442°, −10.01788°) are thus younger than the outcrops of diamictite in the cliff. The cobbles (termed "stones") at this locality are nearly entirely comprised of granitic crystalline basement and are most easily seen in the loose boulders on the foreshore. No stones of carbonate rock (characteristic of the lower portions of the Port Askaig Tillite Formation) are seen, and the base of the formation is inaccessible.

Fig. 26 Highly deformed granitic "stones" in a loose boulder at Doogort (Loc. 2.9). Hand lens is 8 cm long

5.12 Locality 2.10: Minaun Heights View Point (53.95743°, −10.02723°)

Access and other information This is primarily a scenic stop. Park at the car park beside the transmitter (53.95750°, −10.02651°) and walk *c.* 50 *m* west to the sparse outcrops (53.95743°, −10.02723°). A path past the transmitter leads southwest towards the cairn (a distance of *c.* 1250 m) and affords nice views of South Achill and Clare Island in fair weather.

At this locality, green-white quartzites, green meta-sandstones and chlorite-albite schists strike NE-SW and dip 55° SE. Heavy mineral laminae in the quartzites define foresets which imply bedding is right way-up and younging south. Further to the south (localities 2.11 onwards), bedding steepens as the major, recumbent, southward-facing F2 fold is progressively "downbent" approaching South Achill.

5.13 Locality 2.11: Dooega (53.91795°, −10.01905°)

Access and other information Park by the north end of inlet (53.92122°, −10.01823°). Walk to the east end of the beach and then walk south-southwest for 300 m along the small cliff top above the wavecut platform (53.91795°, −10.01905°). The plant growing at the base of the cliff is Chilean rhubarb (a species of *Gunnera*), which along with Rhododendron and Fuchsia are invasive species that thrive in the humid climate and peat-rich soils of western Ireland.

This locality is important for a number of reasons. Way up is abundant (Fig. 27) in these Appin Group (Fig. 9) quartzites and meta-sandstones. Bedding is right way-up and striking NE-SW and dipping 50° SE similar to the previous locality (2.10), but the metasedimentary rocks are typically too quartz-rich to pick up a schistosity. The laminae which define the sedimentary structures (both planar and trough cross bedding) are composed of heavy minerals including zircon and magnetite. Modern sands immediately offshore are also rich in heavy minerals and locally form placer deposits.

Exercise
Determine way-up at this locality. Is it consistently to the south-east?

Fig. 27 Cross-bedding in quartzites at Dooega (Loc. 2.11) indicating way-up to the southeast. Photo taken looking southeast; compass clinometer is 10 cm long

5.14 Locality 2.12: Ashleam Bay (53.90408°, -9.99333°)

Access and other information Park at the head land above the cove (53.90408°, -9.99333°), then head south to the picnic tables by the steel fence (53.90327°, -9.99253°) and follow the path down. Where the path crosses the culverted stream, walk *c.* 50 m NE along the stream bank. Quaternary varve deposits are visible high up on the SE bank (53.90275°, -9.99096°). Then follow the stream down to the beach.

A detailed 1:2,500 map of South Achill and Achill Beg (Chew 2005) can be used to accompany visits to localities 2.12–2.24. A summary map of South Achill and Achillbeg is presented in Fig. 28 and of Achillbeg in Fig. 38. The headlands which surround Ashleam Bay are composed of quartzite, which are resistant to erosion compared to the schists that form the core of the bay and which define a prominent valley extending to the eastern coast of South Achill. The Quaternary varve deposits seen in the stream above the beach presumably formed in a proglacial lake within this topographic depression. Pelitic layers within the Appin Group quartzites on the north side of the bay contain abundant tourmaline. The central portion of the bay is comprised of thin units of quartzite, grit, schist and sulphurous graphitic schist. A curious limonite-stained breccia (Fig. 29) crops out at the southern end of the bay (53.90192°, -9.99334°). The quartzites at the southern end of the bay are interpreted as a correlative of the Islay Quartzite (Fig. 9), meaning that the Port Askaig Tillite Formation equivalent is absent. However late shearing and faulting are pervasive within Ashleam Bay, and it is possible the Port Askaig Tillite Formation equivalent has been tectonically excised. At localities 2.10 and 2.11, bedding strikes NE-SW and dips at a moderate angle to the southeast. In Ashleam Bay, bedding has swung

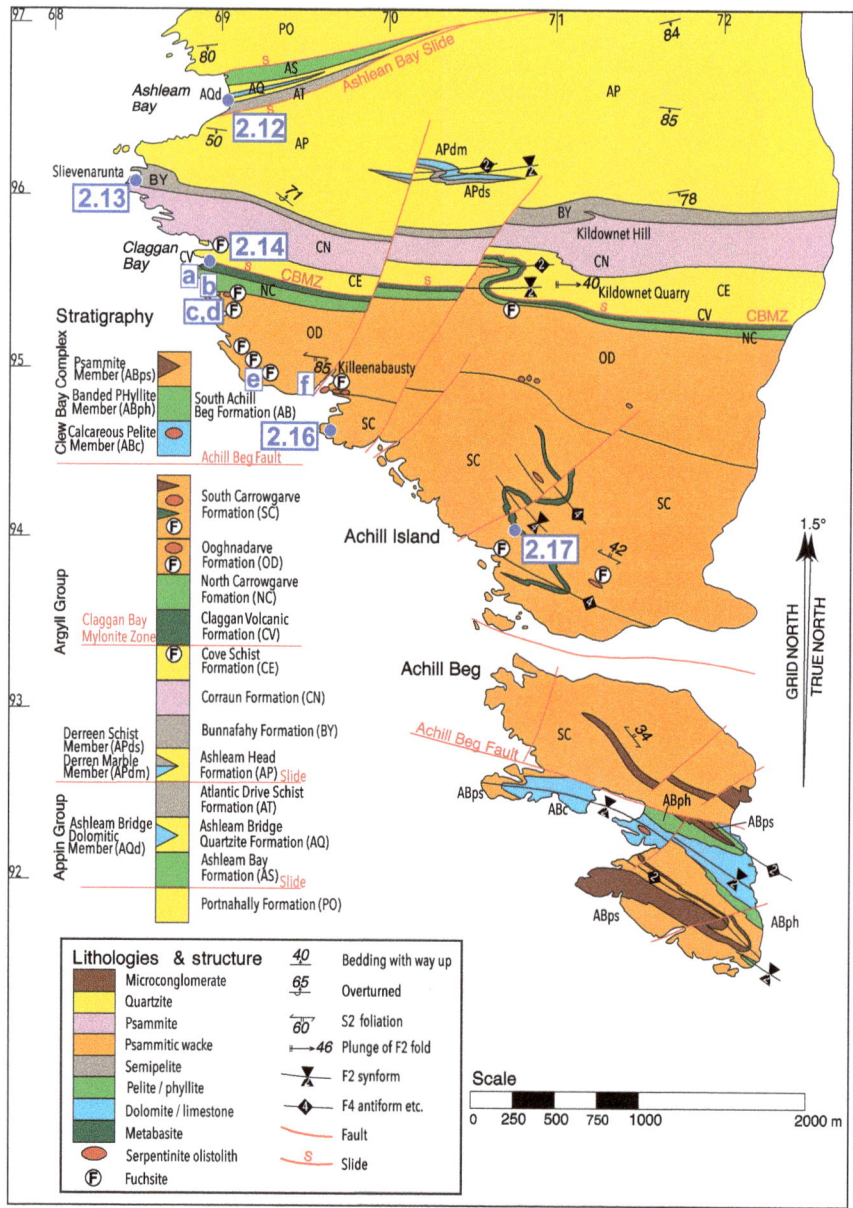

Fig. 28 Geology of south Achill and Achillbeg (from Chew 2005). Boxed blue letters (a-f) refer to locality 2.15

Fig. 29 Limonite-stained breccia at Loc. 2.12 (53.90192°, -9.99334°). Photo taken looking south; compass clinometer is 10 cm long

around to an E-W orientation and dips steeply to the south. This regional change in strike and dip at the southern margin of the North Mayo inlier in the vicinity of the north shore of Clew Bay is evident on the map and cross section in Figs. 1 and 2, and is due to late shearing.

Exercise
Speculate on the origin of the limonite-stained breccia. Is the brecciation sedimentary or tectonic?

5.15 Locality 2.13: Bunnafahy Conglomerate and Nearby F2 Folding (53.89541°, -9.99889°)

Access and other information Follow the road south from Ashleam Bay, and park at the layby (53.89867°, -9.99702°). Walk *c.* 350 m west-southwest to the cliff top exposures (53.89746°, −10.00181°). The cliffs at locality 2.13 are dangerous and extreme caution is required.

This unit consists of fragmented quartzite blocks embedded in a pebbly, psammitic (and often calcareous) matrix. A thin, pure-white quartzite bed occurs at the top of the formation and can be followed from the west coast eastwards. The fragmentation of the quartzite blocks within the psammitic matrix has been regarded as either tectonic (Morley 1966; Phillips 1974) or soft-sedimentary (Max 1974) in origin. A tectonic origin is thought to be more likely, as the elongate quartzite blocks (which commonly have aspect ratios greater than 10:1) embedded in the psammitic matrix may represent D2 boudins (Fig. 30). In addition, sedimentary lamination is well preserved in some of the larger quartzite blocks suggesting the blocks were lithified prior to disruption and arguing against a soft-sedimentary origin. A thin (30 cm) chlorite-albite schist a few metres to the south may represent the first episode of basic volcanism seen in the Argyll Group rocks of South Achill.

Walk east-southeast along the cliff path to the iron age hill top fort of Dun Bunnafahy (53.89541°, -9.99889°). *En route*, a spectacular vertical rib of white quartzite is visible, and likely formed as a result of the surrounding graphitic schists being eroded away. The diverse suite of lithologies at this outcrop (quartzites, meta-sandstones, graphitic schists) exhibit high competence contrasts and are heavily folded. In places, "pods" of metasandstone are embedded in a pelitic matrix, which plunge 45° E and are parallel to the F2 fold axes. However, the plunge is remarkably consistent at this locality, with no evidence of any plunge reversals implying that non-cylindrical ("sheath") folding is unlikely. Closer inspection suggests that these elongate pods are thickened F2 fold hinges, with extremely attenuated limbs and

Fig. 30 East-plunging quartzite rods in the weathered meta-sandstone matrix of the Bunnafahy Conglomerate (Loc. 2.13; 53.89746°, -10.00181°). Small, asymmetric isoclinal fold closures are visible above the compass clinometer. Photo taken looking east; compass clinometer is 10 cm long

may be considered as D2 boudins - the eastward-plunging quartzite blocks of the Bunnafahy Formation may well be of a similar origin.

> **Exercise**
> Discuss the origin of the rod-like quartzite blocks in the Bunnafahy Conglomerate. Is it pre-tectonic or tectonic in origin?

5.16 Locality 2.14: Claggan Bay (53.89290°, -9.99453°) - the Achill Blueschist

Access and other information Park at the end of the stone-metalled track by the sea (53.89290°, -9.99453°). Note there is only space for about four cars. Claggan Bay is a tide dependent locality (ideally 90 min either side of low tide).

The headland at the north end of the bay is formed by a vertical wall of orange-weathering quartzites. Interbedded with these quartzites at the northern end of the shore are black graphitic schists. Some of these graphitic schists on the wavecut platform contain a bright green mineral known as fuchsite, which is a chromium-rich mica which grows when sediment is rich in material derived from ultramafic igneous rocks. The middle of the bay is formed by highly strained quartzites. This shear zone (known as the Claggan Bay Mylonite Zone) extends from west to east across South Achill. The sheared quartzites can also be examined where the cliff of glacial till has been eroded back (53.89376°, -9.99366°). Sparse 1–3 mm pebbles of blue quartz can be seen, and serial sectioning of pebble-bearing blocks shows that the quartzites have experienced oblate strains. On the east side of the island, the quartzite is quarried to use as paving stones due its thin, flaggy appearance. South of the shear zone, and best seen at low tide, is a package of basic metavolcanic rocks. Large round pods (53.89297°, -9.99506°), comprised mainly of epidote (Fig. 31), have been interpreted as being volcanic bombs, and contain relict igneous textures. Inside these pods of epidote, glaucophane has been found (Gray and Yardley 1979) and is indicative of high-pressure, low-temperature metamorphic conditions. South Achill is the only occurrence of blueschist metamorphism within the Dalradian Supergroup.

> **Exercise**
> Discuss the origin of the epidote-rich blocks.

Fig. 31 Epidote-rich pod wrapped by a basic metavolcanic matrix at Loc. 2.14 (53.89297, -9.99506). Photo taken looking west; field notebook is 16 cm long

5.17 Locality 2.15: Headland South of Claggan Bay to the Winch West of Killeenabausty

Access and other information The following itinerary involves walking approximately 900 m SW along the coast, starting at the Claggan Bay parking stop (53.89290°, -9.99453°) and encompasses several important outcrops *en route*. Return to Claggan Bay after locality (f).

5.17.1 Locality 2.15a: Chloritoid Schists (53.89279°, -9.99490°)

The headland immediately south of Claggan Bay is comprised of mica schists with a vertical foliation. Close to the contact with the Claggan Volcanics, these mica schists contain a conspicuous dark green mineral known as chloritoid. This mineral grows only in rocks with a high aluminium content. This horizon is useful for correlation within the Irish Dalradian and has been identified by Geological Survey Ireland as occurring stratigraphically above basic metavolcanics in the Easdale Subgroup rocks (Argyll Group) in several other Dalradian inliers in Ireland.

5.17.2 Locality 2.15b: The Refolded Fold (53.89233°, -9.99393°)

Here is a beautiful example of a refolded fold (Fig. 32). This second generation of folds has an axial planar foliation, which is the main S2 foliation seen in this region. These second-generation folds also fold the flaggy strained quartzites, suggesting the Claggan Bay Mylonite Zone is an earlier structure. Also visible at this locality are late S3 crenulations.

5.17.3 Locality 2.15c: Talc Sheet (53.89181°, -9.99299°)

This white soapy sheet of rock is made of talc (Fig. 33). Talc, like fuchsite, grows in metamorphic rocks due to the breakdown of ultramafic minerals. Throughout the black graphitic schists south of this locality, fuchsite is relatively common. Their origin is discussed in locality 2.15f.

Exercise
Identify fuchsite in the black graphitic schists at this cove.

5.17.4 Locality 2.15d: Late Shearing (53.89177°, -9.99324°)

Access and other information This small cove is protected by a wavecut platform, and the talc sheet at (c) is found on the north side of this cobble strewn beach. Underneath the cobbles are a series of wave-polished exposures with excellent examples of late shearing, but the exposure varies from year to year as a result of Atlantic winter storms.

This cove is the best place to see the structures associated with late shearing in South Achill. Unlike the Claggan Bay Mylonite zone (CBMZ) which is earlier than the main foliation, this shearing is later than the main foliation. Chew (2003) attributes the gradual steepening of bedding in south Achill (localities 2.11 onwards) to late shearing which progressively "downbends" the major, recumbent, southward-facing F2 fold (Fig. 2).

When a phase of ductile shear is superimposed on an already foliated rock, to preserve a simple shear deformation path, crenulations develop to compensate for the displacement component of foliation slip normal to the shear zone wall (Dennis and Secor 1987). In a dextral shear zone, when the pre-existing foliation is at an acute, clockwise angle to the shear zone wall, movement away from the shear zone wall due to foliation slip is compensated by reverse-slip crenulations (RSC) which transfer slip up to "higher" foliation planes (Fig. 34a); when the slipping foliation is

Fig. 32 Refold of an early F1 fold by an F2 fold at Loc. 2.15b (53.89233°, -9.99393°). Photo taken looking east; compass clinometer is 10 cm long

Fig. 33 Talc pod (Loc. 2.15c; 53.89181°, -9.99299°), partly covered by cobbles of red, unmetamorphosed sandstone. Photo taken looking south; compass clinometer is 10 cm long

at an acute anticlockwise angle to the shear zone wall movement normal to the shear zone wall is compensated by normal-slip crenulations (NSC) (Fig. 34b; Dennis and Secor 1987). At this locality, the late shearing clearly affects the main S2 nappe fabric (Fig. 34). Two sets of structures cut the main foliation—F3 crenulations (RSC) and shear bands (NSC) (Chew 2003). This dextral shear zone affects all the rocks from Ashleam Bay southwards, but is particularly well developed in the graphitic schists.

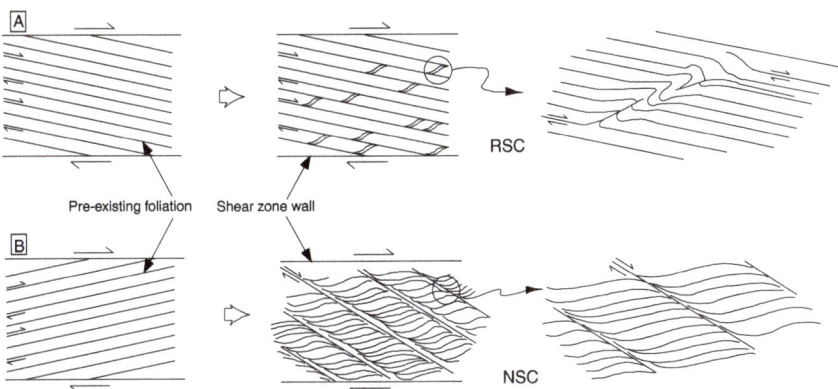

Fig. 34 Angular relationships predicted between the shear zone wall and the foliation-slip and crenulation-slip surfaces, from Dennis and Secor (1987). **a** Reverse-slip crenulations. **b** Normal-slip crenulations

Fig. 35 Normal sense (NSC—top panel) and reverse sense (RSC—bottom panel) affecting the main E-W trending, vertical S2 nappe fabric (Loc. 2.15d; 53.89177°, -9.99324°). The normal sense crenulations are dextral shear bands; the reverse sense crenulations are asymmetric F3 micro-folds with an associated S3 crenulation cleavage. Compass clinometer in both images is 10 cm long

Exercise
The cobbles in this cove are dominated by red, unmetamorphosed sandstone. By consulting a regional geological map, determine a plausible geological source and a likely transport direction.

5.17.5 Locality 2.15e: More Late Shearing (53.88757°, - 9.98852°)

Here is another good example of the late F3 crenulations being affected by dextral shear. The F3 axial planes strike consistently anticlockwise of bedding, and have a consistent 'Z' asymmetry. F3 fold axes plunge moderately to the northeast. The largest F3 folds have wavelengths of only a few metres, and typically the smaller F3 folds can be observed to "root" in the S2 foliation surface. These features are typical of the reverse-slip crenulations of Dennis and Secor (1987). With progressive dextral shear, the S3 foliation which initiates anticlockwise to S2 is progressively brought into parallelism and is particularly evident where there is a large ductility contrast across a bedding surface as seen at this outcrop. Within a relatively rigid psammite bed, S3 is usually oblique to S2. If the S3 foliation continues into an adjacent graphitic pelite layer, then commonly the S3 foliation swings clockwise into parallelism with S2, and hence the ductile graphitic pelite layers are accommodating the bulk of the displacement. The main foliation has been dated from several localities along this coastline section; both Rb–Sr and Ar–Ar techniques give 460 Ma ages for this foliation. The overall disposition of the meta-greywacke and graphitic schist beds seen on this coastal walk are controlled by isoclinal F2 folds (see the 1:2,500 map of Chew 2005).

Exercise Glacial striae trend 140–320° along this coastline and are prominent at this outcrop. Identify the glacial striae, and based on the 'roche moutonnée' geometry of the outcrops along the coastline determine the likely ice transport direction, and link it into your observations of where the cobbles of red sandstone were derived from at the previous locality

5.17.6 Locality 2.15f: Winch and the Serpentinite Mélange (53.88721°, -9.98489°)

Access and other information This winch was used to salvage a boat (the 4,330 ton Aghia Eirini SS cargo steamer) which foundered on the cliffs below on the 10th December 1940 when her steering gear failed *en route* from Cardiff to Buenos Aires with a cargo of coal. From here, looking east is a spectacular view of the serpentinite mélange, with large blocks of serpentinite clearly visible. The path down to the mélange is very dangerous as it is on the edge of a steep cliff and is not recommended.

Like the fuchsite and talc seen in the graphitic schists, serpentinite is another product of metamorphosed ultramafic material. It can be demonstrated (Kennedy 1980; Chew 2001) that the large blocks of serpentinite (Fig. 36) are intruded by the enclosing graphitic schists, and both serpentinite and enclosing schist possess the same set of fabrics. This suggests the serpentinite mélange is a sedimentary mélange that was deformed later in the Grampian orogeny. Serpentinite and related ultramafic

Fig. 36 Serpentinite mélange at Kileenabausty looking east from the winch (Loc. 2.15f; 53.88721°, -9.98489°)

detritus is common throughout the Upper Argyll Group Dalradian of Scotland and Ireland, and hence South Achill is not atypical. A model invoking serpentinization of a zone of upper mantle during a phase of major rifting (hyperextension) in Easdale Subgroup times was presented in Chew (2001) and Chew and van Staal (2014) using the ocean-continent transitions of the present North Atlantic rifted margins as an analogue. Serpentinite protrusions were interpreted to have tapped the zone of inferred serpentinized mantle at shallow depth, generating a horizon rich in ultramafic detritus in the deep marine Easdale Subgroup basins.

5.18 Locality 2.16: Ooghnadarve (53.88495°, -9.98306°)

Access and other information Park by the road (53.88634°, -9.98108°) and walk 200 m south (53.88495°, -9.998306°).

This is probably the best place in South Achill to see the mineral garnet; some of the examples seen in the semi-pelitic schists at this locality are up to 1 cm in diameter. Although garnet is a very common metamorphic mineral, these garnets are unusual in that they contain pyrope-rich cores. These were interpreted as relicts of early eclogite-facies metamorphism (Harris 1993a, b), but later investigation (Yardley et al. 1996) demonstrated that they are more likely to be detrital in origin as in the pyrope-rich cores are unzoned but vary significantly in composition from core to core and ii) have higher X_{Mg} values than any other phase present and were likely never in equilibrium with the bulk rock composition.

5.19 Locality 2.17: Carrowgarve (53.88005°, -9.96616°)

Access and other information Park in the lay-by (53.87889°, -9.996773°) and enter the gate in the field across the road. Walk 160 m northeast up the field; the outcrop straddles the wall on the NW side of the field (53.88005°, -9.96616°).

The hillside is composed almost entirely of grey mica schists (meta-greywackes), but this is one of a few scattered outcrops of metabasite (Fig. 37). The metabasite contains glaucophane, along with albite, epidote and garnet. This assemblage is far more useful for pressure–temperature estimates than the assemblage at Claggan Bay where the glaucophane is isolated within the epidosite pods. Here the high-pressure assemblage can be texturally related to the main deformation fabrics (it is syn-D1 to early D2 in age), and PT estimates suggest a pressure of 10 kbars and a temperature of 450 °C (Chew et al. 2003) which is approximately on the blueschist / greenschist boundary. The conspicuous banding of glaucophane and garnet seen in thin section suggests a volcanic origin (similar to Claggan Bay), although we are now several hundred metres up section.

Fig. 37 Isoclinal F2 fold closure in a laminated metabasite (Loc. 2.17; 53.88005°, -9.96616°) containing glaucophane, garnet, epidote and albite. Photo taken looking northeast; compass clinometer is 10 cm long

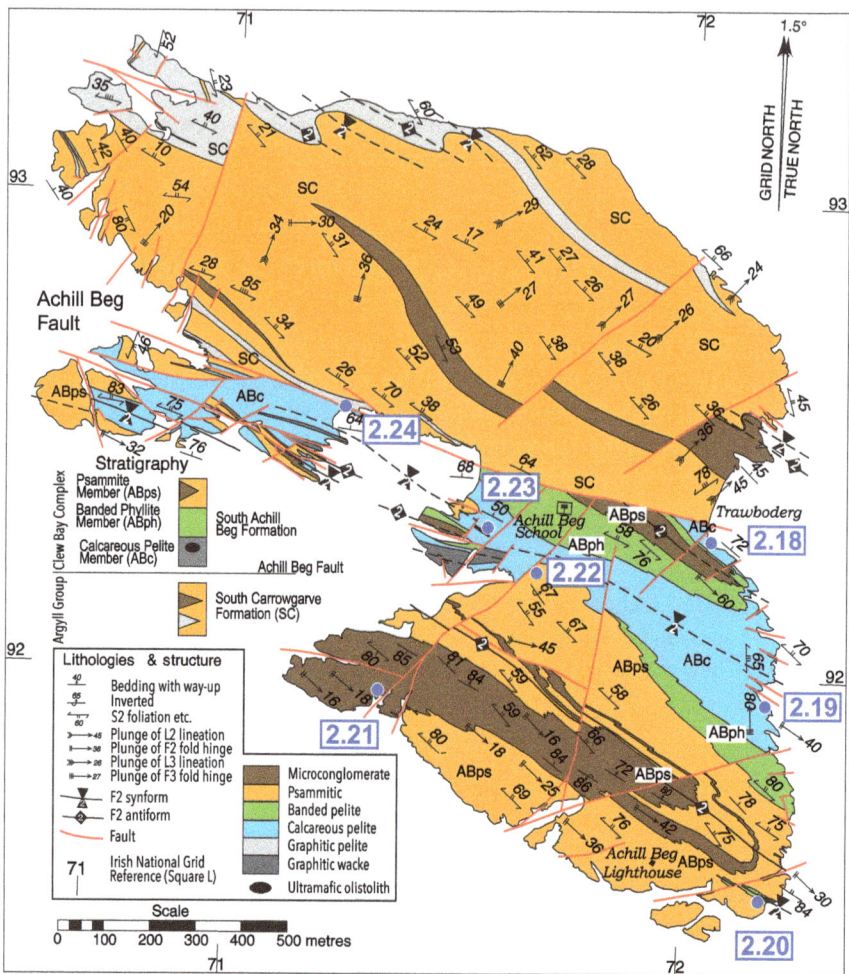

Fig. 38 Geological map of Achillbeg adapted from Chew (2003) and Chew (2005)

5.20 Achillbeg

Access and other information Achillbeg is accessible by hiring a boat from Cloughmore Pier in South Achill to Trawboderg pier (53.86144°, -9.94387°), south of the beach on the east coast of the island. There are no shops or other public facilities on Achillbeg - the island is uninhabited apart from holiday homes in summer although it boasted a population of 178 in 1841. Including the boat journey, this itinerary will take approximately 5 h. It is exclusively devoted to the Clew Bay Complex on the southern part of the island, which offers excellent exposure of both major fold structures and the Achillbeg Fault. These localities were last visited in 2006.

Fig. 39 The Banded Phyllite Member of south Achillbeg at Loc. 2.18. Compass clinometer when open is 18 cm long

Fig. 40 Late crenulations affecting the Calcareous Pelite Member of south Achillbeg at Loc. 2.19. Hammer is 40 cm long

5.21 *Locality 2.18: Trawboderg (53.86450°, -9.94698°)*

This locality shows the three formations which make up the Clew Bay Complex on south Achillbeg (Fig. 38). The youngest rocks are meta-sandstones which in places (such as here) contain beds of microconglomerate. These outcrops are well seen on the south side of the beach and are part of the Psammite member of the South Achillbeg Formation. Also visible are two foliations. The main S2 foliation is nearly vertical, and crenulates an earlier S1 foliation. Structurally below this formation lies the Banded Phyllite Member of the South Achillbeg Formation. Also visible are late F3 folds crenulating the main foliation. These may be related to the late crenulations also seen in South Achill.

5.22 Locality 2.19: Trawboderg Pier (53.86144°, -9.94387°)

This locality is comprised of black slates with interbedded thin orange layers. They are the youngest rocks of the South Achillbeg Formation, and are termed the Calcareous Pelite Member. They are intensely folded, often by late crenulations. This locality is in the core of a major synform. Walking along the coastline, Fig. 38 can be used to see how the folds affect the outcrop pattern of the various members of the Achillbeg Formation.

5.23 Locality 2.20: Southeast Achillbeg (53.85782°, -9.94390°)

A very tight synform is exposed at this locality. In the centre of the gully, the Calcareous Pelite Member is exposed, which is flanked on either side by the banded phyllites. Note how thin the members are here—they have been attenuated on the fold limbs by the intense folding. Then walk 900 m west-northwest parallel to strike along the spectacularly exposed cliffs on the south side of the island, passing by Achillbeg lighthouse (53.85815°, -9.94730°) which was built because the Clare Island lighthouse was often obscured by fog. The south coast is formed of psammite, with occasional microconglomerate bands.

5.24 Locality 2.21: Southwest Achillbeg (53.86120°, -9.95670°)

This region is probably the best locality in South Achillbeg for way-up criteria and grading can be observed in the pebble beds (Fig. 41).

5.25 Locality 2.22: 175 m SSW of Achillbeg School (53.86335°, -9.95179°)

Looking east, there is a well-exposed synform in the rock face.

Fig. 41 Grading in coarse pebbly sandstone of the Psammite Member of south Achillbeg facing down on the S2 cleavage. Hammer is 40 cm long, photo taken 190 m east of locality 2.21 (53.86112°, -9.95386°)

5.26 Locality 2.23: 185 m WSW of Achillbeg School (53.86431°, -9.95350°)

This long (40 m approximately) outcrop is composed entirely of talc—once again the product of metamorphism of ultramafic material. However, unlike South Achill, the field relationships of this are unclear. Its contact with the black pelites is unseen, and there is no ultramafic material in any of the surrounding rocks.

5.27 Locality 2.24: East Side of Dun Kilmore (53.86656°, -9.95848°)

The Achillbeg Fault zone runs through this locality. It is impossible to mark the exact location of the fault, which separates the Dalradian from the Clew Bay Complex, so the term fault zone is more appropriate. Walking southwest for about 40 m, the stratigraphy is recognisable once more as the calcareous pelites of the Clew Bay Complex. Fine grained mica-albite schists from this locality have yielded a 460 Ma Rb–Sr age, identical to the ages obtained from the main S2 fabric in the South Achill Dalradian (Chew et al. 2003).

References

Aftalion M, Max MD (1987) U-Pb zircon geochronology from the Precambrian Annagh Division gneisses and the Termon Granite, NW County Mayo, Ireland. J Geol Soc Lond 144:401–406

Anderton R (1985) Sedimentation and tectonics in the Scottish Dalradian. Scot J Geol 21:407–436

Bluck BJ, Dempster TJ (1991) Exotic metamorphic terranes in the Caledonides: tectonic history of the Dalradian block, Scotland. Geology 19:1133–1136

Cadman AC, Heaman L, Tarney J, Wardle R, Krogh TE (1993) U–Pb geochronology and geochemical variation within two Proterozoic mafic dyke swarms, Labrador. Can J Earth Sci 30:1490–1504

Chew DM (2001) Basement protrusion origin of serpentinite in the Dalradian. Irish J Earth Sci 19:23–35

Chew DM (2003) Structural and stratigraphic relationships across the continuation of the Highland Boundary Fault in western Ireland. Geol Mag 140:73–85

Chew DM (2005) 1:2,500 geological map of south Achill island and Achill beg, western Ireland. J Maps 1:18–29. https://doi.org/10.4113/jom.2005.25

Chew DM, Daly JS, Page LM, Kennedy MJ (2003) Grampian orogenesis and the development of blueschist-facies metamorphism in western Ireland. J Geol Soc Lond 160:911–924

Chew DM, Daly JS, Flowerdew MJ, Kennedy MJ, Page LM (2004) Crenulation-slip development in a Caledonian shear zone in NW Ireland: evidence for a multi-stage movement history. In: Alsop GI, Holdsworth RE, McCaffrey KJW, Hand M (ed.) Flow Processes in Faults and Shear Zones. Geol Soc Lond Spec Pub, 224:337–352

Chew DM, Daly JS, Magna T, Page LM, Kirkland CL, Whitehouse MJ, Lam R (2010) Timing of ophiolite obduction in the Grampian Orogen. Geol Soc Am Bull 122:1787–1799

Chew D, Drost K, Petrus JA (2019) Ultrafast, >50 Hz LA-ICP-MS Spot Analysis Applied to U-Pb Dating of Zircon and other U-bearing minerals. Geostandard Geoanaly Res 43:39–60

Clift PD, Carter A, Draut AE, Van Long H, Chew DM, Schouten HA (2009) Detrital U-Pb zircon dating of lower Ordovician syn-arc-continent collision conglomerates in the Irish Caledonides. Tectonophysics 479:165–174

Chew DM, van Staal CR (2014) The ocean-continent transition zones along the Appalachian-Caledonian margin of Laurentia: examples of large-scale hyperextension during the opening of the Iapetus Ocean. Geosci Can 41:165–184

Daly JS (1996) Pre-Caledonian history of the Annagh Gneiss Complex, north-western Ireland, and correlation with Laurentia-Baltica. Irish J Earth Sci 15:1–21

Daly JS (2009) Precambrian. In Holland CH, Sanders IS (eds). The Geology of Ireland, 2nd edition. Edinburgh: Dunedin Academic Press, pp 7–42

Daly JS, Flowerdew MJ (2005) Grampian and late-Grenville events recorded by mineral geochronology near a basement-cover contact in N. Mayo. Ireland. J Geol Soc Lond 162:163–174

Daly JS, Heaman LM, Fitzgerald RC, Menuge JF, Brewer TS, Morton AC (1995) Age and crustal evolution of crystalline basement in western Ireland and Rockall. In: Croker, PF and Shannon, PM (eds) The Petroleum Geology of Ireland's Offshore Basins. Geol Soc Lond Spec Pub, 93:433–434

Daly JS, Muir RJ, Cliff RA (1991) A precise U-Pb zircon age for the Inishtrahull syenitic gneiss, County Donegal, Ireland. J Geol Soc Lond 148:639–642

Dempster TJ (1985) Uplift patterns and orogenic evolution in the Scottish Dalradian. J Geol Soc Lond 142:111–128

Dennis AJ, Secor DT (1987) A model for the development of crenulation cleavages in shear zones with applications from the southern Appalachian Piedmont. J Struct Geol 9:809–817

Dewey JF, Mange M (1999) Petrography of Ordovician and Silurian sediments in the western Ireland Caledonides: tracers of a short-lived Ordovician continent-arc collision orogeny and the evolution of the Laurentian Appalachian-Caledonian margin. In: Continental Tectonics (edited by MacNiocaill, C. & Ryan, P.D.). Geol Soc Lond Spec Pub 164:55–107

Dewey JF, Ryan PD (2017) Storm, rogue wave, or tsunami origin for megaclast deposits in western Ireland and North Island, New Zealand? PNAS 114:E10639–E10647

Dewey JF, Shackleton RM (1984) A model for the evolution of the Grampian tract in the early Caledonides and Appalachians. Nature 312:115–121
DePaolo DJ (1981) Neodymium isotopes in the Colorado front range and crust-mantle evolution in the Proterozoic. Nature 291:193–196
Fitzgerald RC, Daly JS, Menuge JF, Brewer TS (1996) Petrogenesis of the Annagh Gneiss Complex, NW County Mayo. Abstracts, Annual Irish Geology Research Meeting, University College Dublin, Ireland, p 23
Flowerdew MJ (2000) The thermal history of Proterozoic rocks in the Caledonides of NW Ireland and the response of mineral dating systems to deformation. Unpublished PhD thesis, University College Dublin
Flowerdew MJ, Chew DM, Daly JS, Millar IL (2009) Hidden Archaean and Palaeoproterozoic crust in NW Ireland: evidence from zircon Hf isotope data from granitoid intrusions. Geol Mag 146:903–906
Flowerdew MJ, Daly JS, Guise PG, Rex DC (2000) Isotopic dating of overthrusting, collapse and related granitoid intrusion in the Grampian orogenic belt, NW Ireland. Geol Mag 137:419–435
Gower CF (1996) The evolution of the Grenville Province in eastern Labrador, Canada. In: Brewer, TS (ed.). Precambrian crustal evolution in the North Atlantic Region. Geol Soc Lond Spec Pub, 112:197–218
Gower CF, Krogh TE (2002) A U–Pb geochronological review of the Proterozoic history of the eastern Grenville Province. Can J Earth Sci 39:795–829
Graham JR, Wrafter JP, Daly JS, Menuge JF (1991) A local source for the Ordovician Derryveeny Formation, western Ireland: implications for the Connemara Dalradian. In: Morton, AC, Todd, SP, Haughton, PDW (eds). Developments in Sedimentary Provenance Studies. Geol Soc Lond Spec Pub, 57:199–213
Gray JR, Yardley BWD (1979) A Caledonian blueschist from the Irish Dalradian. Nature 278:736–737
Griffith R (1839) A general map of Ireland to accompany the Report of the Railway Commissioners shewing the principal physical features and geological structure of the country. Scale One Inch to Four Miles. Hodges and Smith, Dublin; James Gardner, London
Halliday AN, Graham CR, Aftalion M, Dymoke P (1989) The depositional age of the Dalradian Supergroup: U-Pb and Sm-Nd isotopic studies of the Tayvallich Volcanics, Scotland. J Geol Soc Lond 146:3–6
Harris AL, Pitcher WS (1975) The Dalradian supergroup. In: Harris AL, Shackleton RM, Watson J, Downie C, Harland WB, Moorbath S (eds) A correlation of Precambrian rocks in the British Isles. Geol Soc Lond Spec Rep 6:52–75
Harris AL, Baldwin CT, Bradbury HJ, Johnson HD, Smith RA (1978) Ensialic basin sedimentation: the Dalradian Supergroup. In: Bowes DR, Leake BE (eds) Crustal evolution in north-western Britain, Seel House Press, Liverpool, pp. 115–138
Harris AL, Haselock PJ, Kennedy MJ, Mendum JR, Long CB, Winchester JA, Tanner PWG (1994) The Dalradian Supergroup in Scotland, Shetland and Ireland. In Gibbons W, Harris AL (eds) A revised correlation of Precambrian rocks in the British Isles. Geol Soc Lond Spec Rep, 22:33–53
Harris DHM (1993a) Structure, metamorphism and stratigraphy of Achill Island, Co Mayo, Ireland.: Ph.D. thesis, University of Keele
Harris DHM (1993b) The Caledonian evolution of the Laurentian margin in western Ireland. J Geol Soc Lond 150:669–672
Harris DHM (1995) Caledonian transpressional terrane accretion along the Laurentian margin in Co. Mayo, Ireland. J Geol Soc Lond 152:797–806
Heaman L, Parrish R (1991) U–Pb geochronology of accessory minerals. In Heaman L, Ludden JN (eds) Short course handbook on applications of radiogenic isotope systems to problems in geology. Mineralogical Association of Canada, Toronto, 19:59–102
Henrichs IA, O'Sullivan G, Chew DM, Mark C, Babechuk MG, McKenna C, Emo R (2018) The trace element and U-Pb systematics of metamorphic apatite. Chem Geol 483:218–238

Highton AJ, Hyslop EK, Noble SR (1999) U–Pb geochronology of migmatisation in the Northern Central Highlands: evidence for pre-Caledonian (Neoproterozoic) tectonometamorphism in the Grampian block, Scotland. J Geol Soc Lond 156:1195–1204

Johnston JD (1995) Major northwest-directed Caledonian thrusting and folding in Precambrian rocks, North West Mayo, Ireland. Geol Mag 132:91–112

Johnston JD, Phillips WEA (1995) Terrane amalgamation in the Clew Bay region, west of Ireland. Geol Mag 132:485–501

Kennedy MJ (1966) The geology of North Achill Island, Co. Mayo, Ireland: Ph.D. thesis, Trinity College Dublin

Kennedy MJ (1969) The structure and stratigraphy of the Dalradian rocks of north Achill Island, Co., Mayo, Ireland. Quarterly J Geol Soc Lond 125:47–81

Kennedy MJ (1980) Serpentinite-bearing melange in the Dalradian of County Mayo and its significance in the development of the Dalradian basin. J Earth Sci-Dublin 3:117–126

Kennedy MJ, Menuge JF (1992) The Inishkea Division of northwest Mayo: Dalradian cover rather than pre-Caledonian basement. J Geol Soc Lond 149:167–170

Kirkland CL, Alsop I, Daly JS, Whitehouse MJ, Lam R, Clark C (2013) Constraints on the timing of Scandian deformation and the nature of a buried Grampian terrane under the Caledonides of northwestern Ireland. J Geol Soc Lond 170:615–625. https://doi.org/10.1144/jgs2012-106

Kirwan PJ, Daly JS, Menuge JF (1989) A minimum age for the deposition of the Dalradian Supergroup sediments in Ireland. Terra Abs 1:16

Krogh TE (1982) Improved accuracy of U–Pb zircon ages by the creation of more concordant systems using the air abrasion technique. Geochim. Cosmochim. Acta 46:637–49

Leake, BE, Tanner, PWG (1994) The Geology of the Dalradian and Associated Rocks of Connemara, Western Ireland. Royal Irish Academy. Dublin, 96pp

Long CB, MacDermot CV, Morris JH, Sleeman AG, Tietzsch-Tyler D, Aldwell CR, Daly D, Flegg AM, McArdle PM, Warren WP (1992) Geology of North Mayo. A geological Description to accompany The Bedrock Geology 1: 000,000 Map Series Sheet 6, North Mayo. Geological Survey of Ireland

Marcantonio F, Dickin AP, McNutt RH, Heaman LM (1988) A 1800-million-year-old gneiss terrane in Islay with implications for the crustal structure and evolution of Britain. Nature 335:62–64

Max MD (1970) Metamorphism of Caledonian metadolerites in north-west county Mayo, Ireland. Geol Mag 107:539–548

Max MD (1974) Reply to discussion on the paper: a note on the stratigraphy and structure of Achill Island. Geol Surv Ireland Bull 1:485–492

Max MD (1989) The clew bay group: a displaced terrane of highland border group rocks (Cambro-Ordovician) in northwest Ireland. Geol J 24:1–17

Max MD, Long CB (1985) Pre-Caledonian basement in Ireland and its cover relationships. Geol J 20:341–366

Max MD, Sonet J (1979) A Grenville age for pre-Caledonian rocks in NW Co., Mayo, Ireland. J Geol Soc Lond 136:379–382

Max MD, Long CB, Sonet J (1976) The geological setting and age of the Ox Mountains Granodiorite. Geol Surv Ireland Bull 2:27–35

Max MD, Long CB, MacDermot CV (1992) Bedrock Geology of North Mayo. 1:100,000 Map Series, Sheet 6. Geological Survey of Ireland

MacLennan S, Park Y, Swanson-Hysell N, Maloof A, Schoene B, Gebreslassie M, Antilla E, Tesema T, Alene M, Haileab B (2018) The arc of the Snowball: U-Pb dates constrain the Islay anomaly and the initiation of the Sturtian glaciation. Geology 46:539–542

McAteer CA, Daly JS, Flowerdew MJ, Whitehouse MJ, Kirkland CL (2010) A Laurentian provenance for the Dalradian rocks of north Mayo, Ireland and evidence for an original basement-cover contact with the underlying Annagh Gneiss complex. J Geol Soc Lond 167:1033–1048. https://doi.org/10.1144/0016-76492009-147

McCay GA, Prave AR, Alsop GI, Fallick AE (2006) Glacial trinity: Neoproterozoic Earth history within the British-Irish Caledonides. Geology 34:909–912

McLelland J, Daly JS, McLelland JM (1996) The Grenville Orogenic Cycle (ca. 1350–1000 Ma): an Adirondack perspective. Tectonophysics 265:1–28

Menuge JF, Daly JS (1994) The Annagh Gneiss Complex. In: Harris AL, Gibbons FA (eds) A correlation of Precambrian rocks in the British Isles, 2nd edition, Geol Soc Lond Spec Rep 22:59–62

Menuge JF, Daly JS (1990) Proterozoic evolution of the Erris Complex, northwest Mayo, Ireland: Neodymium isotope evidence. In Gower CF, Rivers T, Ryan B (eds) Mid-Proterozoic Laurentia-Baltica. Geol Assoc of Canada Spec Pap 38:41–52

Menuge JF, Williams DM, O'Connor PD (1995) Silurian turbidites used to reconstruct a volcanic terrain and its Mesoproterozoic basement in the Irish Caledonides. J Geol Soc Lond 152:269–278

Mohr P (1987) The Cill Alla dike swarm, Cos Sligo and Mayo: physical parameters. Irish Natural J 22:326–334

Morley CT (1966) The geology of South Achill and Achill Beg, Co. Mayo, Ireland. Unpublished Ph.D. thesis, Trinity College Dublin

Muir RJ, Fitches WR, Maltman AJ (1992) Rhinns Complex: a missing link in the Proterozoic basement of the North Atlantic region. Geology 20:1043–1046

Page LM, Stephens MB, Wahlgren C-H (1996) ^{40}Ar-^{39}Ar geochronological constraints on the tectonothermal evolution of the Eastern Segment of the Sveconorwegian Orogen, southcentral Sweden. In: Brewer, TS (ed.). Precambrian crustal evolution in the North Atlantic Region. Geol Soc Lond Spec Pub, 112:315–30

Park RG (1995) Palaeoproterozoic Laurentia-Baltica relationships: a view from the Lewisian. In: Coward MP, Ries AC (eds) Early precambrian processes. Geol Soc Lond Spec Pub, 95:211–224

Phillips WEA (1974) A note on the stratigraphy and structure of Achill Island, a criticism. Geol Surv Ireland Bull 1:479–483

Phillips WEA (1981) The Precaledonian basement and the orthotectonic Caledonides. In: A Geology of Ireland, Holland CH (ed) Chaps 2 and 3. Scottish Academic Press. Edinburgh

Pidgeon RT, Compston W (1992) A SHRIMP ion microprobe study of inherited and magmatic zircons from four Scottish Caledonian granites. T Roy Soc Edin-Earth 83:473–483

Rivers R (2015) Tectonic setting and evolution of the Grenville orogen: an assessment of progress over the last 40 years. Geosci Can 42. http://dx.doi.org/https://doi.org/10.12789/geocanj.2014.41.057

Rivers T, Culshaw N, Hynes A, Indares A, Jamieson R, Martignole J (2012) The Grenville Orogen—A post-LITHOPROBE perspective, Chapter 3, in Percival JA, Cook FA, Clowes RM (eds) Tectonic Styles in Canada: the LITHOPROBE Perspective: Geol Assoc Canada Spec Pap, 49:97–236

Roddick C, Max MD (1983) A Laxfordian age from the Inishtrahull Platform, Co. Donegal, Ireland. Scot J Geol 19:97–102

Rogers G, Dempster TJ, Bluck BJ, Tanner PWG (1989) A high precision U-Pb age for the Ben Vuirich granite: implications for the evolution of the Scottish Dalradian supergroup. J Geol Soc Lond 146:789–798

Rooney AD, Chew DM, Selby D (2011) Re-Os geochronology of the Neoproterozoic—Cambrian Dalradian Supergroup of Scotland and Ireland: implications for Neoproterozoic stratigraphy, glaciations and Re-Os systematics. Precambrian Res 185:202–214

Romer RL (1996) Contiguous Laurentia and Baltica before the Grenvillian–Sveconorwegian orogeny. Terra Nova 8:173–81

Ryan PD, Sawal VK, Rowland AS (1983) Ophiolitic melange separates ortho- and para-tectonic Caledonians in western Ireland. Nature 302:50–52

Sanders IS, Daly JS, Davies GR (1987) Late proterozoic high pressure granulite facies metamorphism in the NE Ox inlier, NW Ireland. J Met Geol 5:69–85

Sanderson DJ, Andrews JR, Phillips WEA, Hutton DHW (1980) Deformation studies in the Irish Caledonides. J Geol Soc Lond 137:289–302

Scanlon RP (1998) Provenance of the Dalradian Keem Conglomerate, Achill Island, Co. Mayo, Ireland: MSc. thesis, University College Dublin

Smith M, Robertson S, Rollin KE (1999) Rift basin architecture and stratigraphical implications for basement–cover relationships in the Neoproterozoic Grampian Group of the Scottish Caledonides. J Geol Soc Lond 156:1163–1173

Soper NJ (1994) Was Scotland a Vendian RRR junction? J Geol Soc Lond 151:579–582

Soper NJ, Anderton R (1984) Did the Dalradian slides originate as extensional faults? Nature 307:357–360

Soper NJ, England RW (1995) Vendian and Riphean rifting in NW Scotland. J Geol Soc Lond 152:11–14

Sutton JS, Max MD (1969) Gneisses in the north-western part of County Mayo, Ireland. Geol Mag 106:284–290

Sutton JS (1972) The pre-caledonian rocks of the Mullet Peninsula, County Mayo, Ireland. Sci Proc R Dub Soc 4:311–329

Tanner PWG (1996) Significance of a pre-intrusion fabric in the contact aureole of the 590 Ma Ben Vuirich Granite, Perthshire, Scotland. Geol Mag 133:683–695

Tanner PWG, Leslie AG (1994) A pre-D2 age for the 590 Ma Ben Vuirich Granite in the Dalradian of Scotland. J Geol Soc Lond 151:209–212

Thompson P (1985) Dating the British Tertiary igneous province in Ireland by the $^{40}Ar/^{39}Ar$ stepwise degassing method. Unpublished Ph.D. Thesis, University of Liverpool

van Breemen O, Halliday AN, Johnson MRW, Bowes DR (1978) Crustal additions in late Precambrian times. In Crustal Evolution in Northwestern Britain and Adjacent Regions, Bowes DR, Leake BE (eds), Geol Jour Spec Iss 10:81–106

Westropp TJ (1912) The promontory forts and early remains of the coasts of county Mayo. Part 2. The Mullet. The Journal of the Royal Society of Antiquaries of Ireland, Sixth Series, 2, No. 3 (Sep. 30, 1912), 185–216

Williams DM, Harkin J, Armstrong HA, Higgs KT (1994) A late Caledonian melange in Ireland: implications for tectonic models. J Geol Soc Lond 151:307–314

Williams DM, Harkin J, Higgs KT (1996) Implications of new microfloral evidence from the Clew Bay Complex for Silurian relationships in the western Irish Caledonides. J Geol Soc Lond 153:771–777

Williams DM, O'Connor PD, Menuge JF (1992) Silurian turbidite provenance and the closure of Iapetus. J Geol Soc Lond 149:349–357

Winchester JA, Max MD (1988) Pre-Dalradian rocks in NW Ireland. In Winchester JA (ed) Late Proterozoic Stratigraphy of the North Atlantic Regions: 131–145. Blackie, London

Winchester JA, Max MD (1984) Geochemistry and origins of the Annagh Division of the Precambrian Erris Complex, N.W. Co. Mayo, Ireland. Precambrian Res 25:397–414

Winchester JA, Max MD (1987a) A displaced and metamorphosed peralkaline granite related to the late Proterozoic Labrador and Gardar suites: the Doolough Granite of County Mayo, northwest Ireland. Can J Earth Sci 24:631–642

Winchester JA, Max MD (1987b) The pre-Caledonian Inishkea Division of northwest Co. Mayo, Ireland: its geochemistry and probable. Geol Jour 22:309–331

Winchester JA, Max MD, Long CB (1988) The Erris Group, Ireland. In Winchester JA (ed) Late Proterozoic Stratigraphy of the North Atlantic Regions: 162–176. Blackie, London

Yardley BWD (1980) Metamorphism and orogeny in the Irish Dalradian. J Geol Soc Lond 137:303–309

Yardley BWD, Barber JP, Gray JR (1987) The metamorphism of the Dalradian rocks of western Ireland and its relation to tectonic setting. Philos T Roy Soc A 321:243–270

Yardley BWD, Condliffe E, Lloyd GE, Harris DHM (1996) Polyphase garnets from western Ireland: two-phase intergrowths in the grossular-almandine series. Eur J Mineral 8:383–392

The Slishwood Division and Its Relationship with the Dalradian Rocks of the Ox Mountains

J. Stephen Daly, Michael J. Flowerdew, David M. Chew, and Ian S. Sanders

Abstract The Slishwood Division of northwestern Ireland is located along the southwestern margin of the Laurentian palaeocontinent. As detailed elsewhere in this guide (e.g., Chapters "The Basement and Dalradian Rocks of the North Mayo Inlier" and "The Central Ox Mountains"), the metamorphism and deformation of this region is generally attributed to the collision of outboard magmatic arcs with the Laurentian continental margin during the $c.$ 470 Ma Grampian Orogeny. The Slishwood Division exhibits several features that are unique within the Grampian orogen. In particular, the metamorphic history of these rocks—early eclogite-facies and later high-pressure-granulite-facies—is different from that in any other region of the Grampian belt in Ireland or Britain. The Slishwood Division is plausibly explained as an exotic terrane that experienced at least part of its tectonic history in an outboard position, originating as a microcontinent rifted from Laurentia that experienced deep tectonic burial and arc magmatism before colliding with the Laurentian margin during the Grampian Orogeny. This chapter provides a series of field locations from which the evidence supporting this model was obtained, including evidence for a tectonic suture, the North Ox Mountains Slide, along which a small remnant ophiolitic fragment may be examined.

J. S. Daly (✉)
UCD School of Earth Sciences, University College Dublin, Belfield, Dublin 4, Ireland
e-mail: stephen.daly@ucd.ie

M. J. Flowerdew
CASP, West Building, Madingley Rise, Madingley Road, Cambridge CB3 0UD, UK

D. M. Chew · I. S. Sanders
Department of Geology, Trinity College Dublin, Dublin 2, Ireland

© The Author(s), under exclusive license to Springer Nature Switzerland AG 2022
P. D. Ryan (ed.), *A Field Guide to the Geology of Western Ireland*, Springer Geology Field Guides https://doi.org/10.1007/978-3-030-97479-4_3

1 Introduction

The Slishwood Division of the Northeast Ox Mountains, Lough Derg and Rosses Point inliers (Max and Long 1985) is located along the southwest margin of the Grampian belt (Fig. 1). It comprises mainly psammitic paragneiss, minor pelite, abundant metabasite intrusives and several occurrences of ultrabasic rocks of diverse origins. The Slishwood Division rocks record a complex metamorphic history involving high-pressure granulite-facies (Lemon 1971; Phillips et al. 1975; Sanders et al. 1987) and earlier eclogite-facies metamorphic assemblages (Sanders et al. 1987; Unitt 1997; Flowerdew and Daly 2005; Sanders 2018). These high-grade metamorphic events occurred either before or during an early stage of the Early Ordovician Grampian Orogeny (Phillips et al. 1975; Sanders et al. 1987; Flowerdew et al. 2000; Flowerdew and Daly 2005), during which the Slishwood Division was juxtaposed with and interleaved within the Neoproterozoic to early Palaeozoic Dalradian Supergroup (Alsop 1991; Flowerdew 1998/9; Flowerdew et al. 2000; Flowerdew et al.

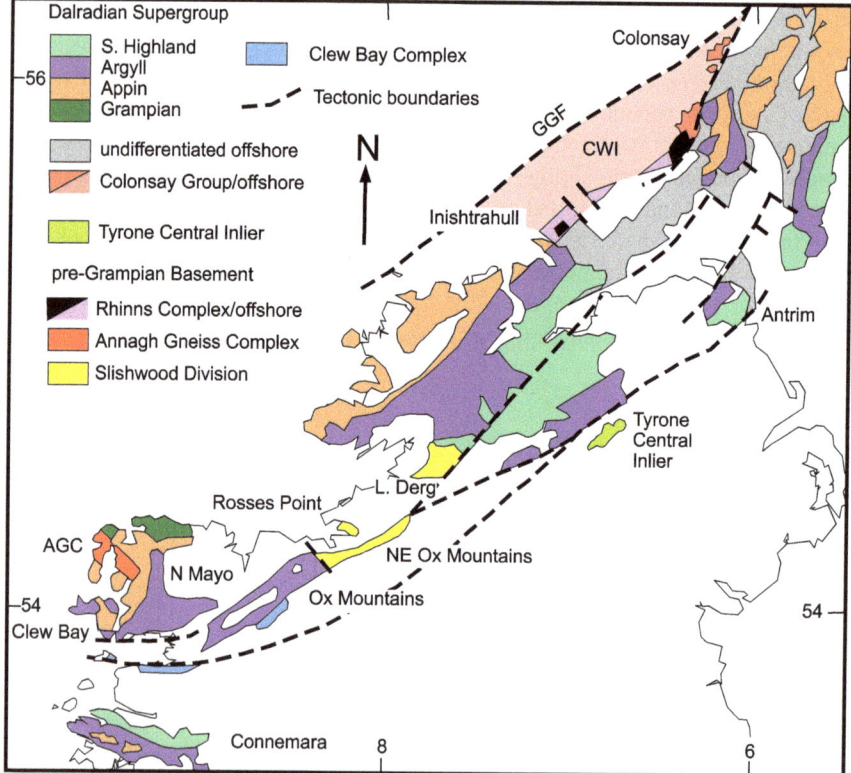

Fig. 1 Location of the Slishwood Division within the Grampian belt of northwestern Ireland (after Daly 2009). Abbreviations: AGC = Annagh Gneiss Complex, GGF = Great Glen Fault, CWI = Colonsay West Islay Block

2005). During the Grampian Orogeny the high grade rocks of the Slishwood Division experienced decompression from eclogite- to high pressure granulite-facies conditions and were then extensively retrogressed under amphibolite-facies conditions (Phillips et al. 1975; Sanders et al. 1987; Flowerdew et al. 2000; Flowerdew and Daly 2005; Table 1).

Table 1 Geological history of the Slishwood Division

Event	Age constraints
Post-Grampian cooling and exhumation	
Deposition of Lower Carboniferous sediments	*c*. 350 Ma
Upright crenulations and open folding	*c*. 400 Ma?
Cooling below 350 °C	*c*. 442 Ma (Rb–Sr biotite [4,5,6])
Extensional shear zones, deformation of pegmatites	455–442 Ma (Rb–Sr muscovite [4])
Grampian Orogeny	
Peak Grampian metamorphism, amphibolite facies (*c*. 9 kbar 600 °C)	*c*. 465 Ma (Sm–Nd mineral isochron [4,5])
Juxtaposition of the Slishwood Division with the Dalradian along the North Ox Mountains Slide	*c*. 476–463 Ma (^{40}Ar-^{39}Ar hornblende, Rb–Sr muscovite [4])
Tonalite/granite intrusion, hydration	
Tonalite, granite and granite pegmatite intrusion	*c*. 472 Ma (U–Pb zircon [6])
Retrograde metamorphism (c. 700 °C)	*c*. 472 Ma (zircon from amphibole-bearing leucosome, supported by ^{40}Ar-^{39}Ar and Rb–Sr data [1,4,5])
High grade metamorphism	
High-temperature shear-zones, open folding	*c*. 472–468 Ma (U–Pb zircon from dry mesoperthitic leucosome and metabasite overgrowths [1])
High-pressure granulite-facies metamorphism (>850 °C)	*c*. 473 Ma (U–Pb zircon from metabasite dry leucosome [1,2,3])
Ductile folds, boudinage of basic bodies, eclogite-facies metamorphism, emplacement of peridotite (>16 kbar)	595–473 Ma (supported by Sm–Nd data [2,3])
Slishwood Division protoliths	
Minor basic intrusions	*c*. 596 Ma (zircon and Sm–Nd isochrons on remnant clinopyroxene and plagioclase [1,2])
Deposition of sandstone with minor mudrock and limestone	< *c*. 926 Ma (detrital zircon populations [1])
[1] Unpublished Daly and Flowerdew data	[4] Flowerdew et al. (2000)
[2] Flowerdew and Daly (2005)	[5] Flowerdew (2000)
[3] Sanders et al. (1987)	[6] Flowerdew et al. (2005)

Fig. 2 Mineralogical layering, probably magmatic in origin in a metabasite pod west of Slishwood Gap. Zircon from the garnet-rich band in this rock yielded a U–Pb age of 473 ± 6 Ma (Table 1), dating the high-pressure granulite-facies metamorphism. Unfortunately, this locality (54.2196°, − 8.4027°) is currently inaccessible due to forestry

Much of the evidence for the metamorphic history comes from a suite of metabasite rocks that are interpreted as intrusive bodies. They are generally podiform (from a decimetre to several metres in size) and often occur in linear strings that likely represent boudinaged sills or dykes (Fig. 8). A few larger bodies occur especially west of Slishwood Gap (Loc. 3.1; Figs. 2 and 7) sometimes exhibiting modal layering (Fig. 2), that is interpreted as an original igneous feature. Magmatic textures are, however, rarely preserved in the Slishwood Division metabasites. Most exhibit a foliation defined by discontinuous aggregates of alternating mafic and felsic minerals. One exception is a suite of metagabbroic bodies at Knader Lough in the Lough Derg Inlier (Loc. 3.8; Fig. 22) that has locally escaped ductile deformation.

The metabasite suite provides a maximum age for the eclogite-facies metamorphism. This follows because their intrusion has been dated at 580 ± 36 Ma using a Sm–Nd isochron from remnant igneous pyroxene and plagioclase from the Knader Lough metagabbro (Flowerdew and Daly 2005; Loc. 3.8). A more precise U–Pb zircon age of 596 ± 6 Ma has been obtained from a layered metabasite pod at Glen (Loc. 3.6, Table 1). The latter is taken as the age of the entire metabasite suite (Table 1). Deposition of the paragneiss protoliths took place onto an unexposed basement likely similar to the Annagh Gneiss Complex (Chapter "The Basement and Dalradian Rocks of the North Mayo Inlier"; Flowerdew et al. 2009) before c. 596 Ma (intrusion of the metabasite protoliths) and after c. 926 Ma—the age of the youngest detrital zircons (Table 1). These constraints rule out any involvement of Slishwood Division in the Grenville Orogeny (cf. Chapter "The Basement and Dalradian Rocks of the North Mayo Inlier") but are consistent with the broad correlation of its pre-metamorphic evolution with the Grampian Group Dalradian of North Mayo (Chapter "The Basement and Dalradian Rocks of the North Mayo Inlier"), as suggested by Sanders (1994).

Varying degrees of metamorphic equilibrium and decompression from eclogite- to high pressure granulite-facies and later retrogression to amphibolite-facies assemblages are displayed in the metabasites. The paragneisses generally display granulite-facies assemblages. Muscovite is only present with quartz as a later retrograde alteration and this, together with the common occurrence of mesoperthitic alkali feldspar with either kyanite or, less commonly, sillimanite, indicates temperatures above the second sillimanite isograd, i.e., above 700 °C (Sanders et al. 1987; Sanders 2018). The high-grade metamorphic history of the Slishwood Division and some of its ultrabasic rocks is likely unique within the Grampian belt. High-temperature, high-pressure-granulite-facies assemblages indicate metamorphic conditions in excess of 850 °C at 14 kbar (Unitt 1997; Flowerdew and Daly 2005). Moreover, robust textural evidence indicates that the granulite-facies event was preceded by eclogite-facies metamorphism at pressures of 14 kbar (Figs. 3 and 4; Sanders et al. 1987) or even as extreme as 20 kbar (Unitt 1997).

The age and tectonic setting of the eclogite-facies event remains uncertain despite decades of research. The eclogite-facies event is bracketed between c. 596 Ma (Table 1, formation of the metabasite protoliths) and c. 470 Ma, when there is abundant evidence for migmatization under granulite-facies conditions (Table 1; Daly 2009). Whilst it remains a possibility that the granulite-facies event affecting the Slishwood Division records decompression of previously over-thickened Laurentian crust, Flowerdew and Daly (2005) and Daly et al. (2012) favour a model where the Slishwood Division acted as an exotic block experiencing its extreme metamorphism outboard of Laurentia after c. 596 *Ma* and colliding with Laurentia during the Grampian orogeny close to c. 470 Ma (see below).

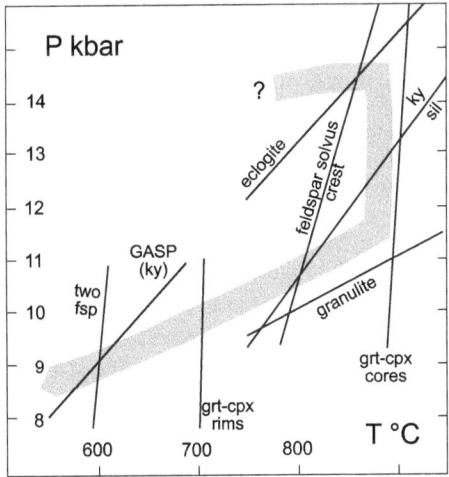

Fig. 3 PT constraints and a suggested PT path for the Slishwood Division applicable to the region south of Lough Gill (Fig. 5) and based on a detailed study from the area around Loc. 3.1 (after Sanders et al. 1987). GASP = garnet-aluminium silicate (ky)-silica-plagioclase barometer; cpx—clinopyroxene, fsp = feldspar, grt = garnet, ky = kyanite; sil = sillimanite

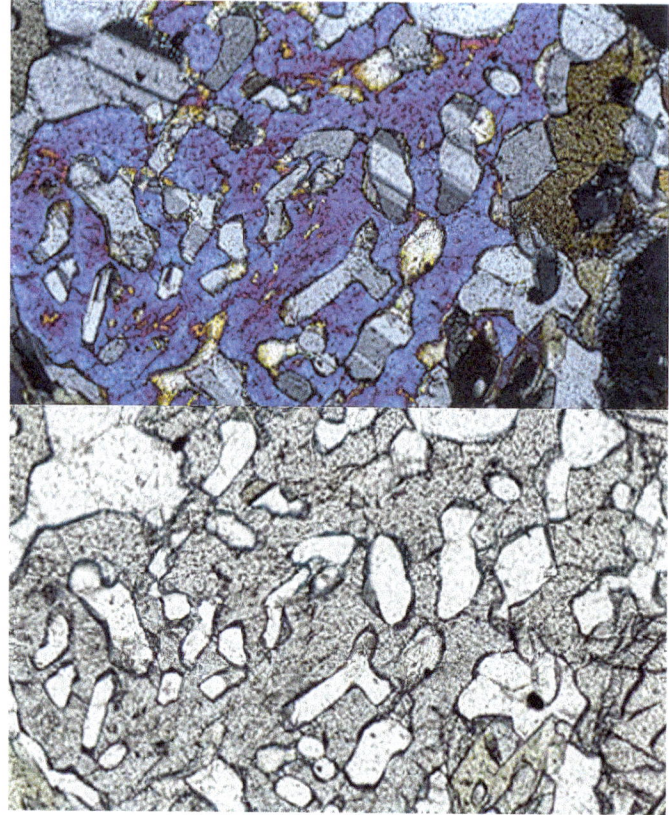

Fig. 4 Plane-polarised (lower) and cross-polarised (upper) photomicrographs of augite-plagioclase intergrowths in metabasite M23 from west of Slishwood Gap (Loc. 3.1). This texture was interpreted by Sanders et al. (1987) to have resulted from the breakdown of original omphacite (high pressure clinopyroxene) during decompression of an earlier eclogite-facies to a later high pressure granulite-facies assemblage. Field of view is c. 1 mm wide. The original image (© I.S. Sanders, Trinity College Dublin) is hosted on The Open University Virtual Microscope (https://virtualmicroscope.org/content/irish-universities-geolab)

A pre-Caledonian age for the high-grade metamorphism was first suggested by Phillips et al. (1975) who tentatively inferred a late Precambrian age from ^{40}Ar-^{39}Ar data on retrograde hornblende from a metabasite body from the NE Ox Mountains inlier. Sanders et al. (1987) interpreted a garnet-clinopyroxene-plagioclase-whole rock Sm–Nd isochron age of 605 ± 37 Ma from another metabasite pod as a cooling age, interpreted as providing a minimum age for the high-grade metamorphism. Later attempts to date the high-grade metamorphism in the metabasites using Sm–Nd mineral isochrons yielded imprecise ages between c. 595–540 Ma (Flowerdew and Daly 2005). Such extreme metamorphism in the late Neoproterozoic in this location is surprising because the geotectonic setting of Laurentia at this time was highly attenuated and undergoing active extension (Soper 1994; Dalziel and Soper 2001; Soper and England 1995; Chew et al. 2010).

Following the eclogite-facies event, whose significance is only partly understood, the subsequent metamorphism and deformation of the Slishwood Division can be confidently attributed to the c. 470 Ma Grampian Orogeny (Table 1), an event generally explained as the result of the collision of a series of outboard magmatic arcs with the Laurentian continental margin (Van Staal et al. 1998) including those at Lough Nafooey (Dewey and Mange 1999) and Tyrone (Chew et al. 2008; Cooper et al. 2011; Hollis et al. 2012).

Decompression during the Grampian Orogeny was accompanied by sinistral shearing and the development of migmatitic leucosomes under high pressure granulite-facies conditions. Zircons in these leucosomes yield U–Pb ages of c. 470 Ma (Table 1; Loc. 3.3, Daly 2009) identical to metamorphic zircons in a metabasite body preserving a pristine dry high-pressure-granulite-facies assemblage, partially decompressed from eclogite (Table 1; Loc. 3.1). Development of the granulite-facies assemblages was followed by slow cooling (Sanders et al. 1987), evidenced by the replacement of sillimanite by kyanite, a highly unusual texture (Fig. 5).

In regions where the high-grade metamorphism is best preserved, a suite of undeformed granite pegmatites cut the foliation (Fig. 6; van Breemen et al. 1978; Molloy and Sanders 1983; Flowerdew 1998/9; Flowerdew et al. 2000) and also facilitated the hydration and retrogression of the high-grade assemblage. A U–Pb zircon age of c. 470 Ma from one such pegmatite indicates that migmatization under granulite-facies conditions, decompression, retrogression and granite pegmatite intrusion all occurred at around 470 Ma. Similar pegmatites are widespread in the Slishwood Division and have yield minimum ages of c. 455 Ma (based on Rb–Sr mineral isochrons from muscovite and feldspar, Flowerdew et al. 2000).

Towards the southwestern end of the NE Ox Inlier a suite of subduction-related tonalites and associated granitic rocks, all of which are dated at c. 470 Ma (Table 1, Flowerdew et al. 2005), intruded into the Slishwood Division after the granulite-facies metamorphism (Flowerdew 1998/9; Flowerdew et al. 2000). They take the form of sheets up to 500 m in thickness southwest of Lough Gill (Ballygawley Tonalite, Fig. 7, Loc. 3.4). These rocks exhibit increasing strain and recrystallization under amphibolite-facies conditions approaching (Fig. 19; Loc. 3.6) and within the North Ox Mountains Slide (Fig. 19; Locs 3.5 and 3.7), a complex zone of deformation and interleaving of the Slishwood Division with the Dalradian rocks (Flowerdew 1998/9; Flowerdew et al. 2000; Flowerdew et al. 2005). Here individual sheets of the Giant's Rock Tonalite are typically c. 50 m in thickness and can be followed along strike for c. 10 km. A similar deformation zone, known as the Lough Derg Slide forms the northern contact of the Slishwood Division with the Dalradian rocks of south Donegal (Alsop 1991).

In summary, we envisage the Slishwood Division as a Neoproterozoic metasedimentary sequence deposited after c. 926 Ma on what became a microcontinental fragment that likely detached from Laurentia during the opening of the Iapetus Ocean. This rifting event produced a suite of metagabbroic intrusives (Locs 3.1, 3.3, 3.6, 3.8, 3.9) at c. 596 ± 6 Ma and likely juxtaposed the metasediment protoliths with fragments of sub-continental lithospheric mantle (Loc. 3.2). This early history

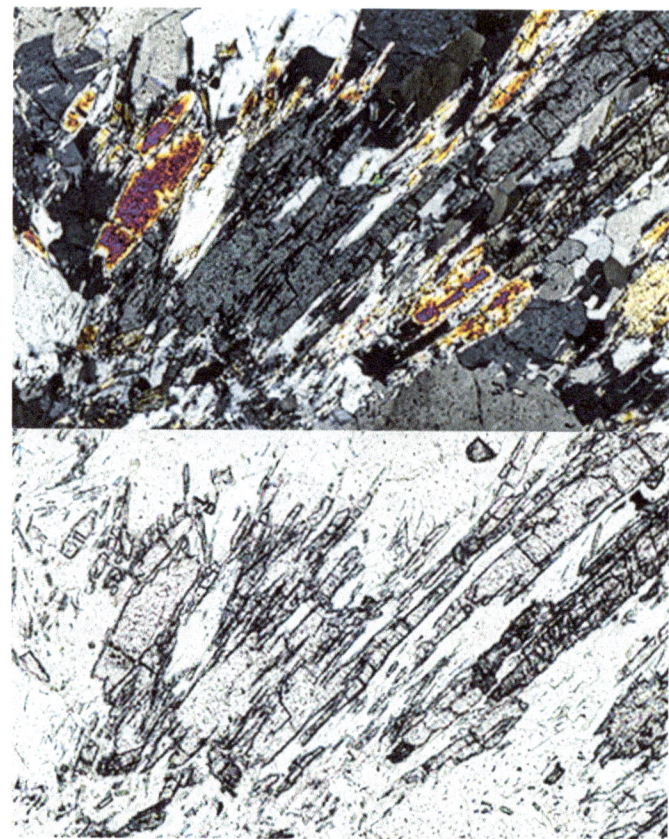

Fig. 5 Plane-polarised (lower) and cross-polarised (upper) photomicrographs showing a single kyanite crystal replacing several sillimanite crystals in metapelitic granulite M22 near Loc. 3.3, east of Slishwood Gap. Replacement of sillimanite by kyanite is very unusual and was interpreted by Sanders et al. (1987) to have resulted from very slow cooling after the development of the high pressure granulite-facies assemblages. Field of view is c. 2.8 mm wide. The original image (© I.S. Sanders, Trinity College Dublin) is hosted on The Open University Virtual Microscope (https://virtualmicroscope.org/content/irish-universities-geolab)

resembles that recorded from the Dalradian Grampian Group rocks of North Mayo (see Chapter "The Basement and Dalradian Rocks of the North Mayo Inlier") with which the Slishwood Division may correlate. Burial of the microcontinent to depths of at least 45 km resulted in eclogite-facies metamorphism. Partial decompression to high-pressure granulite conditions (c. 33 km depth) occurred during or before subduction-related magmatism that produced the tonalites and related granites (Locs 3.4, 3.5, 3.7). Docking of the Slishwood microcontinent back with Laurentia took place during the Grampian collision, as a result of which the Slishwood Division rocks were interleaved with the Dalradian rocks in the Lough Keola area (Loc. 3.7) trapping remnants of the intervening oceanic lower crust or upper mantle represented

Fig. 6 Folded migmatitic leucosomes in psammitic paragneiss cut by discordant dilatant granite pegmatites, east of Slishwood Gap

by the ultrabasic rocks exposed at Lough Keola (Loc. 3.7c). Potentially the entire metamorphic history took place with a few million years close to 470 Ma although it remains plausible that the eclogite-facies event took place significantly earlier.

2 Excursions

2.1 Introduction

Localities 3.1–3.7 in the northeast Ox Mountains Inlier (Fig. 7) are covered by the Ordnance Survey of Ireland Discovery Series map sheet 25 (1:50,000), while localities 3.8 and 3.9 in the Lough Derg Inlier are on sheet 11. The route between these groups of localities is covered by sheet 16.

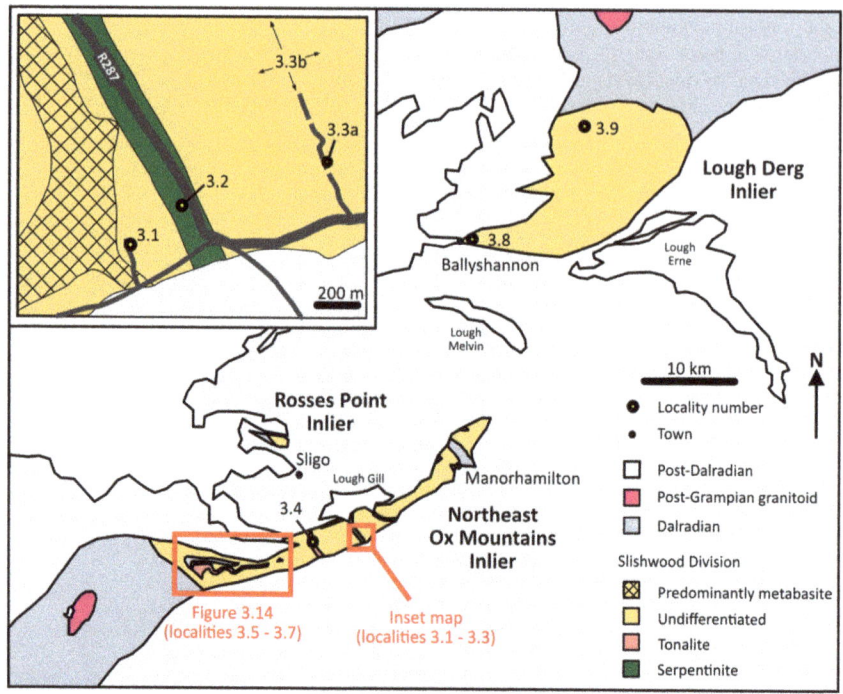

Fig. 7 Sketch geological map of the Slishwood Division inliers after Flowerdew and Daly (2005) showing Locality 3.4 (Ballygawley Quarry) and the locations of Fig. 16 in the northeast Ox Mountains Inlier and localities 3.8 and 3.9 in the Lough Derg Inlier as well as the location of the inset map. The inset map shows the geology at the southern end of Slishwood Gap in more detail together with the locations of localities 3.1–3.3

General directions for localities in the northeast Ox Mountains Inlier (Fig. 7) are given relative to the N4 (Dublin–Sligo road) at Collooney. As far as possible the localities are arranged below in a logical order both to illustrate the sequence of events as currently understood (Table 1) and in geographical sequence. The exception is that localities 3.5 a–c can be reached by two alternative walking routes, so depending on which of these is chosen there may be some re-iterative driving.

Directions for the two localities in the Lough Derg Inlier (Fig. 7) are given relative to the N4 in Sligo.

Convenient access to many localities is over private land. In the authors' experience, permission to cross private land will be granted if politely requested. Fences should not be crossed and gated roads should not be used without first obtaining permission. Great care should be taken on recently harvested forestry land (e.g., the approaches to Loc. 3.5) to avoid timber debris and brambles. Beware also of deep water-filled cuttings and overgrown streams in bogland.

2.2 Excursion Localities in the Northeast Ox Mountains Inlier

"Come away, O human child! To the waters and the wild …"

(W.B. Yeats, *The Stolen Child*).

2.2.1 Locality 3.1: West of Slishwood Gap, Northeast Ox Mountains Inlier (54.210381°, −8.3954930°) *High Grade Paragneiss and Metabasite West of Slishwood Gap*

Access and other information At Collooney roundabout, turn off the N4 and head eastwards along the R290 towards Ballintogher via Ballygawley. Turn left at the church in Ballintogher and drive northwestwards for 1.82 km to the crossroads at the south end of Slishwood Gap at 54.210569°, −8.386090°. Turn left, continue for c. 980 m and park off road at the house on the north side of the road at 54.20781°, −8.39613°. Walk c. 115 m along the road northeastwards to the road junction at 54.20826°, −8.39454°. Leave the road and take the track uphill. Good exposures are available along the track and especially just before and to the east of the gate at 54.21039°, −8.39558°.

The rocks here are typical of the best preserved high-grade psammitic paragneisses and metabasites of the Slishwood Division and provide a good introduction to these rocks. The paragneisses were originally sandstones and siltstones, which have been intruded by basic igneous rocks and later strained (deformed) and metamorphosed to eclogite facies and subsequently under high pressure granulite-facies conditions (Sanders et al. 1987; Flowerdew and Daly 2005; Unitt 1997). Generally, the metabasites are medium-grained but garnet is readily visible in hand specimen. Clinopyroxene is present in most of the metabasite bodies but is at least partially replaced by hornblende. Intergrowths of clinopyroxene with plagioclase are common in the metabasite body closest to the road and were interpreted by Sanders et al. (1987) as evidence for decompression from eclogite facies at temperatures as high as 850–900 °C. A metabasite from here (sample Ox 27, renumbered M23 in Fig. 4) yielded a Sm–Nd isochron age of 605 Ma (Sanders et al. 1987), though its significance is unclear. A leucosome from an anhydrous granulite-facies metabasite pod farther along and near the fence on the western side of the track yielded a U–Pb zircon age of 471 ± 5 Ma (Table 1). The best preserved clinopyroxene is likely to be found in the paler green parts of the metabasite bodies, while the darker coloured metabasite exhibits more extensive replacement of clinopyroxene by hornblende. In some examples large hornblende poikiloblasts are evident in hand specimen but in many cases the petrographic relationships can only be diagnosed in thin section. Paragneisses surrounding the metabasite pods (Fig. 8) contain quartz, mesoperthitic alkali feldspar, garnet, kyanite (sometimes replacing remnant sillimanite, Sanders et al. 1987; Fig. 5), biotite, rutile, zircon and Fe-oxides, the latter two often in heavy mineral bands in

Fig. 8 View to the northeast from Loc. 3.1 towards Killerry Mountain and part of the area examined at Loc. 3.3. In the foreground migmatized psammitic paragneisses and rounded pods of metabasite (weathering back) are well displayed. The tree-covered valley, Slishwood Gap, in the middle distance marks the position of the Slishwood serpentinite (Loc. 3.2)

the more quartzose lithologies. Migmatitic leucosomes are ubiquitous. Muscovite occurs sporadically as an alteration product particularly after kyanite.

To the east, between here and the R287 road (Fig. 7) the rocks become progressively more deformed and the foliation becomes more consistent in orientation. The paragneisses especially become mylonitised. However, they retain their high-pressure granulite-facies mineralogy, so this zone developed under high temperature conditions (c. 850 °C at pressures close to 12 kbar, Sanders et al. 1987). Hence this locality may be typical of regions of thickened metasedimentary lower crust.

Exercises
(1) Carefully examine the foliation within the paragneiss and metabasite pods and determine what features define the foliation within them.
(2) Where the foliation is consistent over a few square metres, measure its dip and strike to compare with Loc. 3.2 and elsewhere.
(3) Discordant granitic pegmatites and granite veins are present at this locality, but they are less common than elsewhere. What do they cut?

2.2.2 Locality 3.2: Slishwood Gap, Northeast Ox Mountains Inlier, (54.212811°, −8.3902073°). *The Slishwood Serpentinite*

Access and other information Return to the crossroads at the south end of Slishwood Gap at 54.210569°, −8.386090° and turn left. After *c.* 380 m turn left again at 54.212981°, −8.3897674°. Stop where the road forks and walk to the house to request permission from the McGarry family both to park on their property and to examine the outcrops. The McGarry family have welcomed geologists on their land for at least 60 years and are well aware of the importance of and interest in these exposures. Return to the vehicles, turn right and park on either the eastern or western side of the concrete bridge.

This locality is listed as a site of national importance by McAteer and Parkes (2004). The Slishwood Serpentinite provides a rare opportunity to examine exposures of the sub-continental lithospheric mantle in somewhat more benign circumstances than those offered by the viewpoint at Locality 2.15f (Chapter "The Basement and Dalradian Rocks of the North Mayo Inlier"). These exposures are all that remains of previously much more extensive outcrops along the R287 northwards for 2.3 km to Slishwood ("Sleuth Wood" of W.B. Yeats in The Stolen Child). Preferential weathering and erosion of the serpentinite here has formed the valley followed by the R287, generally known as Slishwood Gap. The rock here is a highly altered dunite, originally composed mainly of olivine and spinel with subordinate orthopyroxene. Hydration has converted the olivine to serpentine, of which several polymorphs are present including thin (generally just a few mm thick) veins of fibrous chrysotile—white asbestos, for which the locality was prospected by Turner & Newall Ltd in the 1960s. They found a substantial quantity of chrysotile, but the short length of the fibres is not sufficient to spin the asbestos into a yarn and so exploration was abandoned. The serpentinite is well exposed in places along the road and on the eastern

Fig. 9 a Slishwood Serpentinite at Loc. 3.2 displaying very strong mylonitic banding and thin chrysotile veins. The clearest textures such as these are occasionally seen in the outcrops along the driveway to the McGarry house but are also readily seen in the loose but local blocks of serpentinite on either side of the bridge at Loc. 3.2. Lens cap is 5.5 cm in diameter **b** Mylonitised psammitic paragneiss close to the western contact with the Slishwood Serpentinite at Loc. 3.2. Lens cap is 5.5 cm in diameter.

side of the concrete bridge. The best petrographic detail (e.g., Fig. 9a) can be seen in the excavated blocks close to the bridge parapet.

If ground conditions have inhibited investigation of the high-grade shear zone at Loc. 3.1, it may be possible to examine some newly excavated (at the time of writing in 2021) exposures of mylonitised paragneiss (Fig. 9b) near the western contact of the serpentinite at 54.21197°, −8.39141°. Despite the mylonitisation, the paragneiss has retained its high grade assemblage of quartz, mesoperthitic alkali feldspar, garnet, kyanite, biotite and rutile.

Exercises

(1) Carefully examine the rock texture—especially the occurrence of mm-scale pseudomorphs after orthopyroxene. Consider their shape and orientation and the very fact that they have survived. What does this suggest about the timing of the hydration relative to other events affecting these and the adjacent rocks?
(2) How do the chrysotile veins relate to other features of the rock—what age are these?
(3) Compare the orientation of the planar fabric in the serpentinite with that of the paragneiss west of the bridge and between Locs 3.1 and 3.2.

2.2.3 Locality 3.3: Castleore, Northeast Ox Mountains Inlier (54.21485°, −8.37695°). *High Pressure Granulite-Facies Paragneisses, Metabasite Pods and Discordant Granitic Pegmatites*

Locality 3.3a: (54.21485°, −8.37695°). *Migmatised Extensional Shearing at c. 470 Ma*

Access and other information Return to the crossroads at the south end of Slishwood Gap at 54.210569°, −8.386090° and turn left and drive for about 600 m towards Dromahair on the R287. Park off road on the left at 54.21171°, −8.37719° just east of the gates to the large shed. Walk towards Dromahair for about 175 m and turn left into a narrow laneway and walk uphill at least as far as the gate at 54.2151°, −8.37747°. Cross through the gate and head southwards on the east side of the track to Loc. 3.3a at 54.21485°, −8.37695°.

Here, near-vertical NW–SE trending high temperature sinistral shears are occupied by granitic leucosomes. These shears (Fig. 10) are interpreted to have developed under granulite-facies conditions due to the preservation at their margins of the assemblage garnet, quartz, kyanite, K-feldspar, plagioclase and biotite and the absence of coexisting muscovite and quartz. Supersolvus (mesoperthitic) K-feldspar within the leucosome is testament to high temperature crystallization and the lack of hydrous minerals points to melting under dry conditions. Zircons from the leucosomes are mainly inherited detrital grains but they also exhibit very thin zircon overgrowths that have been dated by the U–Pb SIMS method at *c.* 468 Ma (Table 1, Daly et al. 2012). Hence, we interpret the leucosomes to have resulted from melting either due to prograde heating, or, given the dilational nature of the leucosomes and their concentration within the shears, during decompression under high pressure granulite-facies conditions.

Exercises
(1) Over what distance in a NE-SW direction can the zone of shearing be followed?
(2) Estimate the extensional strain at this locality by determining the aggregate width of the leucosomes over several metres in a NE-SW direction a high angle to the shears.

Locality 3.3b: Traverse from (54.2151°, −8.37747°) to (54.22323°, −8.38763°)

Access and other information From Loc. 3.3a, return to the track and follow it northwards for about 2 km until it ends at 54.22323°, −8.38763°. The summit of Killerry Mountain is easily reached beyond the track from which fine views of Lough

Fig. 10 Sinistral shears occupied by granitic leucosome in migmatized Slishwood Division psammite at Loc. 3.3a. Penknife is 11 cm long

Fig. 11 a A planar discordant granite pegmatite cutting polyphase-deformed layering, foliation and migmatitic leucosomes in Slishwood Division paragneiss at Loc. 3.3. Hammer is 40 cm long b A concentrated group of metabasite pods with granitic leucosomes developed in the boudin necks demonstrating how the isolated pods of metabasite so typical of the Slishwood Division likely develop. This photograph was taken near the northern end of the trackway at Loc. 3.3. The paler green interiors of the pods best preserve the high-grade garnet-clinopyroxene-plagioclase assemblages. Note the white lichen (possibly *Pertusaria corallina*), which seems to occur with greater frequency and in larger sizes on the metabasites compared with other lithologies. Lens cap is 5.5 cm in diameter c Heavy mineral bands rich in zircon, rutile and magnetite in relatively quartzose psammite at Loc. 3.3. Under reducing conditions reaction with magnetite has led to the development of biotite mimetic on the original sedimentary laminations. Lens cap is 5.5 cm in diameter

Gill and the Carboniferous hills of north Sligo are available in good weather. Note that in heavy rain the track serves as a drainage channel and is badly eroded in places.

Close to the track and on either side, a variety of interesting features of the Slishwood Division can be freely explored at leisure. The best quality exposure is generally found on the tops and the western side of the numerous roches moutonnés. These included undeformed discordant granite pegmatite dykes (Fig. 11a), metabasite pods (Fig. 11b) and occasional outcrops that display sedimentary structures (Fig. 11c), particularly in the more quartzose paragneiss.

Exercises
(1) Carefully inspect the heavy mineral bands for evidence of way-up. Is the younging direction consistent with the folding?
(2) Carefully evaluate whether apparent instances of cross-laminations are really of sedimentary origin or represent instead the sheared out limbs of isoclinal folds.

2.2.4 Locality 3.4: Ballygawley Quarry, Northeast Ox Mountains Inlier (54.21933°, −8.47127°). *Intrusive Contacts Between Ballygawley Tonalite and Slishwood Division Paragneiss*

Access and other information Return to the R287 from Loc. 3.3 and drive westwards through the crossroads at the south end of Slishwood Gap. From the crossroads drive for about 4.6 km and turn right onto the R284 towards Sligo. From the junction drive about 3 km and turn left into the entrance of Ballygawley Quarry at 54.2202°, −8.46722°. Permission to enter Ballygawley Quarry should be obtained from Sligo County Council (+353 71 911 1111).

A brief stop is worthwhile at this disused quarry locality if permission is obtained from Sligo County Council. Molloy and Sanders (1983) first differentiated a band of "biotite gneiss" several hundred metres wide that traverses the inlier in the vicinity of Ballygawley Lough, noting that it was possibly of igneous origin. Harney et al. (1996) depicted this rock as the Ballygawley Tonalitic Gneiss on the Geological Survey Ireland 1:100,000 map sheet (Sheet 7). It was first described in detail as the Ballygawley Tonalite by Flowerdew (1989/9) and is considered to belong to a suite of subduction-related mainly tonalitic intrusions that includes the Giant's Rock Tonalite (Loc. 3.5) and the Lough Keola Granite (Loc. 3.7), all of which formed at *c*. 470 Ma. In places the well exposed rock faces within the quarry display cross-cutting contacts (Fig. 12) between Slishwood Division paragneiss and tonalite as well as subordinate patches and bands of granite. Late alteration of the tonalite has led to widespread growth of scapolite. A sample from Ballygawley Tonalite taken from the quarry yielded a U–Pb zircon age of 467 ± 6 Ma while a granite sample yielded a U–Pb zircon age of 474 ± 5 Ma and a Rb–Sr biotite-plagioclase isochron age of 449 ± 7 Ma (Flowerdew et al. 2005).

2.2.5 Locality 3.5: Giant's Rock Tonalite, Northeast Ox Mountains Inlier. *Subduction-Related Magmatism at c. 470 Ma*

Access and other information There are two options to reach the outcrops of Giant's Rock Tonalite, a few metres south of Giant's Rock. At the time of writing, both options involve walking over ground recently cleared of mature trees and partially overgrown with brambles.

Fig. 12 The red dashed line outlines the margin of a xenolith of Slishwood Division paragneiss, whose foliation is cut by the Ballygawley Tonalite at 54.21933°, −8.47127° (Loc. 3.4). Brambles provide a scale

Option a Drive northwards on the N4 from Collooney and turn off onto the N59 towards Ballina. Continue past the turnoff for Cooney and turn southwards towards Glen Wood and Coolaney at Lugawarry, 54.201127°, −8.5765958°. Park off road at the Glen Wood picnic site 54.19440°, −8.58379°. Cross the road and walk through a gap into an area of recently cut forest. Walk generally northwestwards but making use of clearer paths made by forestry vehicles or any other clear ground to reach and cross the wall and fence, e.g., at 54.19521°, −8.58147°. Walk first northwards until clear of the birch thicket and at its edge turn eastwards and walk uphill and eastwards towards Loc. 3.5a SSE of Giant's Rock. The total distance from the Glen Wood car park is *c.* 1 km.

Option b Drive northwards on the N4 from Collooney and turn off onto the N59 towards Ballina. Turn left at the turnoff for Cooney, pass localities 3.6a (Scalpnacapple) and 3.6b (Glen), which could be visited first if taking option b. After the Glen locality (large flat outcrops uphill and on the north side of the road), continue past an entrance driveway and a minor road on the right, turning right onto a gated road at 54.186819°, −8.5532695°. Open and close the forestry gate at 54.18692°, −8.55368° and drive northwestwards and then northwards on the stone-metalled track for *c.* 1 km. Park at the forestry loading area at 54.19472°, −8.56266°. From there walk northwards a few metres to the edge of the newly planted trees. Continue westwards for about 200 m along the southern edge of the new plantation and the boundary with what, at the time of writing (August 2021), was old growth forest due to be harvested. This is difficult ground. From the forest edge follow the fence westwards, crossing a small stream and continuing uphill, passing an impressive megalithic tomb on the other side of the fence. Continue westwards, avoiding the highest ground and descend where possible at about 54.195341°, −8.5690664° to Loc. 3.5a. The total distance from the forestry car park is c. 0.5 km. Although option b is shorter, the ground conditions (heathery hummocks and tall bracken concealing boulders) make for slow going. It is noticeable that the field in which the tomb is located offers much easier progress due to effect of grazing on the natural vegetation. However, that field has double fences, about 1 m apart, topped with very effective barbed wire—taking a shortcut across it is not recommended.

The features of interest at localities 3.5a and 3.5b (Figs. 13 and 14) are well exposed on the north-facing escarpments. Unfortunately, access is currently difficult owing to recent overgrowth of vegetation and lack of grazing.

Localities 3.5a and 3.5b display intrusions of the subduction-related Giant's Rock Tonalite into the psammitic paragneisses of the Slishwood Division. The tonalite intruded into the host granulite-facies psammites at 472 ± 6 Ma (U–Pb zircon age, Flowerdew et al. 2005). The cross-cutting nature of the tonalite relative to the earlier fabric within the Slishwood Division paragneiss, the presence within it of high-grade metabasite xenoliths and later deformation affecting the intrusion, demonstrates that the Giant's Rock tonalite was likely emplaced (at c. 472 Ma) after an earlier (eclogite-facies?) event. A Rb–Sr biotite-plagioclase isochron age of 444 ± 7 Ma and a Sm–Nd biotite-zoisite-titanite-plagioclase isochron age of c. 457 Ma obtained from tonalite at this locality indicate that its metamorphism occurred during the Grampian Orogeny (Flowerdew et al. 2005).

Fig. 13 Giant's Rock tonalite at locality 3.5 photographed in 2008 and since extensively overgrown

Fig. 14 Giant's Rock Tonalite at Loc. 3.5a with occasional metabasite xenoliths (to the right of the hammer) as well as numerous smaller paragneiss xenoliths. Here the tonalite (contact marked by the red dashed line) cuts the layering, foliation and migmatitic leucosomes within the Slishwood Division metasediments demonstrating that it intruded after an earlier tectonic event. Hammer is 40 cm long

Locality 3.5a: South-Southeast of Giant's Rock (54.19529°, −8.56903°).
Giant's Rock Tonalite Cutting Early Fabrics in the Slishwood Division Paragneisses

The best tonalite exposures are little deformed and clearly display a variety of inclusions of both paragneiss and metabasite, interpreted as xenoliths, with a history of tectonism that predates the intrusion.

Locality 3.5b: Southeast of Giant's Rock (54.19578°, −8.56830°).
Deformation of the Giant's Rock Tonalite during the Grampian Orogeny

Here sheets of tonalite, which intrude the host paragneisses, have been folded together with the paragneisses about near-recumbent axial planes. Note how the pre-existing foliation in the paragneiss is transposed into the same plane as the later folds. A similar feature is seen in Loc. 3.6a, but here the effect is of greater intensity. Note also that the thin granite veins that cut the tonalite-paragneiss contacts are parallel to the near-recumbent fold axial planes.

> **Exercises**
> (1) Determine the orientation of folds affecting the tonalite. Which limb is extended? Can this be used as a kinematic indicator? If so, do you agree that movement was top to the southeast?
> (2) Compare the shape of the metabasite xenoliths in places where the tonalite is clearly discordant with places affected by the later deformation

Locality 3.5c: Giant's Rock (54.196453°, −8.569652°). *Glacial Erratic and Spectacular View*

This large boulder is the Giant's Rock, after which the Giant's Rock Tonalite was named by Flowerdew (1998/9). It is a spectacular glacial erratic that was deposited by ice that flowed across these rocks *c.* 11,000 years ago.

> **Exercises**
> (1) What is the lithology of Giant's Rock?
> (2) Why is it interpreted to be a glacial erratic?

2.2.6 Locality 3.6 Scalpnacapple and Glen, Northeast Ox Mountains Inlier

Locality 3.6a: Scalpnacapple (54.19735°, −8.54949°). *Extensional Deformation in the Slishwood Division*

Access and other information Drive northwards on the N4 from Collooney and turn off onto the N59 towards Ballina. Turn left at the turnoff for Cooney, stopping on the northwestern side of the road at 54.19753°, −8.54948°, where there is space for several cars near the electricity pylon.

A large cliff exposure comprising mainly psammitic paragneiss, minor pelite and at least one metabasite pod is a conspicuous feature well displayed on the southeastern side of the road (Fig. 15a). A series of asymmetrical F3 shear folds is interpreted as extensional shears that deform all lithologies, except for late discordant granitic pegmatites that cut across them (Fig. 15b). A similar pegmatite from this locality yielded Rb-Sr muscovite-K-feldspar ages of 452 ± 7 Ma and 455 ± 7 Ma (Flowerdew et al. 2000), demonstrating that the late deformation is more likely to be a late phase of the Grampian Orogeny rather than a later Caledonian event, such as the dominant late (Acadian) sinistral deformation seen in the Ox Mountains (cf. Chapter "The Central Ox Mountains"). Metabasite from this locality preserves an early granulite-facies assemblage in which the clinopyroxene is extensively replaced by hornblende (Molloy and Sanders 1983), from which Philips et al. (1975) obtained a ^{40}Ar-^{39}Ar step-heating age. This locality is listed as a site of national importance by McAteer and Parkes (2004).

Locality 3.6b: Glen, Northeast Ox Mountains Inlier (54.189062°, −8.5471934°). *Granulite-Facies Metabasite Pods*

Access and other information Drive south-southeastwards from Scalpnacapple until the road turns through almost 180° and park off road outside the house on the northern side of the road at 54.189664°, −8.5506155°. Walk back to the large flat outcrops on the hillside on the north side of the road, e.g., at 54.189062°, −8.5471934°. Walk uphill over the dip surface of the paragneisses and search for the larger metabasite pods on the lee side of the roches moutonée surfaces.

This locality is listed as a site of national importance by McAteer and Parkes (2004). Metabasites here lack evidence for decompression from eclogite facies. Instead, they contain a granulite-facies assemblage of garnet-plagioclase-quartz-rutile-apatite-zircon. Minor retrogression is represented by hornblende replacing clinopyroxene and titanite replacing rutile. A garnet-rich band from one of the metabasite pods from this locality yielded an unusually high abundance of zircons, the cores of many of which are of early Palaeoproterozoic age and are interpreted as inherited. Almost all exhibit conspicuous overgrowths, dated at 469 ± 5 Ma (Table 1), indicating that they grew during Grampian metamorphism. The core of one grain exhibiting magmatic zoning, also with a rim of Grampian age, yielded a U–Pb zircon age of 595 ± 7 Ma (Table 1) and an ε_{Hf} value of +3.1. Retrograde metamorphism

Fig. 15 a Extensional shears (a distal expression of the North Ox Mountains Slide) affecting Slishwood Division paragneisses at Scalpnacapple (Loc. 3.6a). This photograph illustrates the geological sophistication of Co. Sligo Road engineers **b** An undeformed granitic pegmatite cuts the late (S3) shear fabric affecting Slishwood Division paragneisses at Scalpnacapple (Loc. 3.6a at 54.19735°, −8.54949°). Lens cap is 5.5 cm in diameter

at a later stage of the Grampian Orogeny is confirmed by an ^{40}Ar-^{39}Ar age of 465 ± 2 Ma from poikiloblastic hornblende and a hornblende-plagioclase Rb–Sr isochron age of 482 ± 22 Ma (Flowerdew et al. 2000).

2.2.7 Locality 3.7: Lough Keola, Northeast Ox Mountains Inlier (54.19253°, −8.67123°). *North Ox Mountains Slide—The Suture Zone Between the Slishwood Division and the Dalradian Rocks on the Laurentian Margin*

Drive north from Coolaney and turn westwards following the "Ox Mountains" signpost. After Carrowgavneen, turn northwestwards at the crossroads signposted "Cabragh Lodge". Continue for about 2 km. Note that this road is mainly stone-metalled. It is heavily rutted in places, so caution is advised. A vehicle with manual transmission and high road clearance, ideally four-wheel drive, is recommended. Follow the road until it makes a wide 180° turn towards the southeast (Fig. 16) and continue to the end of the road at Loc. 3.7a at 54.18940°, −8.67484°. It is possible to turn and park several vehicles here.

The main feature of Loc. 3.7 (a–d) is the intimate interleaving of Slishwood Division paragneisses, intrusive sheets of Giant's Rock Tonalite and the Dalradian rocks associated with the development of the North Ox Mountains Slide. Deformation within the North Ox Mountains Slide is a relatively late feature in both the Dalradian and the Slishwood Division rocks and is generally associated with top to the southeast extension. As a result of this deformation the southeastwards dipping layering is parallel in the Slishwood Division, Giant's Rock tonalite and in the underlying Dalradian rocks. This extensive zone of deformation can be followed for at least 11 km gradually becoming more localised into discrete shear planes in the vicinity

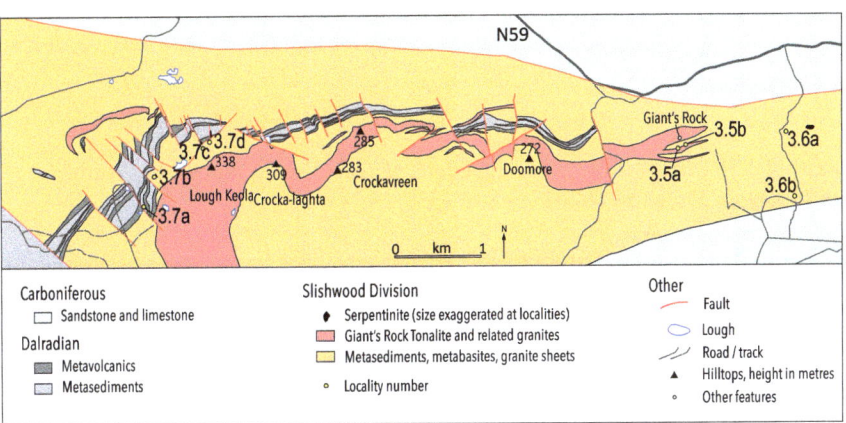

Fig. 16 Geological map (after Flowerdew 1998/9) showing the distribution of Dalradian rocks and the Giant's Rock tonalite within the North Ox Mountains Slide and the location of localities 3.5–3.7

of Scalpnacapple (Loc. 3.6a, Figs. 15 and 16) and results from the tectonic juxtaposition of the Slishwood Division with the Dalradian rocks of the Laurentian margin. (Flowerdew 1998/9, Flowerdew et al. 2005). This suture zone between the Slishwood Division and the Dalradian is marked in places (e.g., at Loc. 3.7d, Fig. 21) by variably serpentinised ultrabasic rocks, interpreted as ophiolitic lower crustal or upper mantle remnants. These were originally described by Lemon (1966) but, in contrast to his depiction of the serpentinites as a continuous sheet, detailed mapping has shown that their exposure is controlled by a series of conjugate faults striking NW–SE and NNW-SSE (Flowerdew 1989/99, Fig. 16). Strongly deformed sheets of a granite (correlated with the Giant's Rock Tonalite) are well exposed within the suture zone at Locs 3.7b and 3.7c, but importantly these rocks are never seen to intrude the Dalradian.

Locality 3.7a: Dalradian Metavolcanics Within the North Ox Mountains Slide (54.18940°, −8.67484°)

Access and other information At the parking place at Loc. 3.7a (Fig. 16), there is a low-lying outcrop of Dalradian metavolcanics intricately interbedded with semi-pelitic and pelitic schist. These rocks probably belong to the Newantrim Member of the Ummoon Formation (Chapter "The Central Ox Mountains") and they can be compared with the same unit exposed at Loc. 4.5. Note the folds plunging shallowly to the ESE and the absence of granitic rocks anywhere intruding the Dalradian rocks both here and at localities 3.7b and c. Samples of Dalradian semi-pelitic schist and a metavolcanic rock taken from a low-lying exposure within the bog, approximately 100 m to the east, have yielded a Rb–Sr muscovite-plagioclase age of 467 ± 7 Ma and a ^{40}Ar-^{39}Ar hornblende age of 457 ± 3 Ma, respectively, indicating that their strong foliation formed during the Grampian Orogeny (Flowerdew et al. 2000). Moreover, the metamorphic assemblage of the semi-pelitic schist (garnet-biotite-muscovite-plagioclase-quartz) is estimated to have crystallised at c. 9 kbar and 550 °C (Flowerdew 2000).

Locality 3.7b: Slishwood Division Paragneiss and Granite Within the North Ox Mountain Slide (54.19263°, −8.67135°)

Access and other information From Loc. 3.7a, walk back along the road northwestwards and cross the drainage ditch on the right at 54.19089°, −8.676036° heading towards the cairn that marks the local summit at Loc. 3.7b.

Leaving the road and heading northeastwards towards the cairn, the first outcrops encountered are Dalradian metasedimentary rocks. These are structurally overlain closer to the cairn by strongly sheared, locally mylonitised, Slishwood Division paragneisses. Beneath the cairn the main features of interest are boudinaged and folded sheets of a granite, correlated with the Giant's Rock Tonalite (Fig. 17a) that

Fig. 17 **a** Foliated granite sheets at Loc. 3.7b (54.19263°, −8.67135°). The cairn at this locality is a useful waypoint in poor weather for the return journey from Loc. 3.7c and d. **b** Mylonitic foliation developed in Slishwood Division paragneiss at Loc. 3.7b, as a result of strain within the North Ox Mountains Slide. This feature resulted from the collision between the Slishwood Division and the Dalradian rocks of the Laurentian margin. Lens cap is 5.5 cm in diameter

Fig. 17 (continued)

intrude the Slishwood Division paragneisses as well as the very strong mylonitic foliation developed in the latter (Fig. 17b).

> **Exercise**
> Determine the apparent extension direction from the boudin necks and compare this with other kinematic indicators—is this consistently NW-SE?

This locality provides a useful waypoint to find Loc. 3.7c and to return safely to the parking place at Loc. 3.7a.

Locality 3.7c: Deformation of the Slishwood Division paragneiss and c. 471 Ma Granite Within the North Ox Mountains Slide (54.19550°, −8.66259°)

A prominent cliff exposure (Fig. 18) of highly deformed, often mylonitic, Slishwood Division paragneiss intruded by a granite sheet related to the Giant's Rock Tonalite allows several features of the North Ox Mountains Slide to be examined. A strongly foliated granitic sheet exposed at the western end of the cliff at 54.19550°, −8.66259° intrudes the paragneiss and has yielded a U–Pb zircon age of 471 ± 5 Ma (Fig. 19; Flowerdew et al. 2005). In several places the mylonitic foliation is folded about shallowly dipping axial planes (Fig. 20). The asymmetry of these folds (Fig. 20) and kinematic indicators based on boudinaged granite pegmatites indicate top to the southeast extensional movements. Refolded folds are also present. In places the mylonitic foliation is so strong that it is hard to recognise the lithology. In places, however, e.g., at 54.19589°, −8.66241° (Fig. 19), high-grade metabasite pods, with remnant granulite-facies assemblages and textures preserved in their interiors are wrapped around by the mylonitic foliation. A Rb–Sr muscovite-plagioclase isochron age of 449 ± 7 Ma (Flowerdew et al. 2000) obtained from the dated granite sheet indicates that extension was likely associated with the Grampian Orogeny. This is

The Slishwood Division and Its Relationship with the Dalradian Rocks …

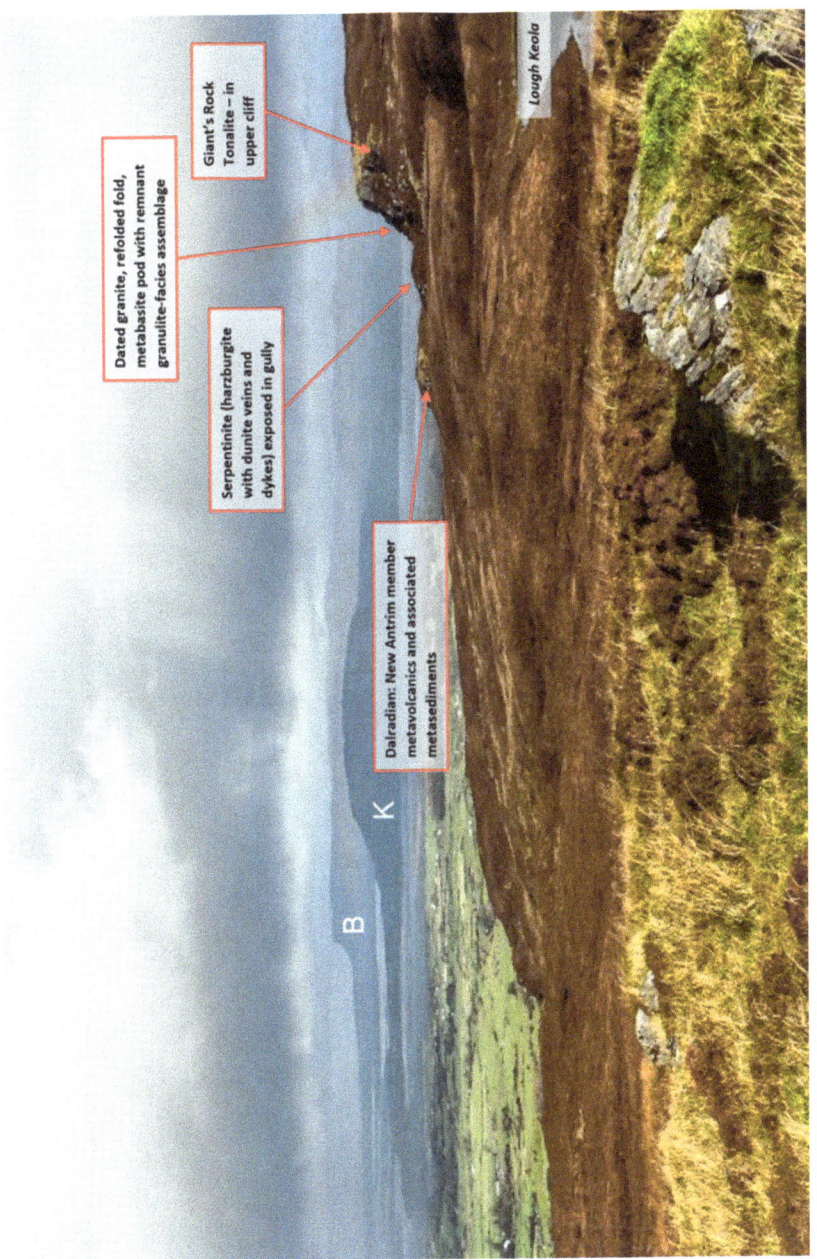

Fig. 18 View from Loc. 3.7b northeastwards towards Locs 3.7c and 3.7d. B = Benbulben, K = Knocknarea

Fig. 19 Overview of the cliff exposure at Loc. 3.7c indicating features discussed in the text. ms = muscovite, pl = plagioclase

Fig. 20 Refolded folds of the mylonitic foliation at Loc. 3.7c indicating top to the southeast extension. Rucksack is 30 cm wide

Fig. 21 **a** Harzburgite (gray) with a more ductile dunite (brown) dyke flowing into a boudin neck in a serpentinite body exposed along a NNW-SSE trending fault (Fig. 19) at Loc. 3.7d. Rucksack is 60 cm long. **b** Harzburgite intruded by a thin dunite vein in a loose but local block at Loc. 3.7d. Lens cap is 5.5 cm in diameter

supported by a Rb–Sr muscovite-plagioclase isochron age of 453 ± 7 Ma from a boudinaged granitic pegmatite vein that cuts the mylonitic foliation but is also deformed within it (Fig. 19).

Exercise
Did the ductile structures here form during a single progressive episode of shearing or as separate discrete events?

Fig. 22 a Late Grampian hydration veins in unfoliated metabasite at Knader Lough (Loc. 3.8). The dark colour corresponds to the presence of hornblende, produced by metasomatic alteration of the high-pressure granulite facies assemblage and has mainly nucleated on and replaced clinopyroxene. **b** Enlarged image of the approximate area outlined by the box in (a) showing how the dark colour of the secondary hornblende reveals the presence and texture of the original randomly orientated plagioclase. (The quartz vein along the main vein is later and a coincidental occupation of the same fracture by later lower temperature fluids). Penknife is 11 cm long

Locality 3.7d: Serpentinite Within the North Ox Mountains Slide (54.19643°, −8.66081°)

Access and other information From the cliff exposure (Fig. 17) walk eastwards across a NNW-SSE-trending fault marked by a conspicuous gully.

Serpentinite is well exposed about 150 m east of Loc. 3.7c along the east-northeastern wall of a deep NNW-SSE trending gully defining a fault. These serpentinised ultrabasic rocks are interpreted as ophiolitic slices emplaced in the suture zone (North Ox Mountains Slide) between the Slishwood Division and the Dalradian rocks. Several textures typical of the lower crustal and upper mantle sections of ophiolites are present including dunite dykes and veins cutting harzburgite on various scales (Figs. 21a, b). In places, strongly foliated dunite is seen to flow into the boudin necks arising from extension associated with the North Ox Mountains Slide (Fig. 21a).

2.3 Excursion Localities in the Lough Derg Inlier

Access and other information From Loc. 3.7 drive back towards Coolaney turning left at 54.177104°, −8.5996109°. Drive northwards past the Glen Wood parking site (one of the routes to Loc. 3.5) and continue to the junction with the N59. Turn right towards Ballysadare and join the N4 there towards Sligo. In Sligo stay on the N4, pass the junction with the N16 and continue on the N15 towards Donegal. In Ballyshannon take the first exit off the N15 at the Erne Roundabout onto Bachelor's Walk. After *c.* 900 *m*, turn left onto Tirconnaill St and continue along Market St and turn left onto College St. After *c.* 470 m, bear right after the Shiel Hospital (on the right) and follow the road to leave the roundabout at the first exit and, staying on College St, drive under the N15 onto Knader Road and continue eastwards along the north shore of Assaroe Lake. Turn left at 54.500721°, −8.1560719° towards Lough Lareen and turn left at 54.501677°, −8.1517786°. It should be possible to drive for *c.* 240 m to park at the hayshed at 54.50264°, −8.15481°. Note that this road is gated so before leaving the vehicle(s), please ask for permission to park. To reach Loc. 3.8, walk on the track past the hayshed for about 50-60 m. The outcrops of interest are on the south side of the track at 54.50570°, −8.15586°. *Please note that localities 3.8 and 3.9 have not been visited by the authors since 2004.*

2.3.1 Locality 3.8: Knader Lough, Lough Derg Inlier (54.50570°, −8.15586°)

The Knader Lough metagabbro is a small (*c.* 100 × 150 m) body wrapped around by the foliation in the enclosing Slishwood Division paragneiss. On its northern side, the body is strongly sheared, while to the south it is virtually free of ductile deformation (Fig. 22). Metamorphic effects are limited to the development of partial coronas of garnet and amphibole around the original igneous minerals, clinopyroxene, plagioclase and ilmenite, which are well preserved. Painstakingly separated aliquots of these igneous minerals have yielded a four-point clinopyroxene-plagioclase Sm–Nd isochron age of 580 ± 36 Ma (Flowerdew and Daly 2005). The low precision of this isochron age is controlled by the naturally limited partitioning of the REE between clinopyroxene and plagioclase. Although the age is imprecise it is likely to be accurate and is considered to date the crystallization of the gabbroic magma. This is supported by the agreement with the more precise U–Pb zircon age of c. 595 Ma obtained from the Glen metabasite at Loc. 3.6b (Table 1).

The lack of ductile deformation and only partial reaction during the metamorphism has resulted in almost complete preservation of the original igneous texture (Fig. 22). Although ductile deformation effects are very minor, the metagabbro shows evidence of solid-state brittle fracture and ingress of a water-bearing fluid, resulting in a series of criss-crossing bands of dark colouration ranging in wide from a few mm to a decimetre and marked by the presence of more extensive development of dark green to black poikiloblastic hornblende, selectively replacing clinopyroxene. As a result of this process, the plagioclase is preserved to a greater extent so that its original size and random grain orientation is readily seen against the dark background (Fig. 22). Outcrops such as this provide a somewhat surprising insight into the flow of water-bearing supercritical fluids in the deep crust, which operate rather differently from their near-surface counterparts in being both highly saline (with specific gravities close to unity) and of very low viscosity.

Closer to the track, the Knader Lough metagabbro is highly sheared over a metre or so in a north–south direction. Here the igneous texture is completely destroyed and a well-equilibrated high pressure granulite-facies assemblage of garnet, clinopyroxene, plagioclase, amphibole, quartz, rutile, ilmenite and biotite yields PT estimates of 810 °C and 15.8 kbar (Flowerdew and Daly 2005).

> **Exercise**
> Why are some metabasites undeformed while others are highly foliated?

2.3.2 Locality 3.9: Oughtdarnid, Lough Derg Inlier (54.60163°, −8.021693°) *Zoned Metabasite Pod*

Access and other information Return to the N15 by the same route and drive northwards towards Donegal. Turn right off the N15 onto the N232 just before the Topaz garage at Laghy and drive southeastwards towards Pettigoe for c. 5 km. Turn left onto a minor road at 54.59849°, −8.038352° and drive for 1.15 km along the southeastern side of Oughtdarnid Hill. Park off road at the gate at 54.601128°, −8.0216914°. Cross the fence and walk uphill to the outcrops around 54.60163°, −8.021693°.

Several metabasite pods are present on the hillside wrapped around by migmatitic leucosomes parallel to the foliation in the Slishwood Division paragneiss. One of these pods (Fig. 23) is spectacularly zoned with a pale foliated interior that preserves the high-pressure granulite-facies assemblage of garnet, clinopyroxene, plagioclase, rutile, quartz, ilmenite, hornblende and biotite. Flowerdew and Daly (2005) reported PT conditions of c 750 °C and 12 kbar for this assemblage. Textural evidence for decompression from an earlier eclogite-facies assemblage is found in symplectitic

Fig. 23 Spectacular boudin of metabasite within migmatized Slishwood Division paragneiss at Loc. 3.9. The paler coloured boudin interior preserves strongly foliated granulite-facies assemblage of garnet-clinopyroxene-plagioclase-quartz-rutile while the darker exterior reflects the replacement of clinopyroxene by poikiloblastic hornblende and associated granitic leucosomes. Penknife is 11 cm long

intergrowths of plagioclase and clinopyroxene as well as plagioclase rims around garnet. Retrograde hydration resulted in the replacement of clinopyroxene by hornblende especially in the darker outer darker region of the pod (Fig. 23), from which Flowerdew et al. (2000) obtained a hornblende ^{40}Ar-^{39}Ar plateau age of 479 ± 6 Ma.

References

Alsop GI (1991) Gravitational collapse and extension along a mid-crustal detachment: the Lough Derg Slide, northwest Ireland. Geol Mag 128:345–354

Chew DM, Flowerdew MJ, Page LM, Crowley QG, Daly JS, Cooper M, Whitehouse MJ (2008) The tectonothermal evolution and provenance of the Tyrone Central Inlier, Ireland. Grampian imbrication of an outboard Laurentian microcontinent? J Geol Soc Lond 165:675–685

Chew DM, Daly JS, Magna T, Page LM, Kirkland CL, Whitehouse MJ, Lam R (2010) Timing of ophiolite obduction in the Grampian Orogen. GSA Bull 122:1787–1799

Cooper MR, Crowley QG, Hollis SP, Noble SR, Roberts S, Chew DM, Merriman RJ, Earls G, Herrington R (2011) Age constraints and geochemistry of the Ordovician Tyrone Igneous Complex, Northern Ireland: implications for the Grampian orogeny. J Geol Soc Lond 168:837–850

Daly JS (2009) Precambrian. In: Holland CH, Sanders IS (eds) The geology of Ireland, 2nd edition. Dunedin Academic Press, Edinburgh, pp 7–42

Daly JS, Flowerdew MJ, Whitehouse MJ (2012) Grampian high-pressure-granulite-facies metamorphism of the Slishwood Division, NW Ireland and its enigmatic eclogite-facies precursor. Geophys Res Abs, 14:13462, EGU General Assembly 2012

Dalziel IWD, Soper NJ (2001) Neoproterozoic extension on the Scottish Promontory of Laurentia: paleogeographic and tectonic implications. J Geol 109:299–317

Dewey JF, Mange MA (1999) Petrography of Ordovician and Silurian sediments in the western Irish Caledonides: tracers of a short-lived Ordovician continent—arc collision orogeny and the evolution of the Laurentian Appalachian–Caledonian margin. In: Mac Niocaill C, Ryan PD (eds) Continental tectonics. Geol Soc Lond Spec Pub, 164:55–107

Flowerdew MJ (1998/9) The Giant's Rock Tonalite and basement-cover relationships in the Northeast Ox Mountains Inlier, northwestern Ireland. Irish J Earth Sci 17:71–82

Flowerdew MJ (2000) The thermal history of Proterozoic rocks in the Caledonides of NW Ireland and the response of mineral dating systems to deformation. Unpublished Ph.D. thesis, National University of Ireland, Dublin

Flowerdew MJ, Daly JS (2005) Sm-Nd mineral ages and PT constraints on the pre-Grampian high grade metamorphism of the Slishwood division, northwest Ireland. Irish J Earth Sci 23:107–123

Flowerdew MJ, Daly JS, Guise PG, Rex DC (2000) Isotopic dating of overthrusting, collapse and related granitoid intrusion in the Grampian orogenic belt, NW Ireland. Geol Mag 137:419–435

Flowerdew MJ, Daly JS, Whitehouse MJ (2005) 470 Ma granitoid magmatism associated with the Grampian Orogeny in the Slishwood Division, NW Ireland. J Geol Soc Lond 162:563–575

Flowerdew MJ, Chew DM, Daly JS, Millar IL (2009) Hidden Archaean and Palaeoproterozoic crust in NW Ireland: evidence from zircon Hf isotope data from granitoid intrusions. Geol Mag 146:903–906

Harney S, Long CB, MacDermot CV (1996) Geology of Sligo-Leitrim. Sheet 7. Bedrock Geology 1:100, 000 Map Series. Geological Survey of Ireland

Hollis SP, Roberts S, Cooper MR, Earls G, Herrington R, Condon DJ, Cooper MJ, Archibald SM, Piercey SJ (2012) Episodic arc-ophiolite emplacement and the growth of continental margins: late accretion in the Northern Irish sector of the Grampian-Taconic orogeny. GSA Bull 124:1702–1723

Lemon GG (1966) Serpentinites in the North-East Ox Mountains, Eire. Geol Mag 103:124–137

Lemon GG (1971) The Pre-Cambrian rocks of the N.E. Ox Mountains, Eire. Geol Mag 108:193–200

Max MD, Long CB (1985) Pre-Caledonian basement in Ireland and its cover relationships. Geol J 20:341–366

McAteer C, Parkes M (2004) The geological heritage of Sligo. An audit of county geological sites in Sligo. Geological Survey of Ireland. Unpublished Report

Molloy MA, Sanders IS (1983) The Northeast Ox Mountains Inlier. In: Archer JB, Ryan PD (eds) A Geological guide to the Caledonides of western Ireland. Geological Survey of Ireland Guide Series 4, 52–55

Phillips WEA, Taylor WEG, Sanders IS (1975) An analysis of the geological history of the Ox Mountains inlier. Sci Proc R Dubl Soc 5B:311–329

Sanders IS (1994) The northeast Ox Mountains Inlier, Ireland. In Gibbons W, Harris AL (eds) A revised correlation of Precambrian rocks in the British Isles. Geol Soc Lond Spec Rep 22:63–64

Sanders IS (2018) Introducing metamorphism. Dunedin Academic Press, Edinburgh and London, p 148

Sanders IS, Daly JS, Davies GR (1987) Late Proterozoic high-pressure granulite facies metamorphism in the NE Ox inlier, NW Ireland. J Met Geol 5:69–85

Soper NJ (1994) Neoproterozoic sedimentation on the northeast margin of Laurentia and the opening of Iapetus. Geol Mag 131:291–299

Soper NJ, England RW (1995) Vendian and Riphean rifting in NW Scotland. J Geol Soc Lond 152:11–14

Unitt RP (1997) The structural and metamorphic evolution of the Lough Derg Complex, Counties Donegal and Fermanagh. Unpublished Ph.D. thesis, National University of Ireland, Cork

van Breemen O, Halliday AN, Johnson MRW, Bowes DR (1978) Crustal additions in late Precambrian times. In Bowes DR, Leake BE (eds) Crustal evolution in northwestern Britain and adjacent regions. Geol J Special Issue 10:81–106

Van Staal CR, Dewey JF, Mac Niocaill C, McKerrow WS (1998) The Cambrian-Silurian tectonic evolution of the northern Appalachians and British Caledonides: history of a complex, west and southwest Pacific-style segment of Iapetus. In Blundell DJ, Scott AC (eds) Lyell: the past is the key to the present. Geol Soc Lond Spec Pub 143:199–242

The Central Ox Mountains

Paul D. Ryan, Ken McCaffrey, David M. Chew, John R. Graham, and Barry Long

Abstract The Central Ox Mountains is a southwest—northeast trending inlier of Dalradian rocks, mostly attributed to the Argyll Group, that lie along a major fault, the Fair Head—Clew Bay Line (FCL). Deposition was associated with rift related magmatism. These rocks were deformed, metamorphosed (up to kyanite zone) and subsequently exhumed during the mid-Ordovician Grampian Orogeny. Subsequent sinistral transpression along the FCL was associated with the emplacement of the granitoids of the Ox Mountains Igneous Complex and the development of major syn-metamorphic shear zones or 'slides' during the Early Devonian Acadian Orogeny. This chapter examines: evidence for early rifting along the FCL coeval with the opening of the Iapetus Ocean; Grampian deformation and metamorphism attributed to mid-Ordovician arc-continent collision; and the emplacement of the OMIC and the later development of the slides during early Devonian Acadian transpression attributed to the closure of the Iapetus Ocean.

1 Introduction

The Central Ox Mountains, County Mayo (Fig. 1), is a low northeast-southwest trending topographic ridge which sits on the trace of a significant Caledonian deformation zone (the Fair Head—Clew Bay line) that defines the southern edge of the Laurentian continental margin in eastern Ireland (Soper and Hutton 1984; Hutton and

P. D. Ryan (✉)
School of Natural Sciences, Earth and Ocean Sciences, National University of Ireland Galway, University Road, Galway, Ireland
e-mail: paul.ryan@nuigalway.ie

K. McCaffrey
Department of Earth Sciences, Durham University, Stockton Road, Durham D1 3LE, UK

D. M. Chew · J. R. Graham
Department of Geology, Trinity College, Museum BuildingDublin 2, Dublin, Ireland

B. Long
Geological Survey Ireland, Beggars Bush, Haddington Road, Dublin 04 K7X4, Ireland

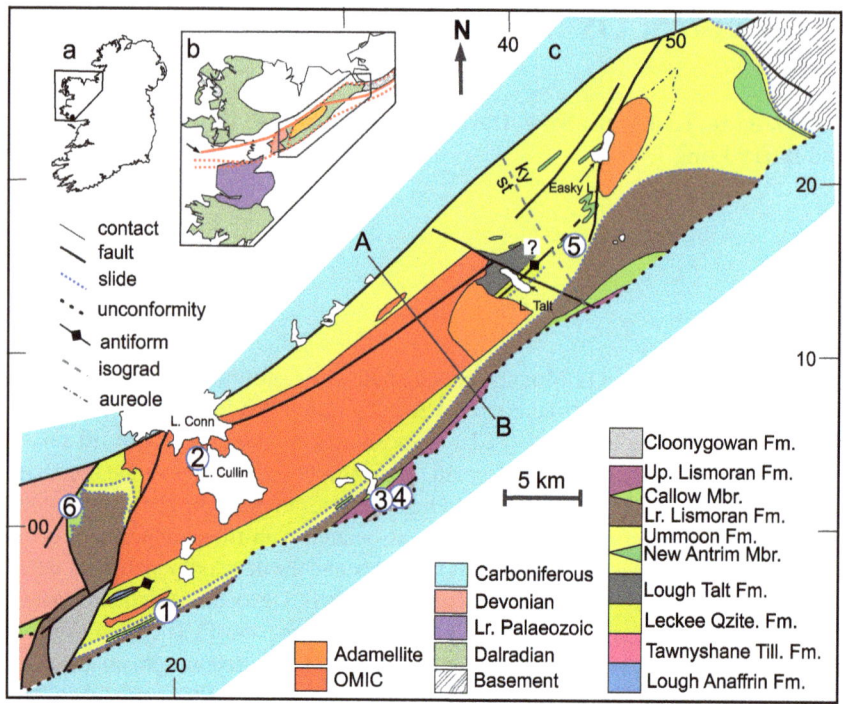

Fig. 1 a Ireland with inset showing location of Fig. 1b **b** summary geology of northwestern Ireland showing major faults in red. The Fair Head—Clew Bay Line is shown as a continuous red line (arrowed). Lr.Palaeozoic refers to the Cambrian to Silurian of South Mayo **c** Geological map of the Central Ox Mountains after Alsop and Jones (1991) and Daly (2009). Abbreviations: Fm. = Formation; Lr. = Lower; Mbr = Member; OMIC = Ox Mountains Igneous Complex; Qzite = Quartzite; Till. = Tillite; Up. = Upper. The position of the localities visited are shown

Dewey 1986). The core of the ridge is an uplifted block (an inlier) of Precambrian to Palaeozoic metamorphic and igneous rocks (Fig. 1) unconformably overlain by Carboniferous and Devonian sedimentary rocks (chapter "Devonian and Carboniferous"). The north-eastern sector of the Ox Mountains in County Sligo contains granulite to eclogite facies Neoproterozoic rocks of the Slishwood Division which are described in chapter "The Slishwood Division and Its Relationship with the Dalradian Rocks of the Ox Mountains". The formation and preservation of the Inlier reflects uplift during late Caledonian strike-slip shear events that occurred along the Laurentian margin (Hutton and Dewey 1986). Within the central Ox Mountains inlier (Fig. 1) polyphase-deformed metasedimentary rocks of the Dalradian Supergroup (Alsop and Jones 1991) record two orogenic events. Firstly, early deformation fabrics (D1 and D2) and up to kyanite grade regional metamorphism developed during the Grampian Orogeny, associated with the onset of ocean closure. The later, early Devonian, Acadian orogeny marked final ocean closure, during which time the Ox Mountains Igneous Complex was emplaced (McCaffrey 1992) and the earlier

Grampian fabrics were partially overprinted, reactivated and retrogressed (D3 and later). Deformation and intrusive structures preserved within these units enable the accretion events and, in particular, the strike-slip deformation to be examined and constrained.

2 The Dalradian

The metasediments of the Central Ox Mountains are assigned to the Dalradian Supergroup (Long and Max 1977). Alsop and Jones (1991) revised the stratigraphic scheme of Long and Max (1977) and established a correlation with the Dalradian succession in Donegal. Crucially, they identified a limestone/tillite/quartzite association at Lough Anaffrin (3 km northwest of Locality 4.1, Fig. 1c) where Appin Group limestones lie in the core of a D2 antiform (Long 1974). Alsop and Jones (1991) recognise the following formations—in ascending stratigraphic order: Lough Anaffrin; Tawnyshane Tillite; Leckee Quartzite; Lough Talt; Ummoon; Lower and Upper Lismoran; and the Cloonygowan (Table 1 and Fig. 1). A further discussion of the Dalradian stratigraphy in Ireland is given in Chapter "The Basement and Dalradian Rocks of the North Mayo Inlier", Sect. 2.4.

The Lough Anaffrin Formation belongs to the Appin Group and all others to the Argyll Group with the exception of the Cloonygowan Formation which is attributed to either the Southern Highland Group (Alsop and Jones 1991) or the Clew Bay Complex (Chew 2001). The metavolcanic rocks of the New Antrim Member (Ummoon Formation, Locality 4.5) have continental tholeiite affinities and can be correlated with metavolcanic horizons in the Dalradian of North Mayo (Winchester et al. 1987; see Chapter "The Basement and Dalradian Rocks of the North Mayo Inlier"). The lavas of the Callow Member (Lismoran Formations) were probably erupted in an oceanic (MORB) setting (Winchester et al. 1987). The aim of this excursion is to investigate the complex tectonic evolution of this inlier rather than study Dalradian stratigraphy in detail. Only the Ummoon Formation (Locality 4.1 and 4.5), the Upper Lismoran Formation (Locality 4.3) and the Cloonygowan Formation (Locality 4.4) are visited.

3 The Ox Mountains Igneous Complex

The Ox Mountains Granodiorite is volumetrically the most important part of the Igneous Complex and was intruded as a composite tabular body c. 25 km long × 5 km wide, during regional transpressional deformation (McCaffrey 1992). Evidence for the timing of the intrusion events relative to the wall rock deformation history can be examined at Cloonkesh (Locality 4.1). The internal composite nature of the intrusion and a range of spectacular sinistral strike-slip shear zone related structures are displayed at Pontoon (Locality 4.2). The accurate U–Pb dating of the Ox Mountains Granodiorite at 412.3 ± 0.8 Ma by Chew and Schaltegger (2005) is important

Table 1 Stratigraphy of the Ox mountains Dalradian succession (Alsop and Jones 1991)

Group	Sub-Group	Formation	Lithology	Thickness
Southern highland		Cloonygowan	Green/grey phyllites, psammites with quartz pebbles. Poorly sorted and beds up to 3 m in thickness	750 m
Argyll	Easdale	Upper Lismoran	Pale green semipelite interlayered with psammite	>700 m
		Callow member	Dark schistose amphibolite preserving some volcanic features	150 m
		Lower Lismoran	Grey-green semipelite interlayered with psammite	400 m
		Ummoon including New Antrim Member (NAM)	Mainly semipelite with persistent schistose amphibolite of the NAM	<600 m NAM <50 m
		Lough Talt	Thin interbeds (< 1 m) of pelites, semipelites, calc-silicates, feldspathic psammite and quartzite	0–300 m
	Islay	Leckee Quartzite	Massively bedded, mostly impure quartzite	580–2300 m
		Tawnyshane Tillite	Pelites, semipelites and psammites with local coarse feldspathic detritus	45 m
Appin	Blair Atholl	Lough Anaffrin	Interbedded tremolite and serpentinite marbles and calc-silicates	>80 m

in constraining sinistral shear events along the Laurentian margin in early Devonian (Lockhovian) times (Becker et al. 2020).

4 Structure and Metamorphism

The rocks of the Central Ox Mountains preserve a long history of metamorphism and deformation spanning the mid-Ordovician Grampian to late-Devonian Acadian Orogenies (Table 2). They sit astride a major fault zone, the Fair Head—Clew Bay Line (Fig. 1b), that was active from the Neoproterozoic to the Carboniferous and marks the upper crustal boundary, in eastern Ireland, between the deformed Laurentian margin and outboard terranes (Chew and Van Staal 2014). There are currently no precise dates for the Grampian metamorphic peak outside of Connemara. However, studies (Flowerdew et al. 2000) in the Slishwood Division (see chapter "The Slishwood Division and Its Relationship with the Dalradian Rocks of the Ox Mountains") suggest the main Grampian event took place between 470 and 459 Ma after the juxtaposition of this division with the Dalradian of the Central Ox Mountains. Peak metamorphism at ~ 460 Ma (Locality 4.5) was followed by rapid uplift and cooling at 460–450 Ma. The later Acadian sinistrally transpressive event is dated by the syn-tectonic emplacement of the Ox Mountains Igneous Complex (OMIC) at 412.3 ± 0.8 Ma (Chew and Schaltegger 2005) which controlled the development of Devonian pull-apart basins (McCaffrey 1997; see Chapter "Devonian and Carboniferous").

4.1 Structure

The earliest Grampian deformation (D1) probably took place under low greenschist-facies conditions and is only evidenced as a fabric wrapped around by D2 fold closures (Jones 1989; McCaffrey 1992). A single large southwest-northeast trending D2 antiformal anticline preserves the only Appin Group sediments in its core (Fig. 3). Limited structural evidence suggests vergence towards the northeast and no major inversion of the stratigraphy. A penetrative fabric formed during D2 folding and the Grampian metamorphic maximum was achieved late in this phase (MP2 of McCaffrey 1992, Locality 4.5). Structural correlations with the Clew Bay region (see Chapters The Basement and Dalradian Rocks of the North Mayo Inlier and The Ordovician Fore-Arc and Arc Complex) suggest that this deformation had a dextral component until 448 Ma (Chew et al. 2004). The Cloonygowan Formation (Southern Highlands Group) was thrust over the Upper Lismoran Formation (Argyll Group) during late D2 (McCaffrey 1992). However, the younger, lower metamorphic grade rocks still occupy the hanging wall suggesting that this thrusting did not fully reverse the original extensional geometry. Rapid uplift evidenced by common ~450 Ma cooling ages (Chew et al. 2003; Flowerdew et al. 2000) took place immediately after this event. A major southwest–northeast upright antiform formed during D3 with an axial planar crenulation cleavage (Figs. 2 and 3). This was associated with regional retrogression under greenschist facies conditions of the Grampian amphibolite facies assemblages (Fig. 3). The OMIC was emplaced late during D3 sinistral transpression (see Sect. 4.3 below) probably associated with the final closure of the Iapetus Ocean. High strain

Table 2 Deformation and metamorphic history of the Central Ox Mountains after Jones (1989), McCaffrey (1992, 1994, 1997), Flowerdew et al. (2000), Chew and Schaltegger (2005)

Time	Framework	Event	Kinematics	Nature of Deformation
Lower Carboniferous (Tournaisian) ~359–347 Ma	Basin margin normal faulting		Normal/Dextral	Syn-sedimentary faulting on NW and SE flanks of inlier
Late Devonian ~383–359 Ma	Transcurrent faulting, basin inversion		Dextral	Brittle reactivation along the Knockaskibbole fault
Middle Devonian ~393–383 Ma	Pull-apart Devonian basin formation. Continued unroofing		Sinistral	Chevron folds and kink bands. Retrogression
Early Devonian ~407–393 Ma	Late- to post-Acadian pull-apart Devonian basin formation. Continued unroofing. Emplacement of late granites	D4	Sinistral	Brittle reactivation of shear zones to form pull-apart basin
Early Devonian ~410 Ma	Acadian sinistral transpression		Sinistral	Final juxtaposition of greenschist-facies Southern Highlands Group against amphibolite-facies Argyll Group
Early Devonian (Lochkovian) 412.3 ± 0.8 Ma	Acadian Syntectonic granite emplacement and transcurrent shear, unroofing central Ox Mountains	D3 Post-peak metamorphic maximum conditions. Locally sillimanite in aureoles	Sinistral	Major F3 upright antiform. Ductile–brittle deformation progressively localized into shear zones within plutons and slides in wall rocks
Ordovician (Sandbian—Katian?) ~455–450 Ma	Late-Grampian orogenic collapse		Down-dip extension	

(continued)

Table 2 (continued)

Time	Framework	Event	Kinematics	Nature of Deformation
Ordovician (Darriwilian—Dapingian?) ~460 Ma	Grampian arc collision and obduction with ophiolites	D2 Penetrative fabric. Late MP2 regional metamorphic maximum (kyanite)	Dextral?	Major thrusting and pervasive fabric formation at mid-crustal levels. West directed overthrusting at upper-crustal levels
Ordovician (Tremadoc—Floian?) ~470–460 Ma	Initiation of Grampian arc-continent collision	D1 (low greenschist?), not present in southwest		
Ordovician (Tremadoc—Darriwilian) ~476–463 Ma				Juxtaposition of Central Ox Mountains with the Slishwood Division

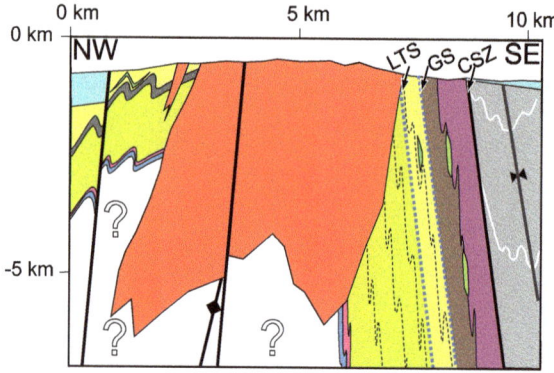

Fig. 2 Speculative cross-section along line A-B of Fig. 1 after McDermott et al. (1996), modified. Abbreviations: CSZ = Callow shear zone; GS = Glennawoo slide; LTS = Lough Talt slide. Colours as for Fig. 1c

zones were developed within the OMIC and the country rock during this time. Those in the country rock likely reactivated earlier Grampian structures. These high strain zones or faults which formed during regional deformation and metamorphism are referred to as slides (Hutton 1981). In the Dalradian these zones form the Glennawoo, Lough Talt slides and the Callow shear zone which disrupt the stratigraphy but define formation boundaries (Fig. 3). Later sinistral reactivation of these slides controlled the development of the Devonian pull-apart basins described in Chapter "Devonian and Carboniferous". These basins were inverted during latest Devonian, Acadian, reversal of slip sense along the Knockaskibbole Fault (Fig. 3). Such reversal of shear sense is reported in multiply deformed terranes globally (Dutta and Mukherjee 2019).

4.2 Regional Metamorphism

Yardley et al. (1979) report progressive Barrovian style metamorphism ranging from the garnet zone to the kyanite zone in the Appin and Argyll Group rocks of the Central Ox Mountains Dalradian. The grade increases from southwest to northeast (Fig. 3) with staurolite occurring in all but the extreme southwest and kyanite in the northeast. Yardley et al. (1979) suggest the peak of Grampian metamorphism occurred at 600–620 °C and at 0.6–0.7 GPa. Amphibolite assemblages were subsequently retrogressed to greenschist facies, especially in the Acadian D3 shear zones (Long and Max 1977; Yardley et al. 1979; Fig. 3). The Cloonygowan Formation, attributed to the Southern Highland Group, have sinistral shear fabrics and do not exceed low greenschist facies grade (Yardley et al. 1979).

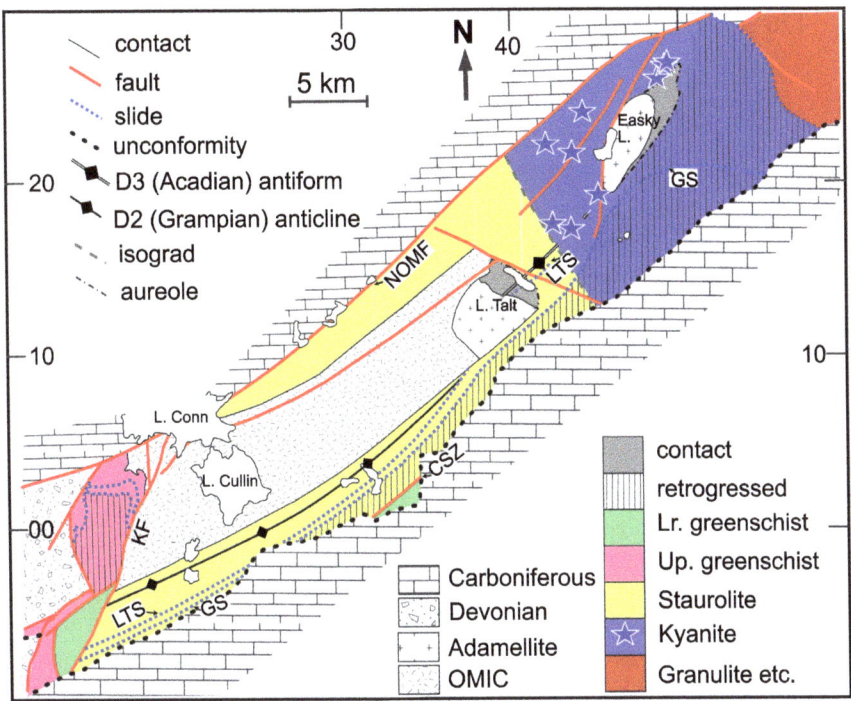

Fig. 3 Structural and metamorphic map of the Central Ox Mountains (after Yardley and Long 1981; McCaffrey 1992, 1997; Yardley et al. 1979). Abbreviations: CSZ = Callow Shear Zone; GS = Glennawoo Slide; KF = Knockaskibbole Fault; Lr. = lower; LTS = Lough Talt Slide; NOMF = North Ox Mountains fault; OMIC = Ox Mountains Igneous Complex; Up. = upper

4.3 Contact Metamorphism

The thermal aureole around the main granodiorite body of the OMIC is limited to the local development of fibrolite and sillimanite. This may reflect the gradual emplacement of the granodiorite as a series of sheets (McCaffrey 1992). In the Lough Talt area a late- to post-Acadian, poorly exposed, adamellite has a well developed aureole where sillimanite and andalusite overprint regional staurolite assemblages (Fig. 3). Similar contact relationships are found in the aureole of the coeval Easky adamellite some 15 km northeast (Fig. 3) for which Yardley and Long (1981) report PT conditions of 595 ± 30 °C and 0.25 ± 0.05 GPa. This suggests ~13–17 km of exhumation during the Acadian and before the ~400 Ma (Long, Max and O'Connor, 1984) emplacement of the Easky adamellite.

5 The Excursion

5.1 Aim

This excursion has two aims. Firstly, to demonstrate the emplacement of a small Early Devonian granitic igneous complex, the Ox Mountains Igneous Complex, during sinistral transpression (Locs 4.1, 4.2). Secondly, to establish that this took place along a major long-lived structure, the Fair Head—Clew Bay Line, that was active from the Neoproterozoic to the Carboniferous (Locs 4.3, 4.4, 4.5 and 4.6).

5.2 Locality 4.1: Cloonkesh (53.893614°, −9.242484°)

Access and other information Limited parking is available on the roadside next to a series of low relief exposures on the northern side of the access road. The access road (53.893668°, −9.258042°) is a right turn onto the L5888, 150 m north of the paired bridges over the Clydagh River on the R130 Castlebar to Pontoon road. Turn left after 350 m then continue a further 800 m.

Evidence for the synkinematic contact relationships for the Ox Mountains Igneous Complex and the Dalradian country rocks can be examined just outside the main body of the intrusion at this locality (Fig. 4a). Here, extensive exposures of psammitic to semi-pelitic lithologies of the Ummoon Formation (Table 1) contain metre to decametre-scale intrusive sheets of Ox Mountains granodiorite and pegmatites. They are part of the metasediments that have been correlated with the Argyll Group of the Dalradian Supergroup (Alsop and Jones 1991) and were deformed and metamorphosed to upper greenschist to lower amphibolite facies metasediments during the Grampian Orogeny.

At this locality, the main structures are tight to isoclinal upright folds with hinges parallel to the main gently NE- or SW-plunging stretching lineation (McCaffrey 1994). These folds formed during sinistral transpressional shearing that characterises the third phase (D3) structures in the Ox Mountains Inlier (Fig. 4b, c). These folds, lineation and accompanying strong steeply dipping planar foliation deform a strong composite bedding, S1 and S2 fabric that formed during Grampian events along the Fair Head—Clew Bay line (Chew et al. 2004).

In these exposures, in cross section view, the relationship between the upright F3 folds and folded sheet intrusions is well displayed (McCaffrey 1994) (Fig. 4c, e). Minor intrusions clearly cross-cut the F3 folds but they themselves are folded with upright more open folds with the same geometry (Fig. 4e) implying they were intruded during the D3 event. In plan view exposures, the intrusive rocks contain a pervasive foliation that is parallel to the fold axial planes. The sheet contacts that were intruded sub-parallel to the host rock fabrics show evidence for sinistral shearing on their margins (Fig. 4d). Minor intrusions intruded at a high angle to the foliation show evidence for anticlockwise rotation and shearing (Fig. 4e). In summary

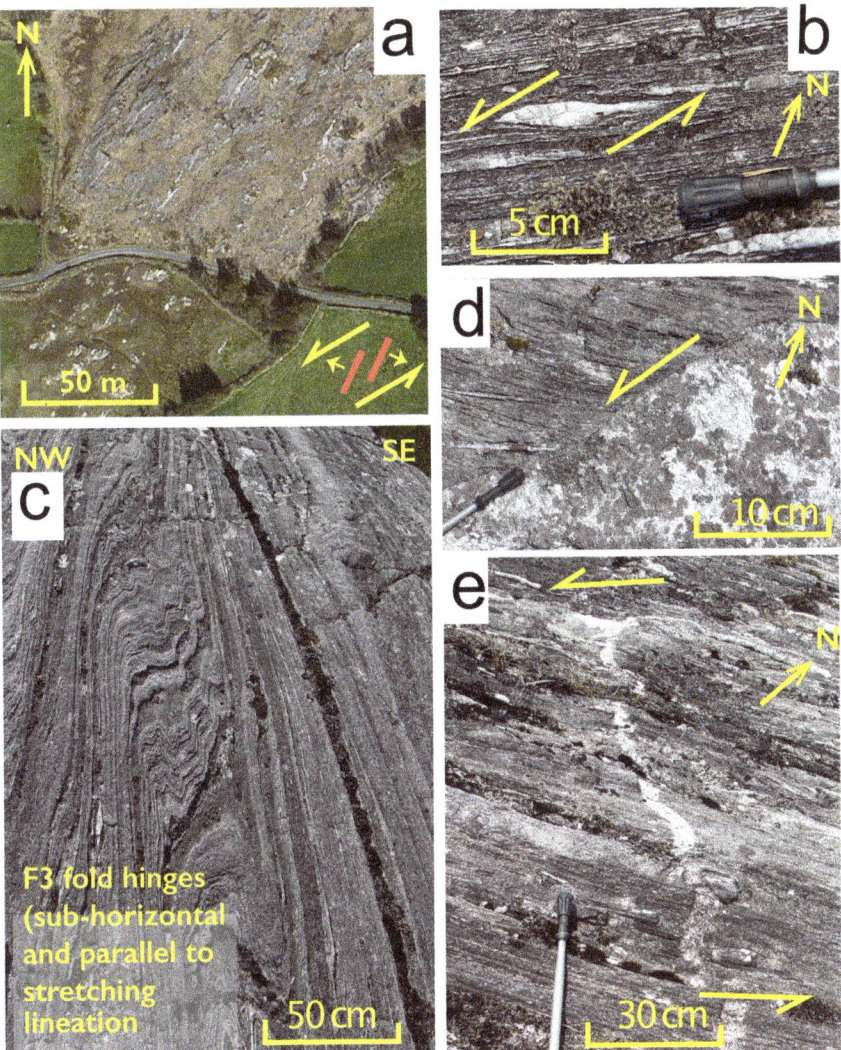

Fig. 4 a Digital Globe image of the Cloonkesh locality shows intrusive structures (light coloured veins). The inset at the bottom right shows the orientation of dilational structures under sinistral transpression **b** sinistral tails to asymmetric quartz boudin **c** subhorizontal F3 fold hinges in Ummoon Formation metasediments which are parallel to the stretching lineation. **d** granodiorite sheets emplaced along a left lateral shear zone **e** granitic vein that cuts across main D3 shear fabric but is itself buckled and sheared by anticlockwise rotation induced by sinistral shear. The photographs in Figs. 4b to 4d are a guide the type of structures that cn be found at this locality

these exposures show convincing evidence that the intrusion of the Ox Mountains granodiorite is synchronous with, and can be used to date, Late Caledonian sinistral shearing along the Highland Boundary (Fair Head Clew Bay line) on the Laurentian margin.

5.3 Locality 4.2: Pontoon Bridge (53.986683° −9.198322°)

Access and other information Parking is available in a lay-by on the east side of the R310 just to the south-west of Pontoon bridge (53.987373°, −9.197352°). Immediately to the south, road cuts in the Ox Mountains igneous complex are now quite weathered and the internal details are much better seen on the lakeside exposures. Walk c. 100 m south to the end of the road cuts and then access the lakeshore by walking east through a small stretch of forest for c.300 m from 53.986683° − 9.198322° (Fig. 5). In summer the undergrowth can be dense. If lake levels are low this locality can be accessed by parking at the small pier 400 m south (53.983692°, −9.198180°) and walking north along the lake shore.

The excellent glaciated rock surfaces by the lakeshore have been described by McCaffrey (1992 Fig. 6; Fig. 6a). In this area the Ox Mountains igneous complex is characterised by tonalite intrusive sheets separated by granodiorite and granitic screens typically 10–15 m thick, although some zones within each sheet show a finer scale (c. 50 cm) sheeting (Fig. 6c, d). The sharp straight contacts indicate intrusion into a crystalline host and the absence of chilled contacts indicate no significant temperature difference between host granodiorite and tonalite intrusions.

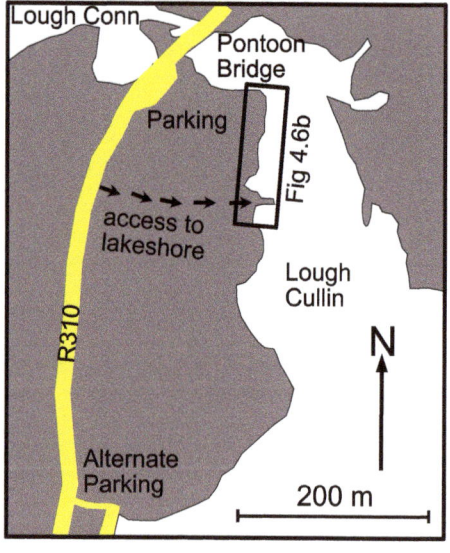

Fig. 5 Location map for locality 4.2

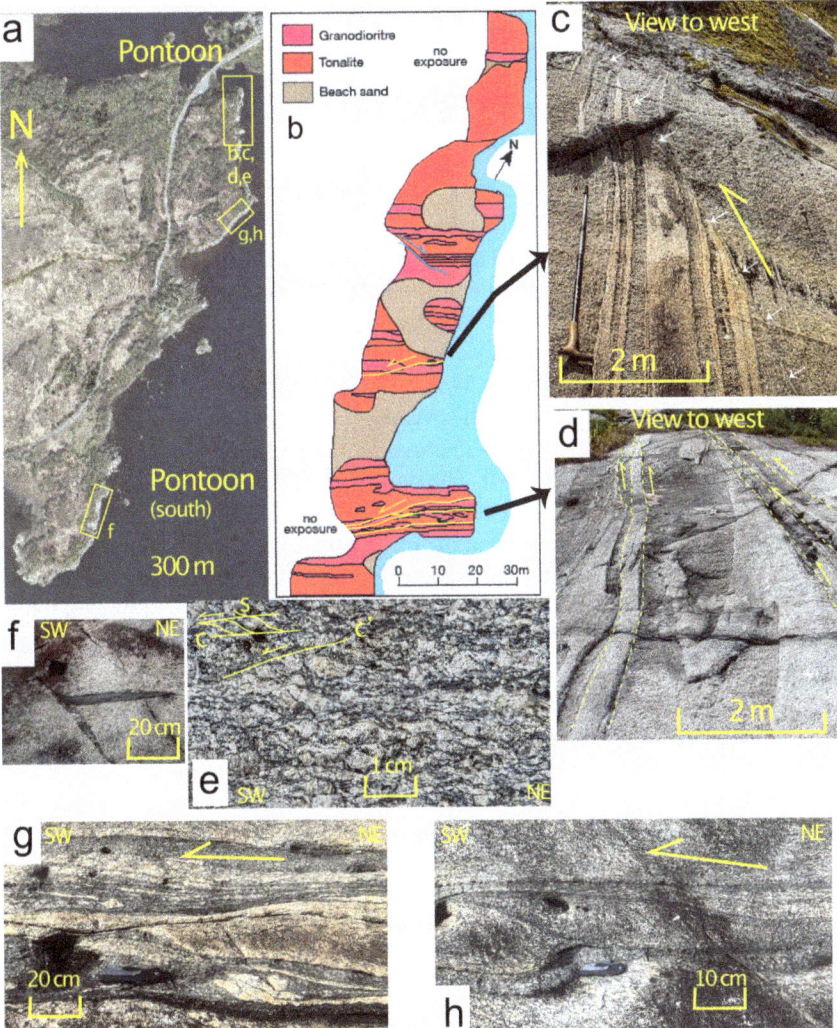

Fig. 6 a Digital globe image for the Pontoon locality. The position of the described features and a supplementary locality (Pontoon south) are shown **b** Sketch geological map showing details of the exposures on the lakeshore (after McCaffrey 1992) **c** Sheeted nature of the tonalite (darker) and the granodiorite (lighter) bodies **d** sharp contacts between the tonalite sheets and the granodiorite. The yellow dashed lines (solid on map) show the location of the early sinistral shears. These are (1 cm wide) fine-grained relict high temperature faults along which the contacts are displace by sinistral shearing with displacements of the order of 1 m or more **e** S-C and Extensional shear bands (C') deform main fabric **f** late dextral shear at Pontoon (on shoreline 500 m west of Healy's Hotel) **g** late shear zone—a sinistral strike slip duplex **h** localisation of late sinistral shearing on internal contacts

The exposures to south of Pontoon bridge (Fig. 6) preserve excellent evidence of the types of structures that can form as a mid-crustal magmatic body cools whilst undergoing bulk transpressional deformation. These include: (1) Early discrete shears, possibly formed during crystallisation; (2) a penetrative solid state foliation with textbook quality shear band fabrics; and (3) late conjugate discrete shear zones that deform the foliation.

(1) *Early discrete shears.* These structures described by McCaffrey (1994) are uncommon in plutonic rocks and their origin is difficult to determine because they are overprinted by the main foliation and in places by the late discrete structures. They are characterised by cm- to m-scale sinistral offsets of the granodiorite and tonalite contacts on thin (1 cm wide) slightly finer-grained bands that form generally clockwise to the main foliation. They are examples of single structures that deform several contacts (Fig. 6c, d) and examples of networks of connected shears that disrupt the multilayer of granodiorite and tonalite contacts (see Fig. 6d). The net effect of this shearing is to extend the contacts in a direction parallel to the intrusive contacts. These early structures possibly represent a form of 'melt enhanced embrittlement' (Davidson et al. 1994). Their preferential preservation in the Ox Mountains Intrusive Complex may be a result of the dynamic tectonic environment in which they formed. They were uplifted and cooled before being completely overprinted by subsequent crystal plastic deformation during ongoing sinistral shearing.

(2) *Main foliation*: The main penetrative foliation is not deflected by the early discrete sinistral shears and is a solid-state fabric, i.e. deformation occurred when all of the mineral phases had crystallised. It is a grain-shape alignment fabric defined by quartz, feldspar and mica which dips steeply to the NW (Fig. 6e). On the foliation planes, a mineral lineation can be detected that gently plunges NE or SW. A textbook sinistral shear band fabric (S-C fabric) is present in many parts of the intrusion and is especially well developed in the Pontoon exposures. The C planes and extensional shear planes (C') are penetratively developed anticlockwise of the main foliation and deflect the foliation and the 'tails' of feldspar grains in an asymmetrical sense that is consistent with sinistral shear (Fig. 6e). Individual shear planes are defined by fine-grained quartz ribbons lined with mica grains. In thin section, McCaffrey (1994) showed that the shear planes were associated with the development of myrmekite. This is a feldspar and quartz intergrowth texture that forms by breakdown of K-feldspar as the plutonic phases cooled and in this case was likely induced by the deformation (cf Simpson and Wintsch 1989).

(3) *Late shear zones*: The main foliation is locally intensified into a series of mylonitic shear zones, up to 50 cm across, that form on internal contacts. Some very good examples of strike-slip imbricates and duplexes (sensu Woodcock and Fischer 1986) are present at Pontoon. They are characterised by dark, fine-grained protomylonites. Also developed are a series of conjugate sinistral and dextral shear zones that modify the main foliation into broad fold

structures by drag (McCaffrey 1992, 1994). Notable examples of the sinistral and dextral shears at Pontoon (53.986683° −9.198322) were illustrated in McCaffrey (1994, Figs. 10 and 11) and shown in Fig. 6f, g and h).

5.4 Locality 4.3: Callow—Lismoran (53.968781°, −9.031283°)

Access and other information Proceed east to Foxford then take the N26 and drive 6.5 km to Callow. Park in the layby at 53.970434°, −9.031375°. Access requires walking, taking appropriate precautions, 190 m south on the N26 to the right hand fork for 'Lismorane'. Cross a small wall to visit the prominent frost shattered outcrops immediately northwest of this junction.

Localities 4.3 and 4.4 aim to show the stratigraphic and structural relationships across a D3 slide zone, the Callow shear zone. There are three such zones on the south limb of the major D3 antiform (Fig. 2): the Lough Talt slide; the Glennawoo slide; and the Callow shear zone (Figs. 2 and 3). These were developed during D3 after the consolidation of the OMIC, which then became a buttress and deformation was concentrated in the country rocks (McCaffrey 1992, 1994). The geology of the Callow area has been described in detail by Jones (1989) and this account is based on this work.

The Lismoran Formation comprising pale green/grey psammites and semipelites is separated into Upper and Lower units separated by the amphibolites of the Callow Member. Locality 4.3 is in the highly sheared Upper Lismoran Formation (Fig. 8a). The main S3 fabric is mylonitic, subvertical and strikes 060°. Quartz segregations are rodded and define a lineation that plunges 015° to the northeast. The F3 folds seen at Locality 4.1 are here transposed by the intensified S3 fabric. Shear band type asymmetric boudins within quartz veins (Fig. 7b) and shear bands, on scales ranging from 1–50 cm, develop as the contact with the Cloonygowan Formation is approached (Fig. 7). All indicators show a sinistral shear sense. Thin section analysis shows that S3 is defined by a strong crystallographic and dimensional orientation of quartz grains with large pressure-shadows filled by muscovite + chlorite + quartz + undeformed feldspar grains. This ductile fabric formed after, perhaps within 1–4 m.y., the crystallisation of the OMIC (McCaffrey 1992). However, sinistral motion on the faults bounding the block continued from the 412 Ma emplacement of the OMIC at least until the Middle Devonian (Table 2; see Chapter "Devonian and Carboniferous").

5.5 Locality 4.4: Cloonygowan (53.961533°, −9.021189°)

Access and other information Drive a further 1.3 km south on the N26 and park at the oblique crossroads with the L5377 signposted for Cloonygowan (53.962342°, −

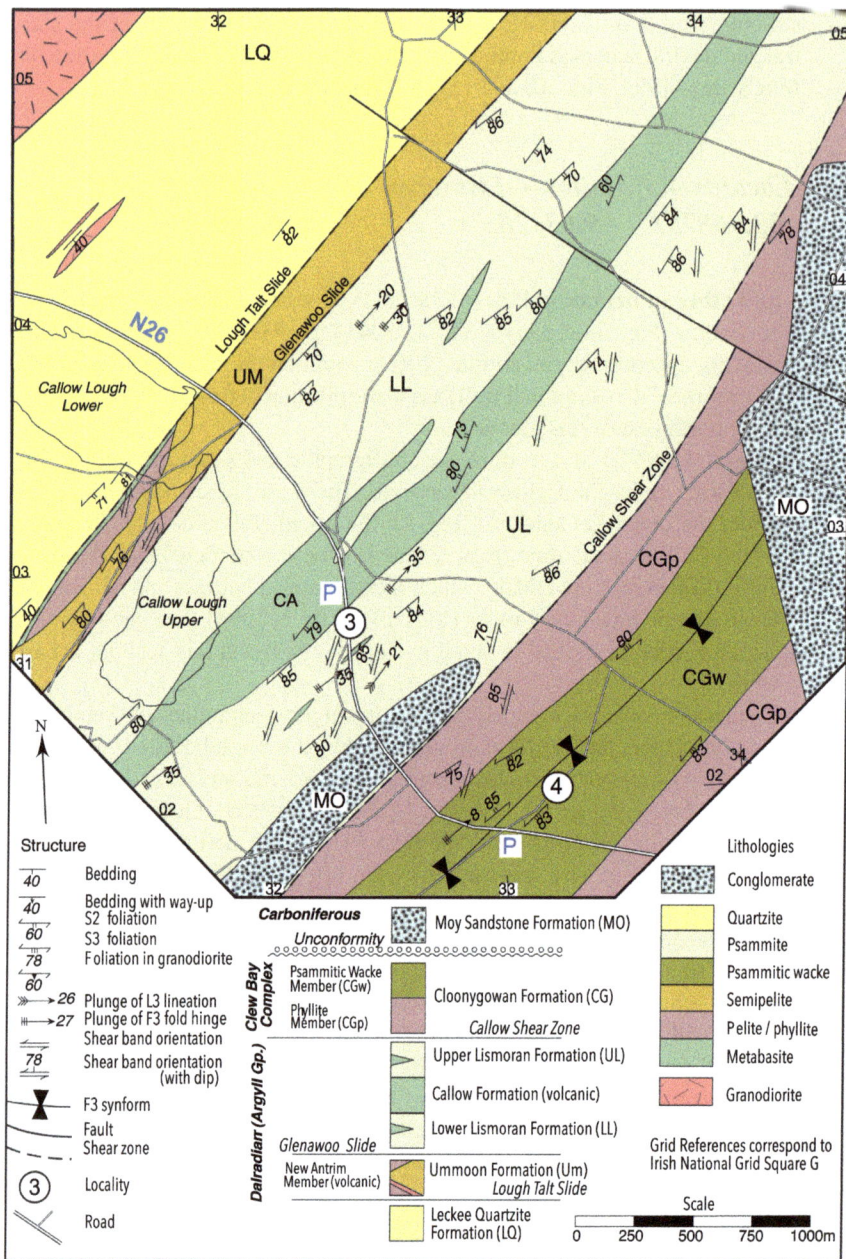

Fig. 7 Sketch geological map of the Callow area showing the locations of stops 4.3 and 4.4 which are represented by the circled '3' and '4' respectively

Fig. 8 a General view of Locality 4.3 **b** Plan view of shear band type asymmetric boudinage of quartz veins. Sense of shear is indicated; however, this section is highly oblique to the local stretching lineation. Both photographs looking west from main N63 (53.968719°, −9.031394°)

9.019105°). Limited parking is available. Walk 150 m northeast on the L5377. The Cloonygowan Formation is exposed in roches moutonnées on either side of the road. The outcrops on the southeastern side are more accessible.

The rocks of the Cloonygowan Formation comprise interbedded green-weathering coarse greywackes with pebbles of plagioclase and opalescent quartz, greywackes with 5–10% pebbles and light greenish/grey phyllites. The matrix contains chlorite, quartz, plagioclase and pyrite. Sorting is poor and coarser beds exhibit normal grading. The Cloonygowan Formation displays two spaced cleavages probably produced by pressure solution where the rock has segregated into quartz and phyllosilicate layers. The early cleavage (S2) is close to bedding in most instances. It is cut and crenulated by a later upright S3 cleavage whose mean strike is 059°. In places, there is a well-developed sinistral S-C fabric. Minor F3 fold vergence gives a major synform running through the grit member (Fig. 7). An associated stretching lineation plunges at about 10° to the northeast. Way-up can be obtained in the grit member where graded bedding and sharp bases are visible. Based on this way-up evidence and employing facing directions on the cleavage, F2 fold closures can be inferred and have a much shorter wavelength than the major F3 synform. Jones (1989) reports sinistral shear bands within 50 m of the contact with the Upper Lismoran Formation.

Fig. 9 a Greywackes of the Cloonygowan Formation with opalescent quartz and plagioclase granules **b** S2 (sub-horizontal and marked by a line) is folded by a minor F3 fold. The D3 stretching lineation, trending 060° is marked by an arrow. An extensional quartz vein at a high angle to this lineation and perpendicular to S2 is at the top of the view. Both photographs are from a locality at southeast side of road (53.962352°, −9.018910°)

The Upper Lismoran and the Cloonygowan Formations thus show a similar structural development but a disparity in maximum metamorphic grade during D2. The main Ox Mountains succession, east of the Knockaskibbole Fault, which includes the Upper Lismoran Formation attained amphibolite facies assemblages (Long and Max 1977; Yardley et al. 1979), whilst the Cloonygowan Formation did not exceed lower greenschist facies (Alsop and Jones 1991). The extent to which the relative uplift of the Ox Mountains Dalradian succession with respect to the Cloonygowan Formation was due to movement along the shear zone as a whole or displacement on the late Callow fault, which now separates these formations, is unresolved (Fig. 7).

5.6 Locality 4.5: Zion Hill (54.100308°, −8.877857°)

Access and other information Proceed to Swinford then take the N5 towards Dublin. Leave the N5 at the Charlestown junction and drive north on the N17 to Tobbercurry. Take the R294 to Annagh and drive 8.9 km to Mullany's Cross. Take the right fork and follow the signposts for 'Mass Rock'. Turn left after 1.7 km then left again after 3.8 km. Drive past the Mass Rock on your right, then turn right after 1.1 km at the 'T' junction with track. Zion Hill is 200 m north on the left. Parking (Fig. 10) is available at the small sandpit (54.102099°, −8.874272°). This locality is on private land and permission should be sought from the nearby farmhouse (54.102817°, −8.874216°). This locality is probably not suitable for large parties.

The aim is to demonstrate: the mafic vulcanism associated with Dalradian sedimentation; the Grampian D2 fabrics and the associated kyanite zone regional metamorphism (620 ± 30 °C and 8 ± 2 kbar); and the transposition of the Grampian fabrics (D2) during the formation of the D3 major antiform (Fig. 3) synchronous with the emplacement of the OMIC in the southwest during the Acadian Orogeny.

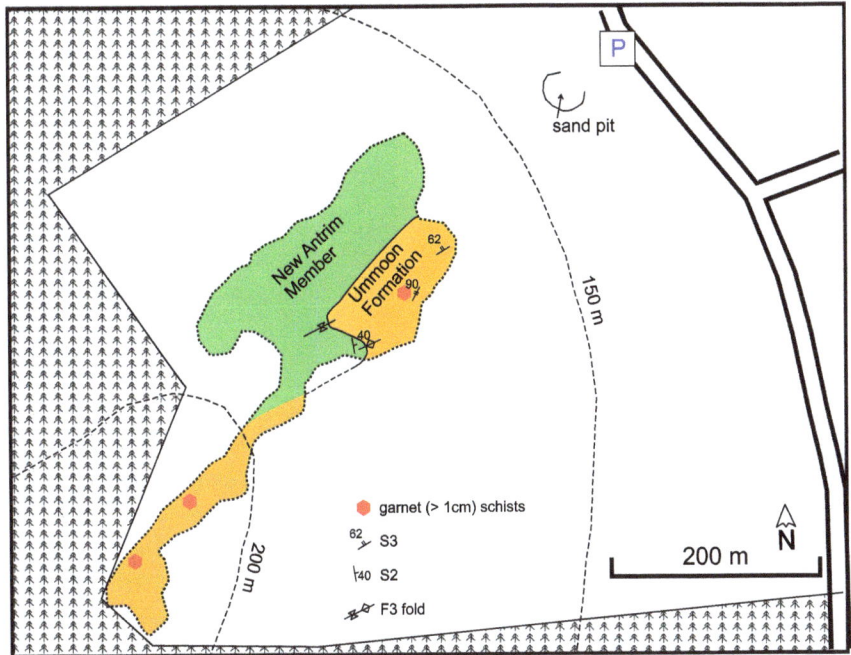

Fig. 10 Sketch geological map of Zion Hill

Proceed southwest climbing the hill behind the small sandpit (Fig. 10). There are low cliffs on the northwest face of the ridge. The main exposures are along the ridge, many of which are frost shattered but largely in place. Pelitic schists of the Ummoon Formation contain centimetric-scale porphyroblasts of garnet, kyanite and staurolite. Zion Hill is just east of the kyanite isograd (Yardley et al. 1979). Almandine garnet can comprise up to 25% of the rock (Fig. 11a). Garnet overgrows the S2 (Grampian) foliation, indicating that the Grampian metamorphic maximum postdated the major S2 fabric development event, but they are wrapped by the Acadian S3 fabric with the development of quartz-filled pressure shadows. The contact with the metadolerites of the New Antrim Member runs approximately along the ridge. The metadolerites are fine grained, sheared and locally have apparent relic ophitic texture. The contact with the pelites is sharp and highly sheared. However, the metadolerites share the same deformation history as the Ummoon Formation, which is consistent with them being erupted/intruded during sedimentation (Long and Max 1977). An earlier (S2) fabric within the metadolerites is preserved within the mesoscale F3 closures (Fig. 10). This earlier fabric is at a high angle to and is folded by F3 (Fig. 11b). The vergence of the asymmetric mesoscale F3 folds indicates that they lie on the southern limb of the main Ox Mountains antiform (Fig. 1). Loose blocks of psammite within the Ummoon Formation, located within the fold closures, show that S3 is a spaced cleavage with pressure solution that crenulates S2 (Fig. 11c).

Fig. 11 a garnetiferous mica schist of the Ummoon Formation (54.099963°, −8.877060°) **b** F3 folds (dashed white lines mark F3 axial surfaces) developed in S2 in New Antrim Member (yellow line) above a small thrust (orange dashed line) (54.099724°, −8.878137°) **c** S2 (yellow line) crenulated by S3 (dotted white line) (54.100398°, −8.876661°) **d** Kyanite (arrowed) in Ummoon Formation 54.098614°, −8.879675°)

The best examples of large (>1 cm) kyanite (Fig. 11d) and staurolite porphyroblasts occur on the ridge near to the summit.

5.7 Locality 4.6: Windy Gap (53.954158°, −9.315519°)

Access and other information From Castlebar take the R310 for Pontoon; after c.1 km take the left fork onto the L1721 towards Burren and proceed to the highest point of the road, called Windy Gap, and park in the large car park at 53.954232°, −9.315546° with fine views over Glen Nephin to the north. This locality (Fig. 12) examines the Dalradian of the Raheen Barr succession (Long and Max 1977) that

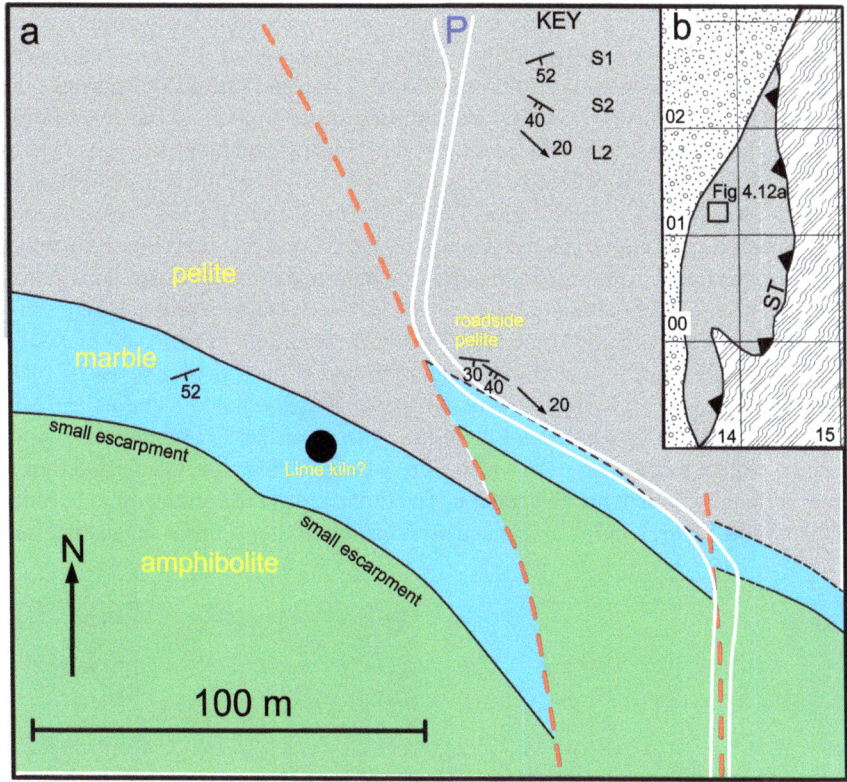

Fig. 12 a Sketch geological map of the Windy Gap region after Long and Max (1977). The 'P' in the north is located at the south end of the lay by **b** Outline map of the Sheeans region showing the Glenisland Formation (grey) in the hanging wall of the Sheeans Thrust (ST). Devonian sediments are shown with a dot—circle ornament and the main Ox Mountains Succession as wavy lines. Irish National Grid lines, Grid Square G, are shown

underlies but is in faulted contact with the Devonian Kings Hill Formation (see Locality 10.5).

The aim of this traverse is to demonstrate the association of mafic magmatism with Middle Dalradian sedimentation and to establish the lower maximum metamorphic grade typical of the southwestern region, west of the Knockaskibbole Fault (Figs. 1 and 3).

The car park lies on the Dalradian Glenisland Formation which contains pelites, white marbles, ortho-quartzites and massive amphibolites overlain by psammites (Long and Max 1977). The section examined occupies the lower part of this formation, and starts on pelites passing up section through marbles to the overlying amphibolites. The Glenisland formation is attributed to the Easdale Subgroup and correlated with the Nephin Succession of North Mayo (Winchester and Max 1996). The pelites are well exposed at a road cutting on the east side of the first bend south of the car park.

Mica schists have a primary schistosity that is overgrown by idioblastic garnet. Currall (1963) reports a muscovite + chlorite + garnet + albite + quartz ± chloritoid (garnet zone) assemblage with no sign of biotite growth. Long and Max (1977) attribute this fabric, which is locally S1, to the Ordovician Grampian Orogeny and correlate it with S2 in the main Ox Mountains Succession. A retrograde crenulation cleavage, which is locally S2, (Fig. 13b) is believed to be coeval with the Devonian emplacement of the OMIC and correlated with the regional S3 fabric (Long and Max 1977). The traverse then crosses a gully with limited exposures of white to grey marble (Fig. 13c), which have been exploited in the past for their lime content, and ortho-quartzite. This marble was used to make agricultural lime. A small circular depression probably represents the remains of an 'egg cup' lime kiln (Fig. 12a). The ridge to the south marks the base of the meta-volcanic sequence. These are fine-grained poorly schistose rocks (Currall 1963) and in this region are mostly metadolerites which Winchester and Max (1996) characterise as continental tholeiites. The Glenisland Formation lies in the hanging wall of the Sheeans thrust (Fig. 12b) which emplaces the Glenisland formation over the Ox Mountains Succession. The thrust dips to the northwest at between 5-15° (Currall 1963) and is associated with cataclastic fault rocks suggesting post

Fig. 13 a View of locality 4.6 looking south from the Windy Gap view car park **b** Garnet mica schists of the Glenisland Formation. The Grampian S1 fabric is in the plane of the photograph. The lineation associated with the Acadian S2 crenulation cleavage is marked by an arrow (53.953453°, −9.315410°) **c** clean, grey calcite marble of the Glenisland Formation (53.953095°, −9.316139°). The stick is 1 m. **d** Remains of 'egg cup' lime kiln where broken limestone was interlayered with fuel and burned to produce agricultural lime. 'Egg cup' kilns, named after their shape, had a ventillated grate at the base called the 'eye' and were used mainly during the eighteenth and nineteenth centuries. Field of view ~20 m, facing towards the west (53.953122°, −9.315725°)

D3 emplacement (Jones 1989) which may have been associated with the inversion of the overlying Devonian basins (see Chapter "Devonian and Carboniferous").

References

Alsop G, Jones C (1991) A review and correlation of Dalradian stratigraphy in the southwest and central Ox Mountains and southern Donegal, Ireland. Irish J Earth Sci 99–112

Becker R, Marshall J, Da Silva A-C, Agterberg F, Gradstein F, Ogg J (2020) The Devonian period. In: Geologic Time Scale 2020. Elsevier, pp 733–810

Chew D, van Staal C (2014) The ocean–continent transition zones along the Appalachian-Caledonian Margin of Laurentia: examples of large-scale hyperextension during the opening of the Iapetus Ocean. Geosci Can 41:165–185

Chew DM (2001) Basement protrusion origin of serpentinite in the Dalradian. Irish J Earth Sci 19:23–35

Chew DM, Daly J, Flowerdew M, Kennedy M, Page L (2004) Crenulation-slip development in a Caledonian shear zone in NW Ireland: evidence for a multi-stage movement history. Geol Soc Lond Spec Pub 224:337–352

Chew DM, Daly JS, Page L, Kennedy M (2003) Grampian orogenesis and the development of blueschist-facies metamorphism in western Ireland. J Geol Soc Lond 160:911–924

Chew DM, Schaltegger U (2005) Constraining sinistral shearing in NW Ireland: a precise U-Pb zircon crystallisation age for the Ox Mountains Granodiorite. Irish J Earth Sci 55–63

Currall AE (1963) The geology of the south-west end of the Ox Mountains, Co Mayo. P Roy Irish Acad B 63B:131–165

Daly JS (2009) The precambrian. The geology of Ireland, 2nd edn. Dunedin Academic Press, Edinburgh, pp 7–42

Davidson C, Schmid SM, Hollister LS (1994) Role of melt during deformation in the deep crust. Terra Nova 6:133–142

Dutta D, Mukherjee S (2019) Opposite shear senses: geneses, global occurrences, numerical simulations and a case study from the Indian Western Himalaya. J Struct Geol 126:357–392

Flowerdew MJ, Daly JS, Guise PG, Rex DC (2000) Isotopic dating of overthrusting, collapse and related granitoid intrusion in the Grampian orogenic belt, northwestern Ireland. Geol Mag 137:419–435

Hutton D (1981) Tectonic slides in the Caledonides. Geol Soc Lond Spec Pub 9:261–265

Hutton DHW, Dewey JF (1986) Palaeozoic terrane accretion in the Western Irish Caledonides. Tectonics 5:1115–1124

Jones CS (1989) The structure and kinematics of the ox mountains, western Ireland: a mid-crustal transcurrent shear-zone. Unpublished PhD thesis, Durham University

Long C, Max M (1977) Metamorphic rocks in the SW Ox mountains inlier, Ireland; their structural compartmentation and place in the caledonian orogen. J Geol Soc Lond 133:413–432

Long CB (1974) A note on the stratigraphy of the probable Dalradian metasediments north and northeast of Castlebar, Co. Geol Surv Ireland Bull 1:459–469

Long CB, Max MD, O'Connor PJ (1984) Age of the Lough Talt and Easky Adamellites in the central Ox Mountains, NW Ireland, and their structural significance. Geol J 19:389–397

McCaffrey K (1992) Igneous emplacement in a transpressive shear zone: Ox Mountains igneous complex. J Geol Soc Lond 149:221–235

McCaffrey K (1994) Magmatic and solid state deformation partitioning in the Ox Mountains Granodiorite. Geol Mag 131:639–652

McCaffrey K (1997) Controls on reactivation of a major fault zone: the Fair Head-Clew Bay line in Ireland. J Geol Soc Lond 154:129–133

Simpson C, Wintsch RP (1989) Evidence for deformation induced K-feldspar replacement by myrmekite. J Met Geol 7:261–275

Soper N, Hutton D (1984) Late Caledonian sinistral displacements in Britain: implications for a three-plate collision model. Tectonics 3:781–794

Winchester J, Max M (1996) Chemostratigraphic correlation, structure and sedimentary environments in the Dalradian of the NW Co., Mayo inlier. NW Ireland. J Geol Soc Lond 153:779–801

Winchester J, Max M, Long C (1987) Trace element geochemical correlation in the reworked Proterozoic Dalradian metavolcanic suites of the western Ox mountains and NW Mayo Inliers, Ireland. Geol Soc Lond Spec Pub 33:489–502

Woodcock NH, Fischer M (1986) Strike-slip duplexes. J Struct Geol 8:725–735

Yardley BWD, Long CB (1981) Contact metamorphism and fluid movement around the Easky adamellite, Ox mountains, Ireland. Min Mag 44:125–131

Yardley BWD, Long CB, Max MD (1979) Patterns of metamorphism in the Ox mountains and adjacent parts of western Ireland. Geol Soc Lond Spec Pub 8:369–374

The Ordovician Arc Roots of Connemara

Bruce W. D. Yardley and Robert A. Cliff

Abstract The geological history of the rocks of Connemara is summarised, concentrating on the period of folding, metamorphism and magmatism that took place as an arc developed during Ordovician times, and produced the rocks we see today. Arc magmatism began after a period of folding and medium pressure metamorphism, and superimposed higher-temperature, lower-pressure assemblages over much of the region, metamorphism was accompanied by folding throughout. Later in the Ordovician, these rocks were thrust up over a thick mylonite unit to their present structural position. Despite many years of intensive mapping and study with twentieth century techniques, there are still some significant uncertainties in our understanding of the development of the rocks of Connemara, and a few of these are highlighted. The available geochronological data is reviewed in detail and ages recalculated with currently recommended decay constants; however, uncertainties remain which could be resolved with current techniques. Five field excursions are presented, incorporating a range of structural, metamorphic and igneous features. These illustrate both arc magmatism and the accompanying metamorphism, including important aspects of the P–T–t path.

1 Introduction

Connemara is a region of spectacular landscapes and seascapes extending from the north shore of Galway Bay along the Atlantic coast north from Slyne Head to Leenane in the west and to Lough Corrib in the east. In geological terms, it is an inlier of igneous and metamorphic rocks with affinities to the metamorphic rocks of the Ox Mountains, North Mayo and Donegal, as well as to the southern Scottish Highlands, cut by Siluro-Devonian post-orogenic granites. Carboniferous rocks rest on the Connemara basement in the east, Silurian ones in the north. Users of this volume should bear in mind that some of the excursions described here are very close to

---

B. W. D. Yardley (✉) · R. A. Cliff
School of Earth and Environment, University of Leeds, Leeds LS2 9JT, UK
e-mail: B.W.D.Yardley@leeds.ac.uk

© The Author(s), under exclusive license to Springer Nature Switzerland AG 2022
P. D. Ryan (ed.), *A Field Guide to the Geology of Western Ireland*, Springer Geology
Field Guides https://doi.org/10.1007/978-3-030-97479-4_5

excursions in other chapters, and plan accordingly. You should also be aware that speeds on the narrow and winding roads are very slow, especially in summer.

The metamorphic rocks of Connemara are late Precambrian in age and belong to the Dalradian supergroup. The recognition of the distinctive late Precambrian tillite in the Dalradian of Connemara by Kilburn et al. (1965) allowed detailed correlations to be made with Dalradian rocks in western Scotland and Donegal. Connemara is however distinct in that it occurs to the south of all the other areas of Dalradian rocks and is separated from them by the Lower Palaeozoic rocks of the South Mayo trough. Connemara is also unusual in that there are a number of syn-orogenic intrusions ranging from gabbros to granites. In this respect, the area is comparable to Aberdeenshire in Scotland. The general distribution of the major rock types is shown in Fig. 1.

Connemara has had a long and distinguished history of geological research extending back to the work of Patrick Ganly in the mid nineteenth century. The modern body of work began in the late1920s, when L. R. Wager started work on the gabbros near Roundstone, before transferring his attention to Greenland. R. M. Shackleton initiated structural and stratigraphic studies of the metasediments in the 1930s and over the next 30 years his group extended their coverage over the Twelve Bens and much adjacent ground, unravelling the polyphase folding as we know it today. This work is summarised by Tanner and Shackleton (1979). B. E. Leake started work south of Clifden in 1951 before studying the Cashel-Lough Wheelaun intrusion

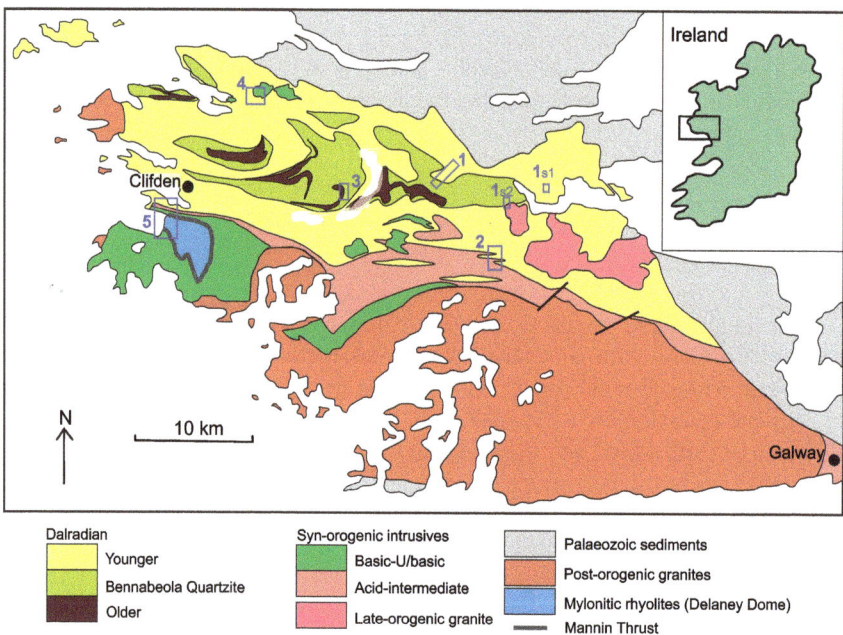

Fig. 1 Outline geological map of Connemara, based on Leake et al. (1981). Numbered boxes denote Excursions

for his Ph.D., and then setting about systematic detailed mapping of the remainder of the metamorphic belt with the help of numerous Ph.D. students. This culminated in the publication of a 1:63,360 geological map (Leake et al. 1981) and an accompanying memoir (Leake and Tanner 1994), updated for the Clifden district by Leake (2021). Much of the geology of the "Leake map" was subsequently incorporated into the Irish Geological Survey's 1:100,000 geological map (Sheet 10, Long et al. 1995). In this chapter, we have used Leake and Tanner (1994) as a reference for terminology and interpretations, except where noted. Their interpretations sometimes contradict the geological map (Leake et al. 1981), especially in the northwest.

Connemara studies have also led to a number of wider advances in geochemistry, petrology and geochronology, and we will point out some of these in what follows. Leake developed improved methods for rapid rock analysis by XRF (Leake et al. 1969) so that he could test his ideas about the evolution of the Connemara gabbros and related rocks. Connemara was also one of the first metamorphic belts to be studied with the newly developing techniques of whole rock geochronology in the 1960s, while at the end of the century the work of Friedrich (1998) brought Connemara to the fore in the emerging field of dating single grains by robust U–Pb methods. In metamorphic petrology, it was one of the first areas for which the metamorphism was described with a P–T–t path (Yardley 1976). With all this intense activity in the second half of the twentieth century it is perhaps not surprising that there has been relatively little work over the past 20-years. The fields of study which were pioneered in Connemara in the last century have now developed new technologies with the potential to give much needed new insights. There remain some significant difficulties with our present understanding of Connemara geology and we will point some of these out in the following pages in the hope that it will encourage new research. Not only are there continuous improvements in laboratory facilities, but also the construction of new road cuttings has opened up new exposures in some crucial areas, although in general there has been catastrophic loss of access to the hillsides since the late twentieth century.

These excursions are largely based on student field trips over many years, and we are indebted to our colleagues and students for many improvements through the years. We are also grateful to Bernard Leake for supplying an advance copy of his in press paper, Bob Wintsch for an extremely perceptive review and to Paul Ryan for his patient and helpful editing.

2 Geological Outline

2.1 Stratigraphy

Connemara has been important for the understanding of orogenic belts in large part because it has a well-defined stratigraphy. This has aided the recognition of large-scale structures as well as making possible geochemical and petrological comparisons between areas with different metamorphic histories. There are distinctive marble and quartzite horizons which trace out structures over kilometres, and mature pelitic sediments which develop a range of metamorphic minerals. The general outline of lithologies has been recognised since the nineteenth century but it was refined in the mid twentieth century through the work of Shackleton and his students in the Twelve Bens, summarised in Tanner and Shackleton (1979). The succession was further refined by B. E. Leake and students in eastern Connemara, presented in Leake and Tanner (1994). Tanner and Shackleton (1979) had already unravelled the complexities of the folding to recognise the stratigraphic sequence illustrated in Fig. 2. Particularly significant was the recognition by Kilburn et al. (1965) of a boulder bed which could be correlated with the glacial tillite in the less-metamorphosed Dalradian sequence of the southwest Highlands of Scotland. Here we are only concerned with the role

Fig. 2 Simplified Dalradian stratigraphy of Connemara, based on Tanner and Shackleton (1979) and Leake and Tanner (1994). For further details of parts of the succession, see Excursions 1 and 3

Ben Levy Formation	pelitic schist semi-pelite psammite
Cornamona Formation	graphitic schists minor marble
Ballynakill Formation	pelitic schist psammite and pebbly bands
Lakes Marble Formation	calcite marble quartzite amphibolte
Streamstown Formation	pelitic schist semi-pelite psammite
Bennabeola Quartzite Fm	quartzite
Cleggan Boulder Bed Fm	psammite boulder beds
Barnanoraun Formation	pelitic schist psammite
Connemara Marble Fm	dolomitic marble quartzite calc-silicate rocks

of stratigraphy in understanding the structure and metamorphism of the region and will not focus on the issues of correlation which are dealt with in existing reviews.

In the following descriptions, we will follow the regional terminology for metamorphosed siliciclastic rocks. Quartzite denotes quartz-rich meta-sandstones which are resistant to weathering and form conspicuous, white outcrops, irrespective of the precise feldspar content. Some quartz-rich meta-sandstones appear to have had coarse quartz grains originally and are referred to as grits, or exceptionally pebble beds. Psammites are brown-weathering meta-sandstones with significant feldspar and some mica. Pelite is used for mica-rich metasediments, usually with garnet and other metamorphic minerals according to grade. The term semi-pelite is used for rocks intermediate between pelite and psammite, having a moderate mica content but lacking index minerals other than garnet.

The oldest Dalradian units in Connemara occur along the central east–west axis of the region particularly at the western end (Fig. 1). They include pelitic schists and other siliceous metasediments and these pass up into the Connemara Marble Formation. The Connemara Marble is undoubtedly the best-known rock type in Connemara and is a formation of originally dolomitic carbonate rocks with variable amounts of silica. As we shall see in Excursion 3, there is considerable variability within the formation but the rock type which is most commonly associated with it is a yellow green serpentine-rich rock or ophicalcite marble (Fig. 3a). Other calc-silicate rocks include a notable bed rich in diopside (Fig. 3b), and there are thick dolomite units with diopside and tremolite. Some of the calc-silicate beds are of metasomatic origin, formed as water released from the adjacent schists penetrated the porosity created by metamorphic decarbonation reactions (Yardley 2009). Later water infiltration produced retrograde serpentine from primary forsterite. Thin interbeds of quartzite are also common.

The Connemara Marble is succeeded by pelitic schists of the Barnanoraun Formation and then by the Cleggan Boulder Bed Formation. This is the key unit for stratigraphic correlation because the association with dolomitic limestone below and quartzite above recurs in Dalradian sequences throughout northwest Ireland and Scotland and allows correlation with the Port Askaig tillite on Islay. Unlike the older formations, it extends almost the entire length of Connemara (Fig. 1) with the most easterly outcrops on Leckavrea Mountain, overlooking the Maam valley (Excursion 1). The Cleggan Boulder Bed Formation contains a range of siliceous metasediments and within it are distinctive purplish weathering psammite units, typically 1–2 m in thickness, which contain clasts of a range of rock types (Fig. 3c). The intense deformation and metamorphism in Connemara mean that none of the characteristic sedimentary features that led to its recognition as a tillite in the southwest Scottish Highlands, can be seen here, but clasts of granite in particular are easy to spot and are weakly flattened parallel to the schistosity. There are also strongly flattened clasts with rims of dark green hornblende and weathered interiors which are likely to represent boulders originally of dolomite.

The younger formations are much more widespread than the older ones (Fig. 1). Quartzite of the Bennabeola Quartzite Formation makes up the main peaks of the central mountains, extending east as far as Derry, near Oughterard, as well as in the

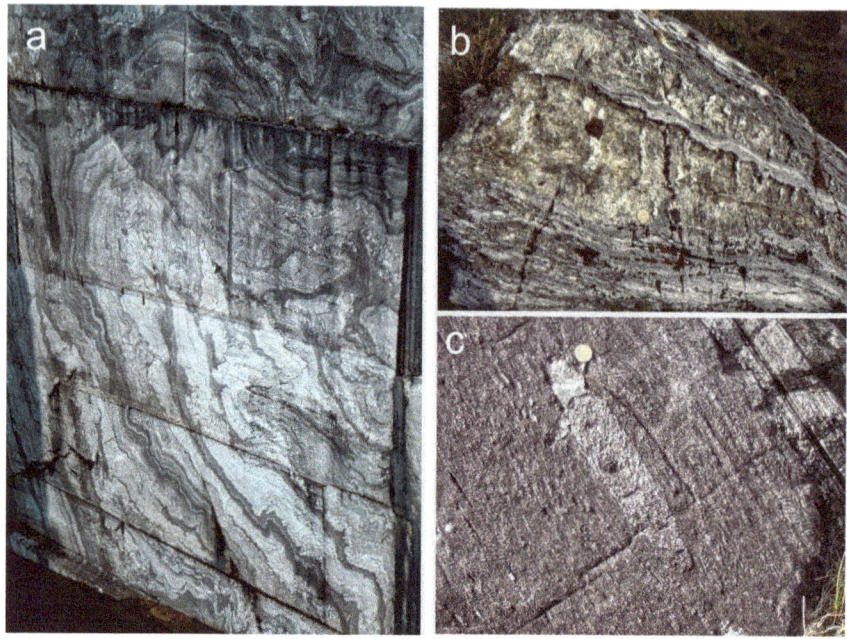

Fig. 3 Distinctive lower Dalradian lithologies **a** Folded Connemara Marble, wire cut surface in the Streamstown Quarry, north of Clifden, **b** Diopside rock interlayered with thin quartzite beds. The diopside rock developed from quartz-bearing dolomitic marble. Creggs Quarry, south of Letterfrack at the foot of Cregg Hill, **c** Cleggan Boulder Bed with granite clast, Glencoaghan (Excursion 3)

northwest at Tully Mountain. It is distinctive because of the high quartz content, although there is some feldspar present also. It is succeeded by the schists of the Streamstown Formation which include muscovite-rich pelites and semi-pelites as well as psammites.

Another very distinctive group of rocks comprise the succeeding Lakes Marble Formation. There are two horizons of calcite marble in this formation, separated by quartzite and grit, often with associated amphibolite and psammite. This is clearly seen on Excursion 1, where the Lakes Marble formation is repeated by folding many times along the excursion route. Even within this walk, the abundance and relative position of quartzite and amphibolite is seen to be very variable. The Lakes Marble Formation is well developed in the north of Connemara, for example in the Maam valley, but the same rocks also occur to the south of the mountains where the bright green grass and fertile fields, seen from the Galway-Clifden Road west of Maam Cross, mark outcrops of marble.

The amphibolites within the Lakes Marble Formation may represent original lava flows or sills in the sedimentary sequence. While it is not possible to confirm which of these possibilities is correct, the variable position and thickness of the amphibolites is probably indicative of an origin as sills for many.

The Lakes Marble Fm. is succeeded by a distinctive pelitic schist, rich in aluminium and iron, which develops a wide range of metamorphic index minerals, notably staurolite. This belongs to the Ballynakill Formation which also contains units of pebble beds in many places, as well as semipelites. There is some uncertainty about the extent of this formation, but Leake and Tanner (1994) assign all the succeeding siliciclastic rocks to it until the distinctive graphitic Cornamona Formation, known mainly from the Maam Valley and Cornamona. The sequence then passes into the Ben Levy Formation, comprising psammites to pelites with some grit bands (Leake and Tanner 1994) and staurolite schists. These rocks are often extensively retrogressed, particularly in the northwest. Uncertainty arises because of the presence of a distinctive tectonic feature, the Renvyle-Bofin slide, which Leake and Tanner (1994) now report as coinciding with the contact of the Benlevy formation. Where there is sufficient exposure, the Renvyle-Bofin slide is often marked by lenses of serpentinite.

2.2 Structure

The distinctive rock units present in many parts of the succession has allowed fold structures to be recognised on all scales. Overall, the central and northern parts of Connemara, including the Twelve Bens, Maamturk and Corcogemore mountains, comprise a fold belt in which complex fold interference patterns can be mapped out, while in the south many of the same rocks occur in a belt with steep dips in which folds are hard to unravel and there are many faulted contacts (Fig. 4). Synorogenic arc intrusive rocks occur mainly in this southern belt. In southwest Connemara, there is a window through the metasediments and associated igneous rocks to underlying metarhyolites, which were thrust under them at the final stage of the arc's history.

Deformation of the Dalradian metasediments in northern and central Connemara has been classically described in terms of 4 phases of deformation, numbered D1 to D4. While this view has not been seriously challenged, it now seems clear that the "D1 phase of deformation" has not in fact produced any structures observable in the field, being based solely on inclusion trails in early garnet. In our experience of bringing students to Connemara, the geology is already sufficiently challenging without having to explain that the first phase of folding is called D2. For this reason, in this chapter we will take a simpler view. We refer to the earliest mappable folds as first phase folds; these are refolded by second phase folds and both sorts are affected by the third phase of folding. Both second and third stage fold axes trend approximately east–west. If following up the existing literature, note therefore that these folds correspond to folds numbered D2 to D4 in most earlier publications.

The first regional deformation phase gave rise to a major isoclinal fold, the Derryclare Fold, now heavily refolded, which runs through the length of central Connemara. It was also responsible for the development of an intense schistosity which remains the pervasive fabric in many rock types, regardless of metamorphic grade. The hinge region of the Derryclare Fold is marked by the occurrence of the

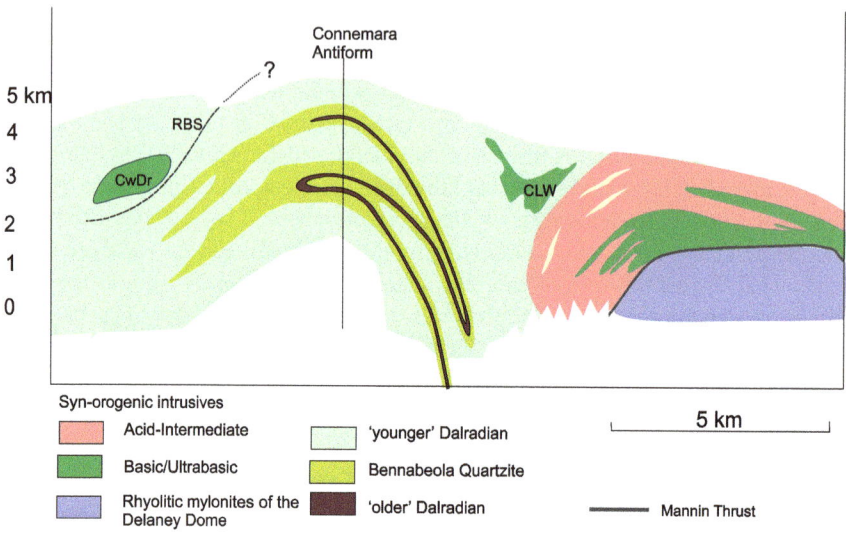

Fig. 4 Generalised north–south cross section across Connemara, based on Friedrich and Hodges (2016), Tanner and Shackleton (1979) and Leake et al. (1981). *Abbreviations* CLW—Cashel-Lough Wheelaun intrusion; CwDr—Currywongaun-Doughrough intrusions; RBS—Renvyle-Bofin Slide

oldest Dalradian formations (Fig. 1) and the refolding is shown in section in Fig. 4. This fold plays a major role in controlling the distribution of the sedimentary rock types in the field and is visited in Excursion 3. Small scale folds linked to this large structure are difficult to spot because they are usually isoclinal and the pervasive axial planar schistosity associated with first phase folding is parallel to bedding except in the hinge itself.

The fabric linked to first phase folds is refolded by later, second stage folds. Usually these do not develop an axial planar fabric of their own. The main exception to this rule is pelitic schists, which commonly have an axial planar fabric defined by minerals which were growing or recrystallizing at the time. At higher grades this is a sillimanite fabric, but at lower grades there is often a mineral segregation fabric developed in mica schists in the hinge regions of second stage folds. The central part of Connemara comprises a stack of second phase folds (Fig. 4), refolding the Derryclare Fold. The northern limit of the fold stack in one interpretation is the Renvyle-Bofin slide, a ductile fault, which appears to have formed at around the same time as the second phase folds and apparently marks a change in stratigraphy. According to Leake and Tanner (1994) the rocks above the slide belong to the Ben Levy formation, and do not occur further south, except possibly in the zone of migmatitic schists interleaved with synorogenic intrusions.

The axial planes of second stage folds have themselves been folded on a large scale in the third stage of folding. This is demonstrated in Excursion 1. The major third phase structure in Connemara is known as the Connemara Antiform. It runs east–west through the central mountains the length of Connemara. In the northeast however there are some additional large-scale third stage folds to the north of the Connemara antiform. Small scale folds related to third stage folds are rare, but minor folds and crenulations are sometimes present, especially in schists near the major hinges.

South of the mountains, Dalradian rocks are often migmatites hosting many small igneous intrusive bodies. Exposure is often poor, particularly to the east. The dips in this area are steep and it is difficult or impossible to recognise continuous fold structures, although strips of rocks such as Lakes Marble formation can be identified. This zone is known as the "steep belt". The steep belt appears to contain some identifiable second phase fold hinges, but the intervening limbs have been disrupted so that the structures and beds cannot be traced through. Much of this disruption seems to be related to the continuation of the second phase deformation and indeed we will see small scale examples of the phenomenon in Excursion 2. The ductile shears which break up the stratigraphy are commonly referred to as slides. Later faults have also contributed to the disruption of the layering within the steep belt.

To the south of the steep belt, the distinctive Lakes Marble Group lithologies are absent, synorogenic intrusive bodies (orthogneisses) are intermingled with migmatites (paragneisses). It is difficult to identify large scale folding, but nevertheless, in the Cashel district, Leake recognised major second phase folds which have folded the synorogenic Cashel gabbro. In the east, such as the area of Excursion 2, steeply dipping migmatised metasediments, probably of the Ballynakill Formation, become interleaved with coarse-grained foliated igneous rocks of the synorogenic intrusive suite on passing southwards. Further west, in the area south of Clifden visited in Excursion 5, the migmatites are followed southwards by metagabbro. Altered gabbros pass down through an amphibolite mylonite (known as the Ballyconneely amphibolite) into a distinct rock unit of pale, rather fine-grained rocks which are interpreted as intensely deformed acid volcanic rocks. The contact between the units is known as the Mannin Thrust (Fig. 4) and the acid metavolcanics are sometimes referred to as the Delaney Dome formation after the region in which they outcrop, which is a broad dome with radial dips. They are included in Excursion 5. The Mannin Thrust marks the boundary between the high-grade rocks of Connemara and acid volcanic rocks which they have over-ridden at the final stage of their emplacement. Deformation in the amphibolites above the thrust suggests hydration at greenschist facies conditions while the acid volcanic rocks contain chlorite, phengitic muscovite, albite and quartz suggestive of low greenschist facies conditions.

2.3 *The Synorogenic Intrusive Suite*

In the southern part of Connemara, a belt of igneous rocks lies between the high grade, migmatitic Dalradian metasediments, with which they interfinger, and the younger Galway granite to the south. This synorogenic intrusive suite comprises a range of intrusive igneous rocks ranging from gabbro to granite in composition and including ultrabasic cumulate rocks. The more acid rocks, commonly referred to as orthogneisses, give rise to low lying ground whereas the gabbros form conspicuous hills. Although most of the rocks of this suite are found in the southern part of Connemara, there are also two gabbro intrusions, Currywongaun (Excursion 4) and Doughrough, in the north west, near Kylemore, with a smaller extension on the coast at Dawros (Leake 1970).

The gabbro bodies are the earliest, largest and most coherent bodies in the intrusive suite. Primary igneous minerals and textures, including layering, are preserved in a number of the gabbros (Fig. 5a, see Excursion 4) but small bodies, and the margins of large ones, are amphibolites and are often foliated. Most early studies were of the intrusions in the south: Cashel-Lough Wheelaun and Errisbeg-Roundstone-Gowla, and are summarised by Leake and Tanner (1994), but the most modern study is of the northern intrusions by Wellings (1998).

Although parts of the gabbro bodies are strongly deformed and all have tectonic fabrics at their margins, the more massive parts of the gabbros still retain many igneous features (Fig. 5b). This contrast will be visited in Excursion 4. While some gabbro bodies have igneous contacts with high temperature country rocks hornfelses (Evans 1964), the northern bodies often sit in strongly foliated country rocks and their boundaries can lie at a high angle to primary layering within them. The northern intrusions are shown as surrounded by paragneiss and hornfels by Leake

Fig. 5 Synorogenic basic intrusions **a** Primary layers of harzburite and anorthosite, Currywongaun (Excursion 4), **b** Lens of gabbro with primary layering and brown pyroxene bounded by sheared gabbro dominated by dark hornblende. Note late stage acid injections. Currywongaun (Excursion 4), **c** Xenolith of schistose country rock in weakly deformed intrusive rock, Kylemore

et al. (1981) but high-grade assemblages are rare in the country rocks, except locally close to gabbro, and the country rocks are generally marked by their strong foliation and retrograde assemblages. Overall, relatively undeformed gabbros sit in a highly deformed matrix of siliceous country rocks. In more detail however, gabbros are increasingly deformed towards their margins, while local patches of high grade hornfelses may adhere to them, with occasional amphibolite apophyses extending into the country rocks. Marginal parts of the gabbro are generally rich in metamorphic hornblende, and quartzofeldspathic veins and lenses are common, sometimes forming an agmatite. These are likely produced by injection of partial melts from the country rocks and are often strongly deformed.

The intermediate to acid intrusions are weakly to strongly foliated and are referred to as orthogneisses. Compositions range from quartz diorite to granite, but these intrusions are notable for their high content of quartz, often greatly in excess of similar rock types elsewhere. It is likely that these magmas were contaminated by siliceous melt derived from the metasediments. This is supported by abundant inherited zircons and their Nd isotope signature (Draut et al. 2002). The dominant components of the orthogneisses are quartz and plagioclase, with subordinate hornblende, biotite and K-feldspar. As we shall see in Excursion 2, secondary chlorite and epidote are abundant. Leake et al. (1981) used the term "K-feldspar gneiss" for the orthogneiss of granitic composition.

The youngest members of this intrusive suite are unfoliated fine grained granites which post-date the deformation (Leake 2021) and are known collectively as the Oughterard granite. They mainly occur east and northeast of Maam Cross as three major bodies extending east to the Carboniferous unconformity near Oughterard. In addition, there are many small, undeformed granite pods, some only a few metres across, and these extend west of Maam Cross, becoming scarcer but persisting towards Clifden.

2.4 *Metamorphism*

Most detailed metamorphic studies in Connemara have concerned the metapelites, and pelite metamorphic zones have been mapped across Connemara (Fig. 6) (Badley 1976; Ferguson and Harvey 1978; Yardley et al. 1980; Leake and Tanner 1994). Nevertheless, only a small part of the area has been subject to modern petrological work, and in large areas, there are no rocks of suitable composition to develop index minerals.

On a regional scale the metamorphic grade in the metasediments decreases northwards, away from the main body of synorogenic intrusions in the south (Fig. 6). However, the basic bodies in both north and south are associated with local high-grade thermal metamorphism, and restites from melting, rich in peritectic garnet and sometimes corundum, occur as xenoliths and in the aureole close to the gabbro contacts. These were extensively studied in the last century but there is little work using modern techniques.

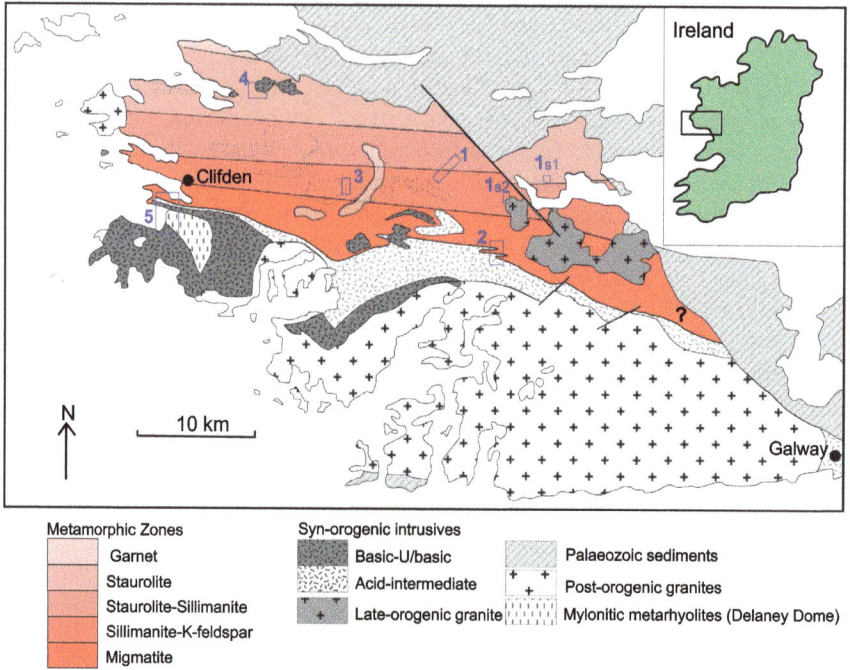

Fig. 6 Summary metamorphic map of Connemara after Yardley et al. (1987). The locations of the excursions are shown

The peak metamorphic grade in pelites ranges from the staurolite zone to upper amphibolite facies migmatites (Yardley and Barber 1991). The extreme northwest of the region may only be at garnet grade, but there has been no modern work and because there is extensive retrograde metamorphism in parts of the area, the peak grade might prove to have been higher. Staurolite is mainly present in rocks of the Ballynakill Formation but is also found in the Benlevy Formation east of Maam (Yardley et al. 1980). The staurolite zone extends the length of Connemara and in the west kyanite is sometimes present also (Ferguson and Harvey 1978). Sillimanite appears as the grade increases southwards, and unusually the reaction from staurolite to sillimanite in Ballynakill Formation pelites first consumed all muscovite in the rock. Staurolite persists with sillimanite but without muscovite over a distinct interval. In the east (Excursion 1, Yardley et al. 1980) andalusite is sometimes the first Al–silicate phase to appear (Fig. 7) and may be replaced by sillimanite.

With increasing grade in the Ballynakill Formation, staurolite-sillimanite schists are succeeded by coarser sillimanite-garnet-biotite schist with plagioclase pophyroblasts that often contain armoured relict inclusions of staurolite (Excursion 1). Yardley (1977) pointed out that the initial breakdown of staurolite is often accompanied by corrosion of garnet, but in the staurolite-sillimanite transition zone garnet

Fig. 7 Photomicrograph of twinned staurolite overprinting a biotite fabric, all included within a large porphyroblast of andalusite. Field of view 0.4 mm, Teernakill

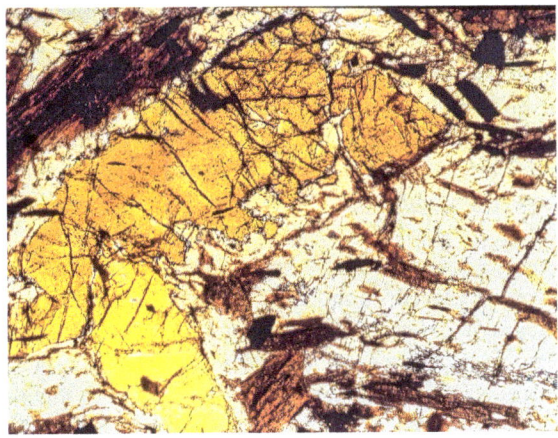

often becomes very cloudy, even to the point of being opaque. This clouding accompanies a change in garnet composition with the cloudy regions having been modified by intracrystalline diffusion while remaining clear cores retain staurolite zone compositions. Garnets in the succeeding garnet-sillimanite schists often have clear new rims overgrowing a clouded interior but there is no compositional change at the textural boundary. At the same grade as the staurolite-quartz assemblage reacts out in the Ballynakill Formation, sillimanite-K-feldspar assemblages appear in most other Formations due to muscovite breakdown and mark the upper sillimanite zone. The first signs of melting are coarse quartz-feldspar veins cutting sillimanite-K-feldspar schists (Excursion 1). These first appear south of the sillimanite-K-feldspar isograd.

There is more widespread development of migmatites in the steep belt and further south (Barber and Yardley 1985) with the development of coarse crystals of garnet, cordierite and sillimanite quite unlike those seen at lower grades and generated by incongruent melting (Yardley and Barber 1991).

Both the breakdown of muscovite before the first signs of melting and the regional andalusite of eastern Connemara (Fig. 7) indicate low-P regional metamorphism during the later part of the metamorphic history. Since kyanite mainly occurs in the west it is tempting to suggest a gradient in pressure, increasing westwards, but this fails to consider the timing of metamorphism. Kyanite and staurolite grew together before the second phase folds, whereas andalusite grew after staurolite. Even if the area as a whole experienced uplift and erosion, parts which reached their maximum temperature before the pressure dropped would not record evidence of low-P metamorphism, while areas which continued to get hotter during uplift would preserve little evidence of earlier, higher pressures. The metamorphic pattern may be related to a shift in the magmatic activity to the east with time, rather than to pressure differences at any one time.

The assemblages of metabasites and other lithologies show little variation, with the exception of the appearance of sillimanite + K-feldspar from muscovite breakdown, noted above. The Connemara Marble Fm contains a range of assemblages, including metasomatic calc silicate rocks, but there has been little modern petrological study, and it is unknown if any of the considerable mineralogical variation reflects different peak temperatures.

In addition to prograde metamorphism, there has also been widespread retrogression, although it has not been investigated in such detail. As already noted, there are distinctive areas of extensively deformed and retrogressed rocks around the northern gabbros. Partial retrogression of biotite to chlorite is widespread throughout Connemara but at Knockaunbaun in the Maam Valley (Yardley 2009), extensive retrogression of staurolite schists is associated with albite–muscovite-chlorite-quartz veins emplaced during the third phase of folding. The absolute age of the vein muscovite is discussed below as locality [13]: Some nearby veins, also apparently emplaced during third phase folding, have vein andalusite and staurolite-enriched margins (Yardley 1986), raising the possibility that the infiltrated fluid was initially hot and derived from amphibolite facies metamorphism of pelites elsewhere. The environs of the Oughterard granite are a possible source for example.

Other lithologies also display evidence of retrogression, although the timing is generally unknown. There is ubiquitous alteration of olivine to serpentine in the ophicarbonate rocks of the Connemara Marble formation, together with alteration of diopside and the formation of talc veins in dolomite. Less dramatically, prehnite is a common joint fill in amphibolite where retrograde chlorite is common. High grade calcite marbles near Maumeen (Excursion 1) exhibit albitisation of plagioclase and alteration of diopside with growth of Mg–Al pumpellyite.

2.5 Relative Timing of Magmatism, Metamorphism and Deformation

The metamorphic zones in Connemara run east–west and are not obviously affected by folding, suggesting that the metamorphic peak was relatively late in the folding history. Garnet inclusion patterns and flattening of the schistosity around the garnet grains (Badley 1976; Yardley 1976) demonstrate that first phase folding took place at garnet grade conditions, with garnets having pre- to syn-tectonic textures with respect to the schistosity associated with first phase folds, which is strongly flattened around them. Staurolite and kyanite (Ferguson and Harvey 1978) grew largely before second phase folding but postdate the first phase fabric (Fig. 7). In contrast, sillimanite often defines a new fabric axial planar to the second phase folds (Excursions 1, 3, Fig. 12). This demonstrates that these folds formed during the peak of metamorphism in the higher-grade zones, but implies that in the staurolite zone the peak metamorphic peak was reached before the second phase folding, while rocks further south were undergoing further heating.

Examining the gabbro contacts in the field (Excursion 4) the different rheology of gabbro and metasediments is often apparent. This is not surprising since in many places away from the immediate contact, the country rocks will have been heated to the point of melting only once the gabbro had already solidified. It appears that the locally intense second phase fabrics only developed once the envelope of country rocks had been heated and melted, by which time the gabbro itself was a strong, massive body which deformed where wet melts and/or water, began to penetrate.

There has been extensive debate about how the intrusions in particular fit into the history of deformation. The gabbro bodies have relatively undeformed interiors but develop extensive tectonic fabrics around their margins. Many earlier studies suggested that the gabbros had been deformed during the first phase of deformation, but Leake and Tanner (1994) ascribed their deformation entirely to the second phase. While the gabbro bodies are deformed by this phase of deformation, it is less clear whether they entirely predate it. Most recently, Leake (2021) has concluded that the gabbros were emplaced late in the first phase of deformation and were subject to later deformation also. Wellings (1998) considered that there was good evidence for the gabbros in the northwest being affected by two deformation phases. He identified these with the regional first and second fold phases, but also showed that they occurred very close in time on thermal grounds.

This focus on timescales as well as classic structural criteria is apposite. Fold events likely take place over timescales of hundreds of thousands of years, but the filling of magma chambers may take significantly shorter time periods, and the basic phase of arc magmatism in Connemara appears to have only involved magma chambers with simple histories since basic rocks never cut intermediate and acid rocks. Thus, the gabbros could have been emplaced within a single phase of regional folding. At the other end of the spectrum, regional heating and cooling takes place over millions of years with rates of the order of tens of degrees per million years, so that large (perhaps >50 °C) changes in metamorphic temperature are unlikely to have occurred within a single regional folding episode.

We think it likely that the gabbro bodies were emplaced within one regional deformation phase, i.e., after the onset of second phase deformation but before it came to an end. In support of this suggestion is the observation that some country rock xenoliths in gabbro have a first phase deformation fabric which has been folded (Fig. 5b). Thus, the two deformation episodes identified by Wellings may both have been part of the same regional second phase of folding, rather than correlating with separate regional phases. The metamorphic textures discussed above appear to preclude a narrow time interval between the first (garnet grade) and second (sillimanite grade in the south) fold phases, because rock volumes on a regional scale are very slow to heat or cool (Yardley and Warren 2021) and the temperature difference between fold phases is at least 50–100 °C, much more at higher grades.

In this interpretation, the second phase of deformation lasted long enough to bracket gabbro emplacement. Further evidence for the complexity of this phase comes from the migmatite zone. Here, it is sometimes apparent that fabric development outlasted the bending of the layering; where folds limbs became too thin, they broke, and the isolated second phase fold hinges sit in a foliated matrix with a fabric which, while axial planar to the folds, evidently continued to evolve after the folding itself was completed (Excursion 2, Fig. 13c).

3 Dating the Magmatism and Metamorphism

3.1 Review of Existing Geochronological Work

Early geochronological work in the region, using mainly Rb–Sr whole rock studies, has been superseded by increasingly precise U–Pb zircon work, culminating in the studies by Friedrich (1998) (Friedrich et al. 1999a, b). Since that time no additional geochronological constraints have been added, despite significant improvements in analytical protocols.

For this article, published ages have been recalculated with current recommended values for decay constants using IsoplotR (Vermeesch 2018) and are summarized in Table 1 and plotted in Fig. 8 with cross references to localities in square brackets. For Rb–Sr (Villa et al. 2015) this has a significant effect and leads to an increase in age of approximately 7 Ma for the relevant age range, compared to ages calculated with the value used in 1990s.

The timing of magmatic activity is documented by U–Pb ages on samples from all the major igneous units other than those within the Dalradian sequence. The metarhyolites of the Delaney Dome are the oldest at 475 ± 3 Ma. The main synorogenic activity within the Connemara massif is dated at 472.0 ± 0.5 Ma and 469 ± 1 Ma (Friedrich et al. 1999a) for two basic members, 467.2 ± 0.9 Ma (Friedrich et al. 1999b) and 461 ± 4 Ma (Cliff et al. 1996) for quartz diorite gneisses, 468 ± 5 Ma (Cliff et al. 1996) for granite gneiss and 466 ± 1 Ma for a granite pegmatite (Friedrich et al. 1999b). Monazite from a cordierite pegmatite is 463 ± 4 Ma and titanite from a calcsilicate pegmatite is 465 ± 4 Ma (Cliff et al. 1996).

Late-orogenic fine-grained unfoliated granites yield ages of 461 ± 1 Ma (Friedrich et al. 1999a) for the Oughterard Granite and 465 ± 5 Ma (Cliff et al. 1996) for a small body at Oorid Lough, indicating that fabric-forming deformation ended before 460 Ma.

The most precise ages directly linked to metamorphism are the 468 ± 2 Ma age for monazite in a pelitic palaeosome in the migmatite zone (Friedrich et al. 1999b) and titanite ages from the metasomatic diopside rock: 462.0 ± 0.5 Ma at Cregg's Quarry and 461.8 ± 0.6 Ma at Glencoaghan (Friedrich et al. 1999b). However, the latter results contrast with older ages published by Cliff et al. (1993, 1996) which although less precise were significantly older. This data has been re-processed with more comprehensive error propagation to check the significance of the older data, resulting in a concordia age for the same Cregg's Quarry sample studied by Friedrich of 476 ± 4 Ma. The samples from Glencoaghan are less concordant but all have $^{207}Pb/^{206}Pb$ ages greater than 470 Ma. Taken together with other titanite data from Friedrich (1998) which was not plotted in Friedrich et al. (1999b), the titanite age pattern is more complicated than can be explained with limited data available at present and does not yet provide a reliable metamorphic age. The hypothesis that the diopside rock itself was generated by fluids from the Oughterard granite of eastern Connemara (Friedrich and Hodges 2016) is not supported by any petrological or field evidence, and is simply an inference from the apparent age, not a support for it.

Table 1 Summary of recalculated mineral ages plotted in Fig. 8 and discussed in the text

Key	Min	Lithology	Locality	Latitude	Longitude	Ma	±	References
1	zr	Basic pegmatite	Currywongaun	Not specified	Not specified	472.0	0.5	Friedrich et al. (1999a)
2	zr	Gabbro	Cashel-L. Wheelaun	53.4330	−9.7420	469	1	Friedrich et al. (1999a)
3a	zr	Granite gneiss	L. Nahasleam	53.4323	−9.5441	468	5	Cliff et al. (1996)
3b		Quartz diorite gneiss	L. Nahasleam	Between 53.4370 and 53.4260	−9.5430 −9.5490	461	4	Cliff et al. (1996)
4a	zr	Quartz diorite gneiss	North of Bunnahown	53.4060	−9.7830	467.2	0.9	Friedrich et al. (1999b)
4b	zr	Pegmatite	North of Bunnahown	53.4060	−9.7830	466	1	Friedrich et al. (1999b)
5	zr	Fine-grained leucogranite[a]	Oorid Lough	53.4558	−9.6160	465	5	Cliff et al. (1996)
6	xe	Granite	Oughterard area	Not specified	Not specified	461	1	Friedrich et al. (1999a)
7	zr	Metarhyolite	Near Ballinaboy	53.4630	−10.0380	475	3	Friedrich (1998), Draut and Clift (2002)
4c	mo	Pelitic palaeosome[b]	North of Bunnahown	53.4060	−9.7830	468	2	Friedrich et al. (1999b)
9	mo	Cordierite pegmatite	Oorid Lough	53.4561	−9.6072	464	3	Cliff et al. (1996)
10a	ti	Diopside rock[c]	Creggs quarry	53.5142	−9.9365	462.0	0.5	Friedrich et al. (1999b)
10b	ti	Diopside rock[c]	Creggs quarry	53.5142	−9.9365	476	4	Cliff et al. (1993, 1996)
11a & b	ti	Diopside rock	Glencoaghan	53.4860	−9.7900	461.8	0.6	Friedrich et al. (1999b)
12	mu	Andalusite pegmatite	Claggan	53.4967	−9.6550	471 461	5 5	Cliff et al. (1996)
13	mu	Muscovite-albite-chlorite vein[d]	Knockan-baun	53.5240	−9.6550	465	3.5	Cliff et al. (1996)
14	mu	Pegmatitic segregation	N59 near junction with R340	53.4592	−9.7700	479	5	Cliff et al. (1996)

(continued)

Table 1 (continued)

Key	Min	Lithology	Locality	Latitude	Longitude	Ma	±	References
15	mu	Pegmatitic segregation	Dawros	53.5680	−9.9650	474	5	Cliff et al. (1996)
16	mu	Pegmatitic segregation[e]	Renvyle village	53.6000	−9.9630	487	5	Cliff et al. (1996)
						446	5	
17	mu	Metarhyolite[f]	Delaney Dome	Near 53.4620	Near −10.0000	457	3	Tanner et al. (1989)

Abbreviations used mo—monazite; mu—muscovite; ti—titanite; xe—xenotime; zr—zircon. 'Key' in column one refers to numbers in Fig. 8
[a]Cross-cutting the dominant schistosity
[b]In migmatite
[c]Subsamples of a specimen 6855
[d]2–5 mm grains
[e]Segregations appear to pre-date 2nd-phase folds
[f]Oriented in mylonitic schistosity

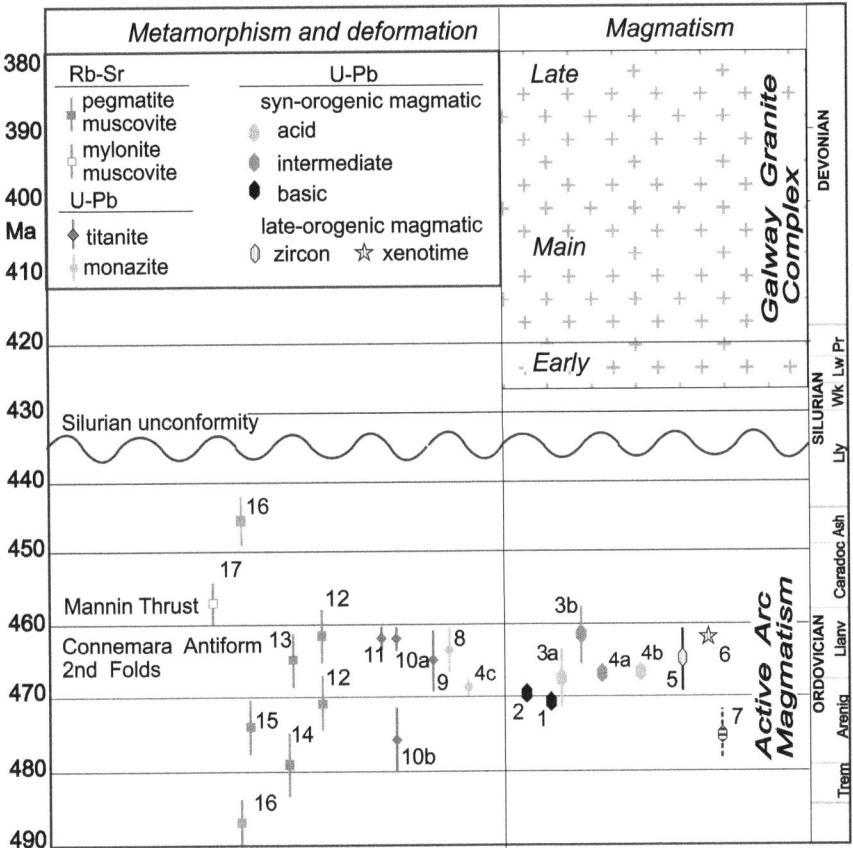

Fig. 8 Compilation of igneous and metamorphic rock ages for Connemara, recalculated as noted in the text. Ages of stratigraphic subdivisions based on the GSA Timescale v.5 (Walker et al. 2018 and Gradstein et al. 2004). Numbers refer to localities listed in Table 1 while symbols denote the type of age. *Abbreviations used* Ash = Ashgill; Lly = Llandovery; Llanv = Llanvirn; Lw = Ludlow; Pr = Pridoli; Trem = Tremadoc; Wk = Wenlock

Additional information is potentially offered by Rb–Sr ages on muscovites, which have been shown to withstand later reheating on regional metamorphic timescales well into the amphibolite facies. Hence provided their microstructural relationships to deformation and their initial Sr ratios are adequately constrained, their ages may be expected to closely approximate the time of crystallization except at the highest grades in the migmatite zone. Present results scatter beyond this expectation but the current data set is small and lacks details on microstructural relationships. Data from Cliff et al. (1996) is plotted in Fig. 8. A small group of samples have ages significantly greater than 470 Ma and may hint at an older stage in the metamorphic history, but others are closer to the 465–470 Ma range suggested by most of the U–Pb data. One much younger age from Renvyle in northern Connemara is unexplained.

Rb–Sr ages on muscovite provide key information on the deformation history. Muscovite from a vein linked to metasomatism and local heating during the third deformation phase on Knockanbaun (Yardley 1986, 2009) has an age of 465 ± 3.5 Ma which suggests this late stage in the deformation history followed closely after the metamorphic peak, possibly overlapping the late-tectonic Oughterard granite. Muscovite from the mylonitic fabric in the metarhyolite of the Delaney Dome Formation dates deformation associated with the Mannin Thrust at 457 ± 3 Ma (Tanner et al. 1989). This emplaced already cool Dalradian metamorphics onto the underlying rhyolites.

Increasingly sophisticated analytical approaches have had limited success in documenting later stages of the cooling history using mica Ar–Ar ages. Only ca. 40% of the results reported by Friedrich and Hodges (2016) are interpreted as dateing cooling to the 300–400 °C range. These are consistent with earlier cooling in the north around 470 Ma and cooling to the same temperature range from 462 to 452 Ma in central and southern Connemara. Other results were tenuously linked to late deformation, considered to be spuriously high due to excess argon, or ascribed to post-Grampian resetting processes.

3.2 Overall Conclusion and Prospect for Future Improvement

The ages discussed here are in good agreement with the field relations and a consistent focus on ages of 462–472 Ma for arc-related magmatism and metamorphism in Connemara has emerged. The regional metamorphism in northern areas, where an effect from arc magmatism is not immediately apparent, appears to have peaked around 470 Ma. Nevertheless, several factors suggest caution is needed:

- The number of dated samples remains small.
- For the U–Pb system in zircon, exhumation to within a few km of the surface by 435 Ma and the presence of large post-orogenic granite plutons nearby suggests the potential for lead loss within ca. 50 Ma, which would result in displacements to younger ages on the concordia diagram without any resolvable offset from the concordia curve. It is common experience for Tertiary age zircons elsewhere to show Pb-loss in excess of 10% in some grains.
- The available single zircon data is not accompanied by the petrographic and geochemical information on individual dated grains that characterize current state-of-the-art studies.
- In many rocks the presence of an inherited zircon component means concordant grains are rare and the use of abrasion to remove more discordant outer parts of grains also enhances the proportion of inherited material.

While the tight clustering of U–Pb ages in a narrow interval is alluring, it remains possible the clustering is the result of similar minor Pb-loss in most grains with a true age somewhat older than the upper bracket on the existing data set. Further improvements in amplifier technology mean that it is now possible to analyse single zircon grains with great precision (e.g., Wotzlaw et al. 2017) but this has yet to be applied to zircons from Connemara. There is also considerable untapped potential to date regional metamorphism using monazite and titanite, once the puzzling discrepancy in diopside rock ages is resolved. The possibility that peak metamorphism was reached at different times in the east and the west could then be investigated.

The main analytical limitation on Rb–Sr muscovite ages hitherto has been error on Rb analysis, which totally dominates uncertainties in the present data set. ICPMS analysis has been shown to allow an order of magnitude improvement in this (e.g., Nebel and Mezger 2006) and, coupled with the ability to measure precise Sr isotopic composition on small samples down to single grains (Cliff et al. 2016), there is the potential to obtain muscovite ages closely tied to microstructural setting.

4 Overview: The Evolution of the Connemara Massif

The rocks we see in Connemara date back to the late Precambrian deposition of the Dalradian series in a basin that extended to the north and northeast. After burial, the earliest record of their metamorphic history is the development of a fabric now preserved only as trails of very fine-grained quartz inclusions in garnet. These garnets are the earliest distinctive metamorphic minerals and display the typical growth zoning of garnet zone pelites. The pressure of garnet zone metamorphism is unknown, but by analogy with other Dalradian rocks in Ireland, they are likely to have formed along a medium-P or even a moderately high-P trajectory (Yardley et al. 1987). Garnet zone metamorphism largely predates the first phase of folding, which resulted in pronounced flattening of the schistosity around garnet. In contrast, there is no such flattening around staurolite grains, which appear to have formed between the first and second folds. Sillimanite growth occurred at the same time as the second phase folds, raising the possibility that the deformation may have been initiated in part by heating, itself linked to the onset of magmatism.

Around the time of initiation of the second fold phase, the gabbros, first of the arc intrusive suite, were emplaced, and we can therefore date this stage to 472–469 Ma, although greater precision is now potentially possible. Synorogenic arc magmatism continued to at least 465 Ma and since by definition these rocks are affected by pervasive deformation, this span effectively dates the duration of the second deformation phase also. In detail, this phase can be broken up into an early phase when the folds initiated, shortly after the gabbros were emplaced, and then continuing deformation of the migmatites and later intrusions which disrupted the folds in the highest grade rocks.

The latest member of the arc suite, the Oughterard granite, was intruded after the final episode of pervasive strain, at around 461 Ma, close to the time that the third phase folds developed.

The earliest preserved metamorphism to garnet grade appears to have been typical medium pressure metamorphism and was followed by staurolite and locally kyanite grade metamorphism which probably predated the basic intrusions. Certainly, the assemblages are similar to those found in Dalradian areas of Scotland and Ireland, which lack synorogenic intrusions. Interestingly however the very fine grain size of the Connemara rocks is notable and may indicate more rapid heating. In contrast, the later metamorphism that took place during and even after the second phase of deformation has features characteristic of low-pressure metamorphism. Not only is andalusite widely developed in the eastern part of Connemara, but there are also local occurrences of cordierite (Leake and Tanner 1994), and muscovite appears to have begun to break down before the first appearance of migmatite leucosomes.

Figure 9, based on Yardley et al. (1987), shows the P–T-t paths for rocks from each end of the traverse visited in Excursion 1, starting from the inferred staurolite zone conditions with pressure at least 550 MPa and temperature near 550 °C. Staurolite-sillimanite schists from Cur Hill developed andalusite which was then partially replaced by sillimanite, implying a drop in pressure but then further heating to progress into the sillimanite field. The significance of this observation is that it rules out a simple clockwise P–T-t path and suggests that uplift and erosion began to affect the rocks before heating from magma emplacement. Garnet-sillimanite schists from Maumeen at the southern end of the excursion did not develop andalusite and are inferred here (Fig. 9) to have followed a similar path to the Cur Hill rocks but lying within the sillimanite stability field. The highest grade migmatites, at Lough Nahasleam (Excursion 2) melted at a pressure near 500 MPa (Barber and Yardley 1985) but were likely staurolite-kyanite schists initially (Leake and Tanner 1994) so must have experienced magmatic heating earlier in the history of uplift and erosion than the localities to the north. They remained hot and partially molten through the

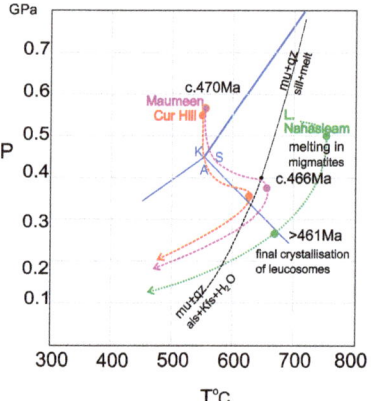

Fig. 9 P–T-t paths for localities included in Excursions 1 and 2. Updated from Yardley et al. (1987) with reference reactions compiled in Yardley and Warren (2021). The round spots on each path are the approximate conditions of staurolite zone assemblages, peak metamorphism and final crystallisation, with approximate ages as discussed above

uplift event however, as some leucosomes crystallised in the andalusite field (Barber and Yardley 1985). Overall, the metamorphic assemblages point to a drop in pressure indicative of the removal of 5–10 km of cover accompanying second phase folding and the continuing deformation.

By the end of the second phase of deformation, most of the region had begun to cool but in the east the emplacement of the Oughterard granite (461 ± 1 Ma) may have led to some further andalusite metamorphism (Leake and Tanner 1994), though this could arguably be considered thermal rather than regional. The third phase folds likely developed close to this time or a little earlier (465 ± 3.5 Ma).

The final stage in the assembly of the Connemara massif came at 457 ± 3 Ma when the arc intrusive sequence and its Dalradian basement were thrust over Ordovician arc volcanics now exposed south of Clifden as the Delaney Dome. These metarhyolites also developed in an Ordovician arc, but in a segment where acid volcanism occurred a few million years before the emplacement of the Connemara gabbros and did not undergo the same metamorphism. The widespread retrogression of many Connemara intrusive rocks likely also dates from around this time.

As the thermal history of Connemara becomes better constrained, it is increasingly valuable for testing tectonic models for western Ireland. The early metamorphic history, before the development of second phase folds, followed a medium-P (Barrovian) trend, similar to most other areas where Dalradian metasediments occur. This was followed in much of the region by low-P metamorphism driven by arc magmatism coupled to progressive uplift and erosion. The earlier medium-P assemblages are preserved in areas that were not heated further during the stage of arc magmatism. The last magmas of the arc suite (which produced the Oughterard granite) were emplaced shortly before Connemara was thrust over low grade metarhyolites on the Mannin Thrust. Present tectonic models for region (Dewey and Ryan 2016, and references therein) do not yet fully accommodate the evolving thermal regime and extended history of uplift and erosion of Connemara, summarised here. We believe that these will provide valuable tests for the tectonic scenarios in the future.

5 The Excursions

5.1 Excursion 1: Dalradian Metamorphism and Structure Across the Mountains. Discovery Series Maps 38, 37 and 44

Access and Other Information: This excursion involves a walk across the central mountains in northeast Connemara, along part of the Western Way. It crosses a number of major second phase folds and illustrates how they are refolded by the third phase Connemara Antiform. It also demonstrates a range of rocks, including quartzite, marble amphibolite and pelite, and their metamorphism. In particular, there is a steady increase in metamorphic grade from staurolite-sillimanite schists at the northern end of the traverse to upper sillimanite zone rocks at the south end. Figure 10 is a map of the route including a detailed breakdown of members of the Lakes Marble Formation along the route. There are 2 supplementary localities (Fig. 1) illustrating complementary features; these may be visited while driving between the ends of the walk or linked to Excursion 2.

Practicalities: ideally this excursion should be done from north to south with a vehicle designated to pick you up at the southern end to shorten the walk. It is however perfectly possible to leave a car on the north side of the mountains and return to it. For large parties it is best to use the car park at the south end. There is no access for coaches to the north end of the route and parties must walk-in on minor roads signposted for the Western Way. They may be dropped off east of Kilmeelikan on the R336 between Leenane and Maam, or at Teernakill, south of Maam on the R336 to Maam Cross. The car park for the southern end is accessed from the N59 Maam

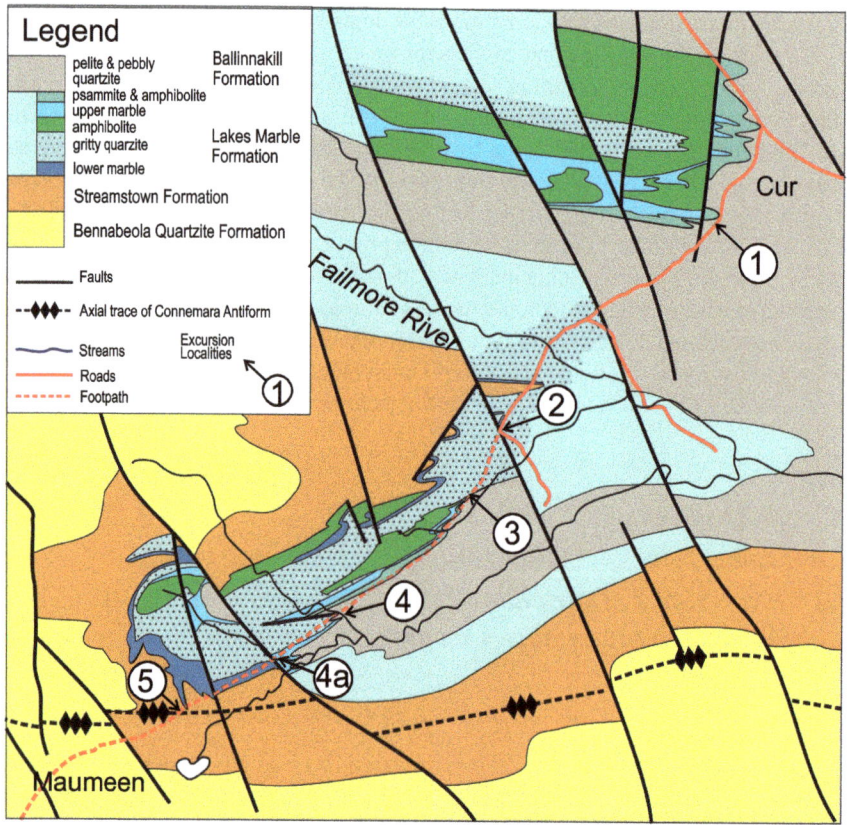

Fig. 10 Geological map of the Western Way between Cur and Maumeen (Excursion 1) to show the localities 5.1–5.5 in the context of roads, tracks and streams, simplified from Yardley (1976)

Fig. 11 Overview of locality 1 looking north to Cur Hill (Knocknagur), showing the areas of outcrop discussed in the text

Cross to Recess road via a minor road 200 m east of Recess church and is signposted for the Maumean Shrine. For smaller groups, a limited number of cars can drop off at the start of the excursion at Cur Hill, and park about 750 m beyond Locality 1. Please never obstruct the road, even momentarily, and ensure all gates are shut. Loc 5.1 is on enclosed land and *you must obtain permission to access it from the nearby farm.*

Cur Hill (Fig. 11) is a conspicuous feature of the Maam valley, shown on maps as Knocknagur. The east end of the hill is cut off by a normal fault running approximately north to south. Conspicuous dark crags are amphibolite. From a distance, these define an antiformal structure just south of the summit, picked out by green grassy slopes marking the presence of calcite marble. In fact, however, the ridge making up Cur Hill is overall a synform and this antiformal structure sits within it. The narrow road in front of the hill passes over pelitic schists which outcrop occasionally by the road. Continue just beyond the field walls to locality 5.1. Note that sillimanite can occur throughout this excursion, but a lower grade staurolite schist is included as supplementary Loc 5.1s, below.

5.1.1 Locality 5.1: Cur Hill (53.515200°, −9.608700°)

Pause at this point to look at the ground to the west. Figure 11 shows the outcrops at this locality. The south side of Cur Hill features dark crags of amphibolite descending to a gully in front of you with bright green grass indicative of marble. Just to the south of this a large knoll marks resistant psammites but further to the left there is relatively little exposure. The underlying bedrock here is schist but most of the outcrops represent sporadic bands of pebbly quartzite.

Go through the gate in front of you fastening it carefully and make your way to the deep green gully at the foot of the main slope. Notice that when you cross the fault line that marks the eastern end of Cur Hill you can see a stream has risen from the fault. Do not visit the spring as the water is used in the local cottages, but it occurs where the fault truncates a marble band on the west side. The gully of bright green grass has cave-collapse features and many outcrops of marble. There are some spectacular examples of folding but please do not be tempted to try and hammer them! These are second phase folds which are almost upright at this locality. The colour banding in the marble is mainly caused by trace amounts of graphite but there are more quartz-rich layers occasionally and these often have rusty weathering due to the presence of pyrite. The purer the marble the more symmetrical the similar style folds. Where there are siliceous bands the folds are no longer of textbook quality!

Upslope on the north side (Fig. 11), the contact between marble and dark green amphibolite can be seen at several places. There is a steep contact above the spring, but further west a small overhang in the cliff has an almost horizontal contact with amphibolite above marble. At first sight this appears to be a thrust, with near vertical layering in the marble, which also appears sheared beneath the gently dipping amphibolite. But the contact maps out as folded bedding and the discontinuity in outcrop probably just reflects the large difference in rheology between amphibolite and marble under amphibolite facies conditions. The flat contact is in a fold closure, the steep one nearby, on the limb. The amphibolite can be studied in loose blocks and has a strong foliation of aligned, often lineated, hornblende with flattened plagioclase. Hornblende also occurs as larger porphyroblasts, possibly replacing an original igneous texture.

The knoll south of the marble gulley is also intensely folded and comprises thinly bedded psammite with minor bands of other lithologies. This unit is impersistent but is often present between the marble and pelites of the Ballynakill Formation. Ballynakill Formation staurolite-sillimanite schist occurs in a few low flat outcrops in the bog a few metres south of the knoll. Proceed south east from here over boggy ground towards the road and further outcrops are mainly of quartzite with quartz veins. The quartzite appears to have had small pebbles originally and these are now rodded along the fold axis direction.

Close to the road the outcrops continue to a small lay-by (with parking for one vehicle), separated from you by a fence. The nearest outcrop to the lay-by consists of staurolite-sillimanite schists with garnet. Provided the weather is dry and sunny it is worth examining this outcrop in detail with a hand lens. Some layers are quite rich in garnets, usually 1–3 mm in diameter. The matrix has abundant biotite and some layers contain golden brown staurolite crystals which may be as much as 30% of the rock. Unfortunately the size of the grains is a lot less than for the garnets and they are extremely difficult to spot. The schist here has occasional cm-sized pink andalusite in veins and these are sometimes pseudomorphed by opaque white sillimanite. These observations are critical to deducing the metamorphic PT path of the region, as discussed in the previous section. Return from this outcrop to the gate where you started. You should not attempt to cross the wire fence.

5.1.2 Locality 5.2: Western Way (53.506100°, −9.624400°)

Continue down the road to the valley bottom. There is a parking area just before the bridge over the Failmore River. As you ascend on the other side, you will notice bright green grass and small outcrops of marble. The Lakes Marble Formation which we left at Cur Hill has been folded and outcrops again in this area. Ahead of you, the route (the Western Way) continues while the farm road turns left. Pause at the gate and look back at Cur Hill; the amphibolite ridge is conspicuous but note the strips of green grass that denote inter-folded marble. The exposure in the valley is not good but the Ballinakill Formation pelites that we saw at Loc 5.1 give way near the river to Lakes Marble Formation rocks that now surround us. From here, you get an excellent insight into the varying scale of second phase folds, including the mesoscopic structures picked out by marble on the ridge and the large scale structure, the Failmore Antiform, you have just traversed.

5.1.3 Locality 5.3: Western Way (53.503100°, −9.626500°)

Continue along the path, noting that much of the rock at your feet is the typical "gritty quartzite" of the Lakes Marble Formation, and continue until there is a conspicuous hawthorn tree on the right of the path. Here, the marble bed seen at Loc 5.1 is present beside the path on the north west side, but a scramble up the outcrop reveals a different sequence from Cur Hill. A few metres of psammite and semi-pelite is succeeded by foliated amphibolite similar to those seen at Locality 1, but this layer is also thin and passes into quartzite at the top of the slope. The quartzite-amphibolite contact can be paced out and is elegantly folded by second stage folds as it runs westwards. The metamorphic grade is distinctly higher than at Locality 5.1. Some of the layers just above the marble contain aligned and flattened lenses of matted sillimanite fibres, 2–5 mm across, weathering up from the outcrop. These layers also contain K-feldspar, indicating the breakdown of muscovite, which is characteristic of the second sillimanite zone (Yardley and Warren 2021).

5.1.4 Localities 5.4, 5.4a: Western Way, Maumeen (53.498000°, −9.635900°)

Continuing along the track, pause at 53.498000, −9.935900 where a stream descends across the path. A few metres downstream, the stream bed has an outcrop of Ballynakill Fm pelitic schist which contains biotite, garnet and sillimanite with staurolite only as inclusions in plagioclase porphyroblasts. These schists are noticeably coarser grained than the pelites at Cur and the biotite has a bronze appearance due to intergrown sillimanite fibres.

As you continue towards the pass at Maumeen, take note of the orientation of the second phase folds in the outcrops along the way. At Locality 5. 3 the folds had a very steep dip to the north, but as you continue southwest you will begin to encounter much more gently dipping layering. A large marble outcrop by a stream crossing at 53.495700, −9.651300 (Loc 5.4a) has very gently north-dipping folds, but they are hard to spot. The stream crossing is close to a major fault; look at the hill to the north west and you will see a sharp break where the fault separates white Bennabeola Quartzite to the right from brown Streamstown Fm outcrops to the left. Continuing on the track, the superficially similar marble outcrops are assigned to the lower marble unit of the Lakes Marble Formation.

5.1.5 Locality 5.5: Western Way, Derryvealawauma (53.493100°, −9.648500°)

As the path rises steeply up to the pass, the last marble bands are left behind. At the top zig-zag, where the path goes right, walk out to the left. The abundant outcrops are pelitic and psammitic schists of the Streamstown Fm. A 1.5 m cliff faces back down the route you have followed, and has conspicuous second phase folds, but these are now recumbent with their axial planes dipping gently to the south. This demonstrates that we have crossed the hinge of the third phase Connemara Antiform structure. Folds in the psammite layers are typical second phase folds with a strong earlier schistosity folded around them, but in the pelitic layers the earlier folded schistosity has been largely destroyed and instead there is an axial planar schistosity picked out by resistant lenses of fibrous sillimanite, known as faserkiesel (Fig. 12). The flat outcrop surfaces are often studded with these white lenses, more nearly circular viewed from above, typically around 5 mm across. They are also cut by occasional veins of coarse feldspar and quartz with minor muscovite, distinct from the quartz veins seen at lower grades. It is likely that these veins are the product of incipient wet melting. The fact that they appear after K-feldspar + sillimanite, demonstrates that the peak metamorphic temperature was achieved at low pressures.

From these, northeast-facing outcrops, return to the path and continue a short distance towards the Maumean Shrine. You rapidly pass on to massive white quartzite of the Bennabeola Quartzite Fm; some layers contain sillimanite in pebble shaped aggregates of quartz with sillimanite, which are picked out by weathering.

If you have a vehicle waiting at the car park on the south side of the mountain, continue down the track, enjoying spectacular views across the migmatites and synorogenic intrusions of southern Connemara. Otherwise return the way you have come.

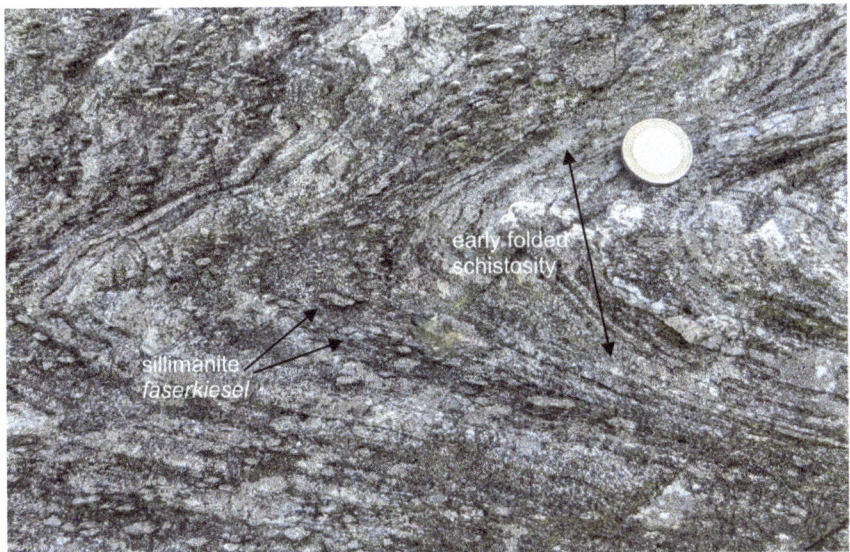

Fig. 12 Second phase fold in Streamstown Formation schists, Loc 5.5, near Maumeen. Sillimanite faserkiesel define a new schistosity axial planar to the folds

5.1.6 Supplementary Localities for Excursion 1

Two localities are suggested which complement those along the traverse shown in Fig. 10. Their approximate locations, to the east and south of Maam respectively, are shown in Fig. 1.

Locality 5.s1: Claggan (53.496688°, −9.491082°)

Access and Other Information: This locality is an overgrown small roadside quarry by the R345 road east of Maam, between Maam Bridge and Cornamona. It lies just before a sharp left hand bend as the road rises to a summit before dropping to Connemara. Parking is possible on the north side of the road in the quarry.

This outcrop is of staurolite zone schists of the Ballynakill Formation, close to the first appearance of sillimanite, and it therefore provides a lower grade addition to the main excursion, as well as good pelite outcrops. Fine grained brown biotite schists with garnet and staurolite are the dominant lithology, and these are cut by short segregation veins of quartz with muscovite and sometimes pink andalusite prisms a few centimetres in size. It can be seen from the assemblage map of Yardley et al. (1980) that in this area andalusite occurs only in veins, but a short distance to the south it appears in the matrix also. There is a distinctive layer at the back of the quarry of a more oxidised pelitic schist with green rather than brown biotite and large (cm-scale) poikiloblasts of pinitised cordierite.

Fig. 13 Geological sketch map of supplementary locality 5.s2 showing the approximate access route to the skarn outcrops

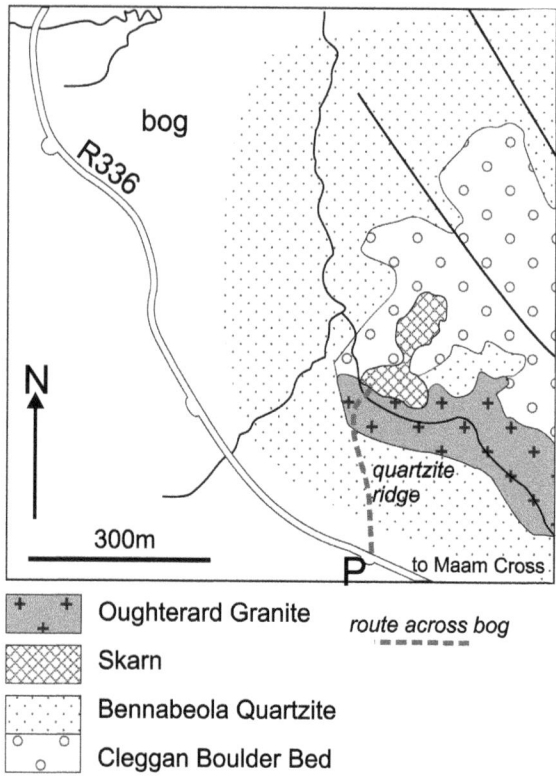

Locality 5.s2: Leckavrea (53.487223°, −9.555977°)

Access and Other Information: Take the R336 from either Maam Cross (to the south) or Maam (to the north) to a long straight section just on the north side of the summit. Park in a lay-by on the south west side of the road, shortly below the summit (53.487223°, −9.555977°) and opposite a gate in the fence on the other side of the road (Fig. 13). At this location at the foot of Leckavrea mountain, there is a small area of the Cleggan Boulder Bed Formation, cut by a skarn of andradite garnet with marginal epidote and by small bodies of Oughterard granite.

The short walk to the outcrop needs care, and luck, to negotiate patches of soft bog. A faint track is present following a ridge of quartzite, but eventually you must cut right and cross a small stream. On the other side is a rectangular digging, probably a trial for copper, in brown skarn (53.489471°, −9.555365°). Immediately behind the digging are further skarn outcrops (Fig. 14) including some showing the skarnification of psammite through selective replacement of feldspathic layers by epidote while quartzite bands are unaffected. Note that the skarn boundaries cut bedding and also cut second phase folds. Away from the epidote margins, the skarn is largely

Fig. 14 Skarn of andradite-rich calcic garnet and epidote, with associated vein quartz, cutting bedding and schistosity in quartz-rich sediments

composed of brown andradite-rich Ca garnet, although small amounts of epidote and pyroxene occur.

Continue north across an old wire fence and a patch of black peat to some outcrops offering excellent views west to Cur Hill and the Maam Valley. A conspicuous granite sheet crosses the outcrop (53.490063°, −9.555006°). These brown-weathering outcrops are of Cleggan Boulder Bed Formation and clasts similar to those described for Excursion 3 can be found in some thick psammite beds. There is a metre scale second phase fold and note that this is cut by the sheet of late orogenic Oughterard Granite.

5.2 Excursion 2: Migmatites and Gneisses Near Maam Cross. Discovery Series Map 45

Access and Other Information: This short excursion involves a walk through high grade migmatites, that can be correlated with the Ballynakill Fm (seen on Excursion 1 at lower grades), and quartz bearing orthogneisses of the synorogenic intrusive suite. Head south from Maam Cross on the R336 for 2 km, where the road crosses an arm of Lough Nahasleam; about 100 m on, over a slight rise, a long straight opens up (Fig. 15). You are by locality 5.6 (Fig. 16a) and should park where it is possible to pull off the road and clear of the bend. There is a small lay-by on the west of the road that is suitable for smaller vehicles (53.436577°, −9.543389°).

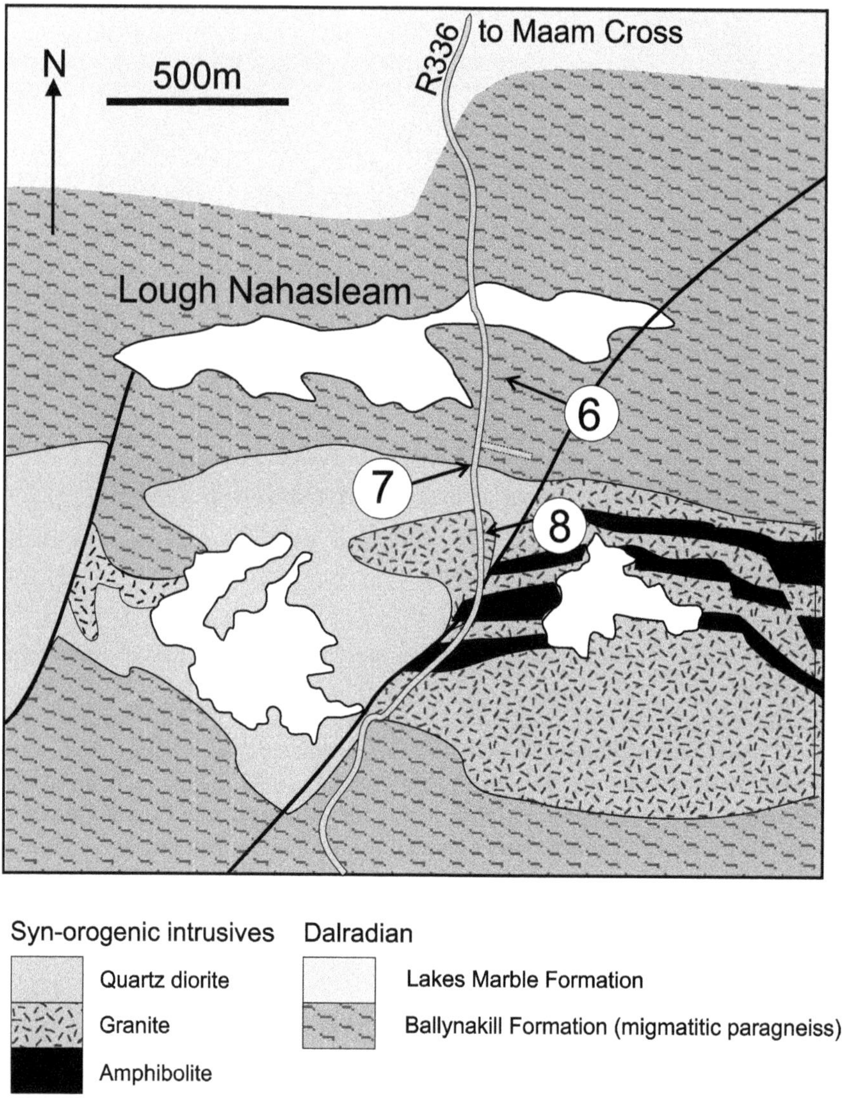

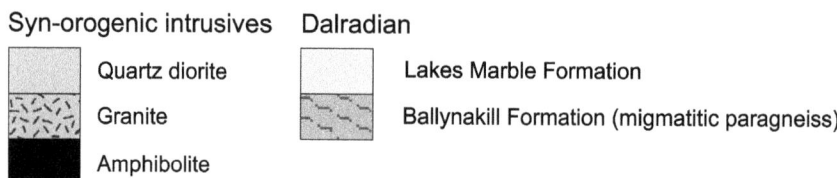

Fig. 15 Sketch map to show the geology and localities 5.6 (6), 5.7 (7) and 5.8 (8) near Lough Nahasleam. Geology based on Leake et al. (1981)

5.2.1 Locality 5.6: Lough Nahasleam (53.436000°, −9.542800°)

On the east side of the road, and separated from it by a deep ditch, is a knoll of migmatite (Fig. 16a) included in the study by Yardley and Barber (1991). A faint path to the south of the outcrop provides access (Note that similar features occur in outcrops on the west side of the road also). The west facing side of the outcrop

Fig. 16 Features of partially melted Ballynakill Fm pelites (paragneiss) near Lough Nahasleam, Locality 5.6. **a** Locality seen from the road, illustrated features are at the front right of the knoll and on top, **b** Nebulitic migmatite with orange-brown weathering peritectic cordierite, **c** Second phase fold of psammite with thickened hinge and attenuated limbs that have become detached from the rest of the bed, immersed in pelitic restite with strong second phase foliation, **d** Leocosome patches hosting peritectic garnet. Intervening restite has ridges of fibrolitic sillimanite replacing biotite

features partially melted pelite with thin bands of psammite. The pelitic migmatite is strongly foliated and contains nebulous patches of fine-grained leucosome. It is also possible to find grains of refractory minerals produced by incongruent melting and at the southwest corner of the outcrop there is abundant centimetre sized cordierite, now heavily altered and weathering to a rusty orange (Fig. 16b). The psammite occurs as disrupted blocks which are now floating in a migmatite matrix. One such block illustrated in Fig. 16c is a second phase fold of which the limbs have been thinned out and broken away leaving the thickened beds in the fold hinge in the midst of pelitic restite with a strong second phase foliation. This demonstrates that deformation continued after the initial folding of layers. Here, and around some other blocks, there are concentrations of migmatite leucosome, which presumably migrated to pressure shadows. On the top of the outcrop to the east there are several drill holes where large diameter core samples have been taken. Here there are large garnet crystals associated with the granitic leucosome material, and sillimanite occurs intergrown with biotite and forming sharp ridges which stick up from the outcrop (Fig. 16d).

5.2.2 Locality 5.7: R336 (53.433929°, −9.543659°)

Return to the road and cross to the west side. Walk south to the vicinity of the 5th telegraph pole. Much of the outcrop is made up of pelitic migmatite, strongly foliated and heterogeneous, with lenses of migmatite leucosome in a matrix of pelitic restite and psammite. On the south side in particular however, a coarse grained rock is present which is foliated but compositionally homogeneous. This is quartz diorite gneiss, part of the synorogenic intrusive suite and with an otherwise intermediate composition contaminated by molten metasediment to have a high quartz content. The mineralogy consists of quartz, plagioclase, now heavily altered to epidote, and hornblende, largely altered to chlorite.

5.2.3 Locality 5.8: R336 (53.431600° −9.544400°)

Continue south towards a bend in the road and cross over. Climb up a large outcrop of pale K-feldspar gneiss. Despite the name, this rock is essentially a weakly foliated granite; it is distinct from the migmatites by being coarser grained and homogeneous. The mineralogy again shows extensive alteration and the epidote alteration of the plagioclase gives the rock a pale green appearance which immediately distinguishes it from the younger Galway granite, present a few kilometres down the road (Chapter "The Late Silurian to Upper Devonian Galway Granite Complex (GGC)"). The outcrop contains a dyke of finer grained granite which lacks any foliation.

5.3 Excursion 3: Early Folds and the Older Dalradian Succession at Glencoaghan. Discovery Series Map 44

Access and Other Information: Glencoaghan is a remote valley in the southeastern corner of the Twelve Bens range. Access is along a minor road which heads north from the N59 at a junction approximately 1 km to the west of the junction with the R341 road to Roundstone. This road is suitable for cars and minibuses but not for large vehicles; there is limited parking available. Most of the excursion is on a steep hill slope and there is a real risk of accidents. As a result, it should not be attempted when the ground is very wet.

Proceed north through a small settlement and continue to the upper part of the valley. Stop where you can park off the road around 150 m before the first cottage in this upper settlement (53.484600°, −9.796300°). The area around the cottages is delimited by a fence and the excursion proceeds on the south and east sides of the fence (Figs. 17 and 18). The fence line provides a convenient route up the hill, but the excursion begins on the low ground by the road south of the fence. The rocks on each side of the road approaching the upper settlement are white quartzites of the Bennabeola Quartzite Formation. They dip gently southeast and a short distance to

Fig. 17 View of the excursion area, looking from Locality 5.9 (9), with blocks of characteristic Boulder Bed psammites in the foreground, towards Locality 5.10 (10) and the route up the fence line to further localities 5.11 (11) and 5.12 (12)

the east of the road pass up into brown weathering psammites of the Cleggan Boulder Bed Formation. Looking east at the hillside you will see green grassy slopes marking the occurrence of the Connemara Marble Formation, and above these are crags of brown psammites passing up into white quartzite again. Thus the valley side at Glencoaghan encompasses much of the lower Dalradian stratigraphy of Connemara, with the distinctive Bennabeola Quartzite occurring both in the valley floor and on the crest of the ridge above, and older rocks occurring in the intervening part of the hillside. The stratigraphy is inverted at road level but right way up in the upper part of the hillside. You will be able to demonstrate that the fold responsible for this repetition of the stratigraphy must predate the abundant parasitic folds which also occur on the hillside.

5.3.1 Locality 5.9: Glencoaghan (53.485000°, −9.784000°)

From the quartzite outcrops near the road, head east towards the base of the slope until you reach browner outcrops of Boulder Bed Formation psammite. Within the Boulder Bed there are units typically 1–3 m thick with a purplish weathered appearance which are of tillite (Fig. 17). You will be able to find clasts of granite and other rock types within these purple weathering beds (Fig. 3c). Thinner beds often show folds and note that both bedding and an earlier, bedding-parallel schistosity are folded. These are second phase folds, with one fold limb gently dipping and the other steep or overturned. Their fold axial surfaces dip southeast, demonstrating that we are south of the Connemara antiform.

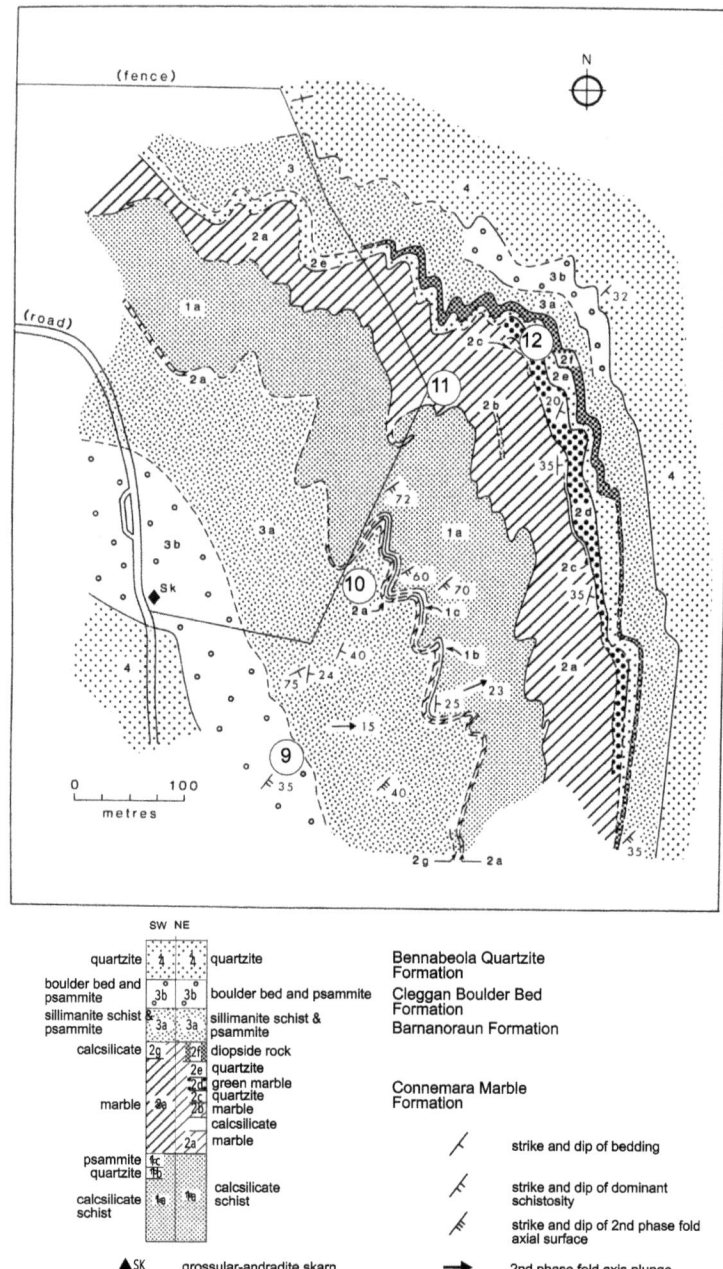

Fig. 18 Geological map of the Derryclare fold at Glencoaghan by BWD Yardley and SH Bottrell. The key for the SW part of the area is for the succession on the lower limb, that for the NW refers to the upper limb. Localities 5.9–5.12 are shown by the circled numbers '9'–'12' respectively

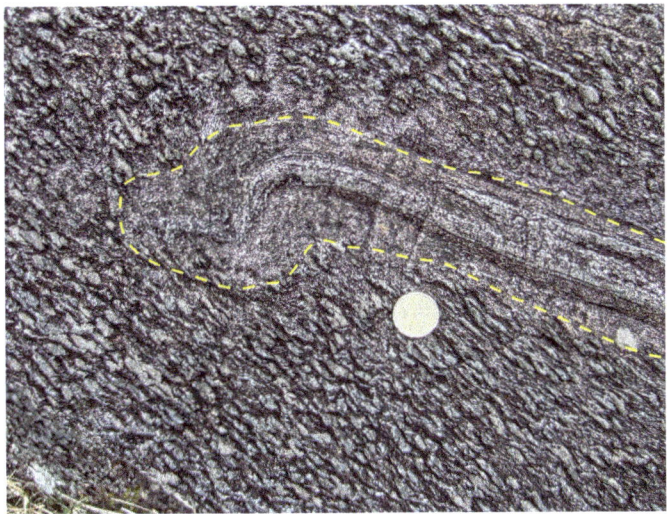

Fig. 19 Refolded fold in rocks of the Barnanoraun Formation. A thin psammite layer picks out a first stage fold which is refolded by a second stage fold with a strong axial planar fabric defined by sillimanite faserkiesel

Continue towards the base of the slope and you will quickly pass from Boulder Bed to a mainly pelitic lithology, the Barnanoraun Formation. Many thin siliceous horizons also display second phase folds, but while these fold the schistosity in the siliceous layers, the pelites develop a new axial planar schistosity marked by the alignment of elliptical lenses (faserkiesel) of fibrous sillimanite a few mm in length, as seen at Excursion 1, Locality 5.5 (Fig. 12). A few blocks have examples of small first phase folds refolded by second phase ones (Fig. 19).

5.3.2 Locality 5.10: Glencoaghan (53.485500°, −9.792900°)

Stunted hawthorn trees at the base of the slope a short distance to the right of the fence mark a line of small marble outcrops in the cores of mesoscale second phase antiformal folds. These mark the top of the Connemara Marble Fm and are structurally overlain by thin beds of psammite and quartzite, they pass up into a massive calc-silicate unit that forms the bluff immediately above.

5.3.3 Locality 5.11: Glencoaghan (53.486600°, −9.791900°)

Proceed with care to climb up the slope next to the fence. When the ground flattens out you have reached an area of dolomitic marble, typically with calcite and tremolite, but with some tremolite pseudomorphs of diopside. It is easily recognised by the lush green grass and gentler slopes. Above this marble are a series of distinctive rocks giving rise to small crags that also pick out further second phase folds. Near the upper contact of the dolomitic marble, a thin quartzite is often present and locally there is a bed of tremolite marble, up to a metre in thickness, with tremolite rosettes marking cm thick layers that formerly contained detrital quartz in the dolomite and developed their present texture when the initial reaction to diopside created porosity which was infiltrated by water (Yardley and Lloyd 1989). This is best found by heading east across the green grass of the marble and inspecting small outcrops.

5.3.4 Locality 5.12: Glencoaghan (53.485100°, −9.790080°)

Upslope from the dolomitic marble, proceeding northeast, are some rather crumbly outcrops of a deeply weathered serpentine marble, an example of the ophicarbonate rocks that characterise the Connemara Marble and are much prized for ornamental stone. This unit is predominantly a serpentine-calcite marble today, but at the peak of metamorphism it contained olivine rather than serpentine; traces remain in some quarries, but none in natural outcrops. Turning to the north, carefully make your way upslope. Above the serpentine marble and separated from it by a few metres of quartzite is a further metasomatic rock unit composed largely of coarse diopside and termed the diopside rock. Both this unit and the serpentine marble are products of metasomatism. Their assemblages result from both the addition of silica in solution and the flushing out of carbon dioxide, released from dolomite breakdown. The pervasive and stratiform nature of the metasomatism suggests that it resulted from the thermally-instigated breakdown of dolomite + quartz at around upper sillimanite zone conditions, which would have generated significant, if transient, porosity throughout the bed. U–Pb ages of titanites from this unit (Table 1) are potentially an important constraint on the timing of metamorphism but, as discussed above, there are large and unexplained discrepancies between different studies. The diopside rock unit is overlain by schists of the Barnanoraun Formation and this was the likely source of silica rich metamorphic fluids.

Proceeding further up through the stratigraphy requires considerable care and good, dry conditions. The intrepid climber will be able to pass on up into the Boulder Bed although the outcrops here have more vegetation and are less easy to study than those by the road. Finally, quartzite is encountered, completing the traverse through the stratigraphy.

Throughout the entire traverse, the vergence of the second phase folds remains constant, indicating an antiformal closure to the north (the Glencoaghan antiform). This demonstrates that the reversal in the way up of the succession is the result of an earlier, first phase fold, the Derryclare fold (Fig. 4).

Although it is clear from the repetition of quartzite and boulder bed that the early fold closure lies somewhere in the middle of the traverse, the exact location is difficult to determine. The distinctive metasomatic calc-silicate horizons seen in the upper part of the section are not repeated lower down. We think it most likely that the early fold is strongly asymmetrical, with the thin marble seen at the very bottom of the slope (Loc 5.10) equivalent to the much thicker units higher up (Loc 5.11) (Yardley et al. 1991a, b). In this interpretation the early fold hinge lies in the calc-silicate unit making up the lower part of the slope. Tanner (1991) considered the fold to be more or less symmetrical, equating the metasomatic beds with the calc-silicate you first climbed up, and treating the lowest marble you met at the bottom of the hill as an anomaly.

5.4 Excursion 4: A Tectonically-Emplaced Gabbro at Currywongaun, Letterfrack. Discovery Series Map 37

A line of gabbro intrusions lies to the north of Letterfrack in northwest Connemara. The smallest of these is at Dawros on the coast. East from the Dawros body across the Dawros River is the Currywongaun body and beyond it the largest of the intrusions forms the hill of Doughruagh above Kylemore Lough. The country rock around the intrusions is shown as paragneiss or hornfels on the map of Leake et al. (1981), but only rarely can the rocks be recognised as hornfelses in the field, and this usually only very close to gabbro or in xenoliths. Often the country rocks are strongly foliated quartz rich schists. This excursion is to Currywongaun with access from a minor road that runs from the N59 just west of Kylemore Abbey, to Tully Cross.

Access and Other Information: From the N59 drive about one-and-a-half km through the small village at the foot of the hill, taking a sharp left hand bend shown in Fig. 20 (lower right corner). Continue beyond the last house where there are several places where it is possible to pull off the road without obstructing gates. Parties with a large vehicle should enquire locally in advance. There is access onto the hillside as shown on the map, although the western gate is of wire and not suitable for parties. Whichever you use, please ensure that it is firmly shut before you move away from it.

5.4.1 Locality 5.13: Currywongaun (53.569000°, −9.932400°)

Whichever gate you use to access the hill you will initially be on country rocks for the most part. Gabbro makes up most of the steep slopes ahead of you, and the contact is near the base of the hill. This locality, which is near the base of the hill, is a small crag in strongly foliated mylonitic country rock, with gabbro above. The contact is exposed here for some metres and is dipping gently into the hillside. Some of the igneous rocks here appear to be hybrid rocks contaminated by the country rock, in which case the original igneous contact must have been quite close to the present tectonised one. Note that the gabbro here is very dark and has the characteristic greenish black colour of hornblende. Further from the contact, where less altered gabbro is encountered, it still contains brown pyroxene. For example, at 53.569000°, −9.932000° there is igneous layering preserved in blocks within foliated hornblende gabbro, shown in Fig. 5b, There are a number of occurrences of pegmatites in the gabbro to the north of

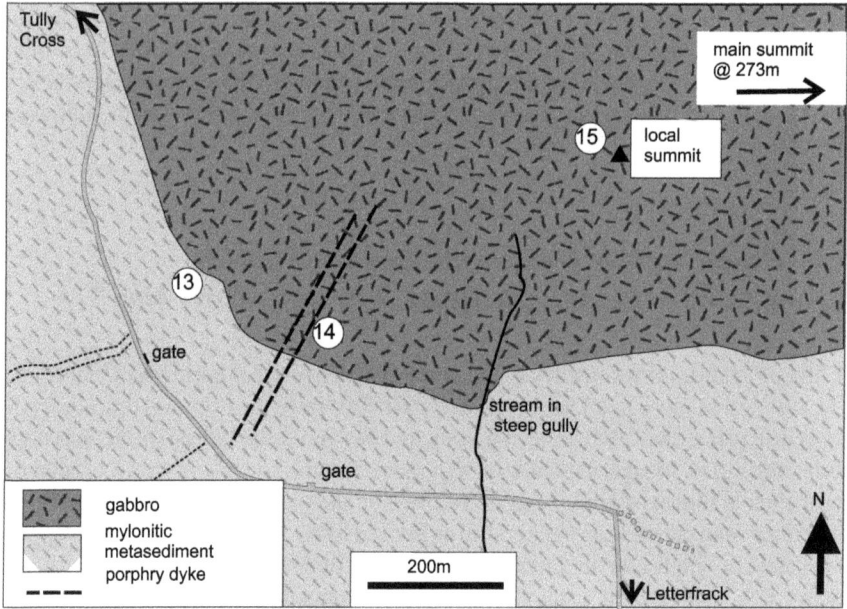

Fig. 20 Map of localities 5.13–5.15 (circled numbers '13'–'15' respectively) in and around the gabbro body at Currywongaun, near Letterfrack

this contact, including examples with tourmaline, muscovite or garnet. Examples are at 53.569800°, −9.930900° (muscovite pegmatite), 53.570500°, −9.930500° (garnet pegmatite) and 53.568200°, −9.929400° (tourmaline-quartz vein).

Follow the contact about 40 m to the east and then head up slope. Crags of gabbro are now partially agmatite, with lenses of quartzofeldspathic leucosome injected into black hornblende gabbro with local shear zones.

5.4.2   Locality 5.14: Currywongaun (53.568200°, −9.929200°)

Continue southeast at a similar height, above the contact. A line of low cliffs near here is in gabbro but contains some large blocks of country rock as xenoliths. From here, work your way up the hill to 53.569200°, −9.930500°, which lies in a wide gully heading to the north northwest. This gully marks the site of a porphyry dyke and occasional scraps of pink dyke rock can be found on the gabbro walls of the gully This dyke has chilled margins, confirming it was emplaced into cooled wall rocks.

5.4.3 Locality 5.15: Currywongaun (53.570600, −9.924100)

Now ascend the hill. As you make your way further from the contact note that xenoliths and agmatite become rarer, and brown pyroxene gabbro becomes more common with the greenish black hornblende restricted to the margins of fractures and small shears. This locality is on a small peak west of the summit of Currywongaun and was the site of the radio transmission masts built by the Marconi company in 1911 as a booster for the signals of the transatlantic radio station near Clifden. All that remains are the concrete bases of the masts and next to one of these is a spectacular outcrop of layered gabbro which features in a number of historic photographs of the transmission station. Alternating layers rich in plagioclase or pyroxene dip steeply west (Fig. 5a).

When you descend from the hill, make sure you return the way you came rather than descending directly down slope.

5.5 Excursion 5: The Mannin Thrust at Mannin Bay. Discovery Sheet 44

The Mannin thrust is arguably the most important tectonic contact in Connemara. The high-grade metamorphic rocks and synorogenic intrusives of the volcanic arc are thrust over low grade meta-volcanic rocks which are exposed in a window south of Clifden. The rocks immediately above the thrust are basic members of the intrusive suite that have been extensively mylonized. These pass up into less highly deformed basic and dioritic intrusions or orthogneiss, followed by pelitic migmatites comparable to those visited in Excursion 2. The first group of outcrops is inland, but the last is on the coast and is best visited when the tide is out.

Access and Other Information: From Clifden head south on the R341 towards Ballyconneely. Turn right for the Alcock and Brown Memorial at 53.461627°, −10.023120° and drive about 800 m to a conspicuous limestone monument with a large parking area. Leave vehicles here while visiting Locs 5.16 and 5.17 (Fig. 21).

5.5.1 Locality 5.16 Alcock and Brown Memorial (53.467000°, −10.032900°)

Although the outcrops are not particularly good, pause at the memorial to take in the geology around. The rocks at the foot of the memorial are migmatite with sillimanite and conspicuous leucosomes. Looking north to Clifden the rocks are mainly Dalradian metasediments with the complex structures of the steep belt. The mountains beyond and away to the northwest are also metasediments but are part of the fold stack of central Connemara. In the opposite direction across Mannin Bay is a

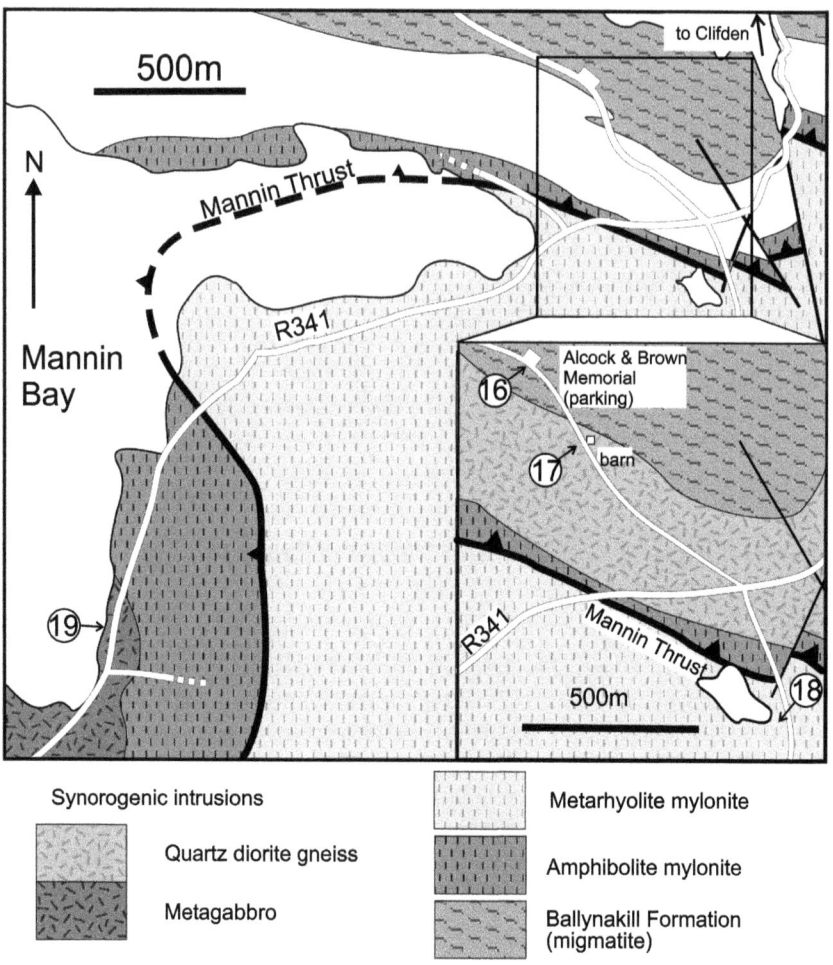

Fig. 21 Geological map of the Mannin Thrust at Mannin Bay after Leake et al. (1981). The inset of the environs of the Alcock and Brown memorial, shows locations 5.16–5.18 marked by circled '16', '17' and '18' respectively. Location 5.19 is marked by the circled '19' in the main diagram

large gabbro body which is part of the basic intrusive suite. An extensive tract of low lying ground to the southeast marks the site of the Delaney Dome, the area where the acid volcanic mylonites outcrop. A distant white pillar sign posted from the monument is the site of the Alcock and Brown landing. The dips in the Delaney Dome are usually gentle, but nevertheless a considerable thickness of mylonite must be present. Furthermore, drilling by the Geological Survey of Ireland in the centre of the dome found no significant change in lithology. It follows that the acid mylonite must be hundreds of metres in thickness.

5.5.2 Locality 5.17: Ballinaboy (53.465000°, −10.030300°)

Walk back down the road to a conspicuous shed on the northeast side with some waste ground opposite. Make your way with care to look at the outcrops on the southwest side of the road, rising to the left of the small valley. Much of the outcrop is quartz diorite orthogneiss with features indicative of medium temperature deformation, including ribbon quartz and fractured plagioclase. At the top of the main outcrop is a southeast facing crag and in places the rocks here are much more basic in composition. The mineralogy is now dominated by hornblende and some hornblendite bodies were probably ultrabasic originally. Where there are flat outcrop surfaces of similar hornblende-rich material at the head of the small valley, it is possible to pick up light reflecting from hornblende cleavages on a sunny day; the hornblende crystals are then seen to be remarkably large, exceeding 10 cm across.

After examining the rocks and their deformation features you can either return to vehicles at the memorial or continue walking and arrange for your transport to pick you up at Loc 5.18.

5.5.3 Locality 5.18: Lough Nagap (53.457600°, −10.021100°)

Access and Other Information: Return to the junction with the R341 and cross, then take the minor road almost opposite, which continues south towards the Alcock and Brown landing site. Continue past farm buildings for about 400 m to a gate in the road. Go through the gate to examine outcrops on both sides of the road (Fig. 21). Cars may be parked at the side of the road.

These outcrops are typical of the mylonites of the Delaney Dome. The rocks are strongly foliated with a greenish silvery appearance. They are fine grained and might be classed as phyllites or phyllonites rather than schists, with small grains of muscovite and chlorite barely visible. Quartz is abundant and on surfaces that cross the foliation it may be possible to see small opaque white albite crystals. Although the fabric is very regular, it is locally affected by kink bands and there are occasional segregations of vein quartz with some chlorite. Rare darker bands are of basic composition. The foliation surfaces carry a pronounced lineation, which indicates the late movement direction in the mylonite is either top to the north (broadly) or the reverse. Shear sense indicators are very rare in the field but overall there is evidence for movement to the south of the upper unit.

5.5.4 Locality 5.19: Mannin Bay (53.445100° −10.062500°)

Access and Other Information: Return to the R341 and head southwest to the road junction at 53.444700°, −10.062100° where there is a small parking area against the sea wall. This locality is in deformed and metamorphosed gabbro and is typical of the highly deformed and metamorphosed gabbro in the vicinity of the Mannin Thrust. It is a very complex outcrop and it is easy to understand that, until modern geochronology was available, some geologists argued that these were in fact older rocks, equivalent to the Lewisian Complex of Scotland, such was the range of cross-cutting relationships.

Descend to the beach, which is of itself of interest as an example of a Connemara "coral strand", a beach made up of fragments of coralline algae from deeper water. There are a wide range of outcrops and these include amphibolites with varying proportions of hornblende and plagioclase, and quartz diorite. The outcrops show variation in both the rock type and the fabric; they are also cut by a few later veins. The compositional differences are most evident in the proportion of dark hornblende and in the presence or absence of quartz. It seems likely that there are both examples of metamorphosed primary layering and of deformed primary agmatite leucosomes such as those seen on Excursion 4. The intensity of the deformation fabrics is very variable and there is also considerable variation in grain size which may reflect an earlier stage in the deformation. Quartz is typically intensely deformed to a ribbon texture, similar to the style of grain scale deformation seen at Loc 5.17. It is not practical to undertake a detailed unravelling of the magmatic and deformational features of these complex outcrops, but what they do provide is an insight into what was happening to the high-grade rocks of the Connemara orogenic belt as they were being cooled and uplifted from depth but before they reached the present tectonic level of the Mannin Thrust. The temperatures at which these rocks developed were lower than their peak metamorphic temperatures, but higher than the temperatures attained by the acid mylonites of the footwall.

Exercises

1. Based on the information provided in this chapter, including Figs. 8 and 9, construct plots showing how the depth of burial of the rocks at Cur Hill (Excursion 1) and at Lough Nahasleam (Excursion 2), has varied through time from 480 to 430 Ma. For both areas, assume that early metamorphism was at kyanite zone conditions and that pressure increased by 0.1 GPa (100 MPa) for every 2800 m burial. You can assume that Cur Hill was about 1 km below the surface when the Silurian unconformity developed, Lough Nahasleam a few kilometres deeper.
2. Gabbros such as the one seen at Currywongaun can contain various other rock types including xenoliths of country rocks, sheets of granitic material originally partial melts of the country rocks, intrusions of later magmas of the synorogenic suite and hydrothermal veins. Suggest how these might differ in appearance in the field.
3. Why is the Connemara synorogenic intrusive sequence interpreted as having been emplaced in an arc?

4. How did the depositional environment of Dalradian sedimentation vary between the deposition of the Connemara Marble Formation and the Ballynakill Formation?
5. Collect measurements of the strike and dip of second phase fold axial surfaces during Excursion 1. Plot your data using an equal area stereographic projection and deduce the plunge of the third phase Connemara Antiform in this area.
6. As part of Excursion 4, make several traverses across the contact between the gabbro and its country rocks. Is the contact always clear cut or are there rocks near the contact with some characteristics of both types? Discuss how you would decide to draw the contact if mapping here.

References

Badley ME (1976) Stratigraphy, structure and metamorphism of Dalradian rocks of the Maumturk Mountains, Connemara, Ireland. J Geol Soc Lond 132:509–520

Barber JP, Yardley BWD (1985) Conditions of high grade metamorphism in the Dalradian of Connemara, Ireland. J Geol Soc Lond 142:87–96. https://doi.org/10.1144/gsjgs.142.1.0087

Cliff RA, Yardley BWD, Bussy F (1993) U-Pb isotopic dating of fluid infiltration and metasomatism during Dalradian regional metamorphism in Connemara, western Ireland. J Metamorph Geol 11:185–191. https://doi.org/10.1111/j.1525-1314.1993.tb00141.x

Cliff RA, Yardley BWD, Bussy F (1996) U-Pb and Rb-Sr geochronology of magmatism and metamorphism in the Dalradian of Connemara, western Ireland. J Geol Soc London 153:109–120

Cliff RA, Bond CE, Butler RWH, Dixon JE (2016) Geochronological challenges posed by continuously developing tectonometamorphic systems: insights from Rb–Sr mica ages from the Cycladic Blueschist Belt, Syros (Greece). J Metamorph Geol 2016. https://doi.org/10.1111/jmg.12228

Dewey JF, Ryan PD (2016) Connemara: its position and role in the Grampian Orogeny. Can J Earth Sci 53:1246–1257

Draut AE, Clift PD (2002) The origin and significance of the Delaney Dome Formation, Connemara, Ireland. J Geol Soc Lond 159:95–103

Draut AE, Clift PD, Hannigan R, Shimizu NGD (2002) A model for continental crust genesis by arc accretion: rare earth element evidence from the Irish Caledonides. Earth Planet Sci Lett 203:861–877

Evans BW (1964) Fractionation of elements in the pelitic hornfelses of the Cashel-Lough Wheelaun intrusion, Connemara, Eire. Geochimica et Cosmochimica Acta 28:127–156

Ferguson CC, Harvey PK (1978) Thermally overprinted Dalradian rocks near Cleggan, Connemara, western Ireland. P Geologist Assoc 90:43–50

Friedrich AM (1998) ^{40}Ar/3'Ar and U-Pb geochronological constraints on the thermal and tectonic evolution of the Connemara Caledonides, Western Ireland. PhD thesis, Massachusetts Institute of Technology

Friedrich AM, Hodges KV (2016) Geological significance of 40Ar/39Ar mica dates across a mid-crustal continental plate margin, Connemara (Grampian orogeny, Irish Caledonides), and implications for the evolution of lithospheric collisions. Can J Earth Sci 53:1258–1278

Friedrich AM, Bowring SA, Martin MW, Hodges KV (1999a) Short-lived continental magmatic arc at Connemara, western Irish Caledonides: implications for the age of the Grampian orogeny. Geol 27:27–30

Friedrich AM, Hodges KV, Bowring SA, Martin MW (1999b) Geochronological constraints on the magmatic, metamorphic, and thermal evolution of the Connemara Caledonides, Western Ireland. J Geol Soc Lond 156:1217–1230

Gradstein FM, Ogg JG, Smith AG (2004) A geologic timescale 2004. Cambridge University Press, 589pp

Kilburn C, Pitcher WS, Shackleton RM (1965) The stratigraphy and origin of the Portaskaig Boulder Bed Series (Dalradian). Geol J 4:343–360

Leake BE (1970) The fragmentation of the Connemara basic and ultrabasic intrusions. In: Mechanisms of igneous intrusion. Seel House Press, Liverpool, pp 103–122

Leake BE (2021) The geology of the Clifden district, Connemara Co. Galway, Ireland and present understanding of Connemara geology. Irish J Earth Sci (in press).

Leake BE, Tanner PG (1994) The geology of the Dalradian and associated rocks of Connemara, Western Ireland: a report to accompany the 1:63,360 geological map and cross sections. Royal Irish Academy, 96pp

Leake BE, Hendry GL, Kemp A, Plant AG, Harvey PK, Wilson JR, Coats JS, Aucott JW, Liinel T, Howarth JR (1969) The chemical analysis of rock powders by automatic X-ray fluorescence. Chem Geol 5:7–86

Leake BE, Tanner PWG, Senior A (1981) The Geology of Connemara. 1:63,360 coloured geological map complete with cross-sections and accompanying map of the fold traces and regional metamorphism in the Dalradian rocks of Connemara. University of Glasgow

Long CB, McConnell B, Archer JB (1995) The geology of Connemara and South Mayo: a geological description of southwest Mayo and adjoining parts of northwest Galway to accompany the bedrock geology 1:100,000 scale map series, sheet 10, Connemara. Geological Survey of Ireland

Nebel O, Mezger K (2006) Reassessment of the NBS SRM-607 K-feldspar as a high precision Rb/Sr and Sr isotope reference. Chem Geol 233:337–345

Tanner PWG (1991) Metamorphic fluid flow. Nature 352:483–484. https://doi.org/10.1038/352483a0

Tanner PWG, Shackleton RM (1979) Structure and stratigraphy of the Dalradian rocks of the Bennabeola area, Connemara, Eire. Geol Soc Lond Spec Pub 8:243–256

Tanner PWG, Dempster TJ, Dickin AP (1989) Time of docking of the Connemara terrane with the Delaney Dome Formation, western Ireland. J Geol Soc Lond 146:389

Vermeesch P (2018) IsoplotR: a free and extendable toolbox for geochronology. Geosci Front 9:1479–1493. https://doi.org/10.1016/j.gsf.2018.04.001

Villa IM, De Bièvre P, Holden NE, Renne PR (2015) IUPAC-IUGS recommendation on the half-life of ^{87}Rb. Geochim Cosmochim Ac. https://doi.org/10.1016/j.gca.2015.05.025

Walker JD, Geissman JW, Bowring SA, Babcock LE (2018) Geologic Time Scale v. 5.0: Geological Society of America. https://doi.org/10.1130/2018.CTS005R3C

Wellings S (1998) Timing of deformation associated with the syn-tectonic Dawros–Currywongaun–Doughruagh Complex, NW Connemara, western Ireland. J Geol Soc Lond 155:25–37

Wotzlaw J-F, Buret Y, Large SJE, Szymanowski D, von Quadt A (2017) ID-TIMS U–Pb geochronology at the 0.1% level using 10^{13} ohm resistors and simultaneous U and $^{18}O/^{16}O$ isotope ratio determination for accurate UO_2 interference correction. J Anal At Spectrom 32:579–586

Yardley BWD (1976) Deformation and metamorphism of Dalradian rocks and the evolution of the Connemara cordillera. J Geol Soc Lond 132:521–542

Yardley BWD (1977) An empirical study of diffusion in garnet. Am Miner 62:793–800

Yardley BWD (1986) Fluid migration and veining in the Connemara Schists, Ireland. In: Walther JV, Wood BJ (eds) Advances in physical geochemistry, vol 5. Springer-Verlag, pp 109–131

Yardley BWD (2009) The role of water in the evolution of the continental crust. J Geol Soc Lond 166:585–600

Yardley BWD, Barber JP (1991) Melting reactions in the Connemara Schists: the role of water infiltration in the formation of amphibolite facies migmatites. Am Miner 76:848–856

Yardley BWD, Lloyd GE (1989) An application of cathodoluminescence microscopy to the study of textures and reactions in high grade marbles from Connemara, Ireland. Geol Mag 126:333–337

Yardley BWD, Warren CE (2021) An introduction to metamorphic petrology, 2nd edn. Cambridge University Press, Cambridge, p 333

Yardley BWD, Leake BE, Farrow CM (1980) The metamorphism of Fe-rich pelites from Connemara, Ireland. J Petrol 21:365–399

Yardley BWD, Barber J, Gray J (1987) The metamorphism of the Dalradian rocks of western Ireland and its relation to tectonic setting. Phil Trans R Soc Lond A 321:243–270. https://doi.org/10.1098/rsta.1987.0013

Yardley BWD, Bottrell SH, Cliff RA (1991a) Evidence for a regional-scale fluid loss event during mid-crustal metamorphism. Nature 349:151–154. https://doi.org/10.038/349151a0

Yardley BWD, Bottrell SH, Cliff RA (1991b) Metamorphic fluid flow. Reply Nat 352:484. https://doi.org/10.1038/352483a0

The Ordovician Fore-Arc and Arc Complex

Paul D. Ryan, John F. Dewey, and John R. Graham

Abstract This chapter first examines the lithologies of the ophiolitic mélange of the Deer Park Complex and the sheared metasediments of the Clew Bay Complex which, together, are interpreted as a Cambrian to mid-Ordovician subduction accretion complex. Two traverses are then made through the Ordovician sediments of the South Mayo Trough, which lie to the south. These are folded into a major syncline, the Mweelrea-Partry Syncline. The first traverse is through the early to late Ordovician sediments of the north limb, which exceed 9 km in thickness. Turbidites at lower stratigraphic levels show derivation from continental and ophiolitic sources. At higher levels, fluviatile sediments contain silicic volcanics and both first cycle metamorphic and silicic igneous detritus. The second traverse is through the volcanic complexes and the overlying, more proximal, mid to upper Ordovician sediments of the south limb. The arc rocks show an evolution from intra-oceanic composition to one influenced by melting of continental material. The sediments record the unroofing of a middle crustal Barrovian metamorphic complex and a continental arc. The Ordovician of South Mayo is interpreted as originating in a fore-arc basin in an oceanic setting that remained below sea-level during the mid-Ordovician arc-continent collision that caused the Grampian Orogeny and consequently recorded the unroofing of this orogen and the subsequent subduction flip.

P. D. Ryan (✉)
School of Natural Sciences, Earth and Ocean Sciences, National University of Ireland Galway, Unversity Road, Galway, Ireland
e-mail: paul.ryan@nuigalway.ie

J. F. Dewey
John Dewey, University College, High Street, Oxford OX1 4BH, UK

J. R. Graham
Department of Geology, John Graham, Trinity College Dublin, Museum Building, Dublin, Ireland

© The Author(s), under exclusive license to Springer Nature Switzerland AG 2022
P. D. Ryan (ed.), *A Field Guide to the Geology of Western Ireland*, Springer Geology Field Guides https://doi.org/10.1007/978-3-030-97479-4_6

1 Introduction

The Cambro–Ordovician sedimentary and volcanic rocks of western Ireland (Fig. 1a) crop out on the shores of Clew Bay (Clew Bay and Deer Park Complexes), in south Mayo (South Mayo Trough), in east Mayo around Charlestown (Charlestown Group), a small inlier at the head of Clew Bay (Letter Formation) and in south Connemara (South Connemara Group). This chapter deals with the rocks of the Clew Bay and Deer Park complexes and the thick lower to middle Ordovician deposits of the South Mayo Trough (Fig. 1a and b). This guide will not visit the Charlestown Group or the Letter Formation. The South Connemara Group is of probable middle Ordovician age and occurs as a roof pendant in the Galway granite and will be dealt with in chapter "The South Connemara Group" for logistical reasons. The Clew Bay and Deer Park complexes (Fig. 1b) make up a Cambro–Ordovician accretionary prism which separates the Ordovician arc-related basins of South Mayo from the Dalradian rocks of the Laurentian margin (Chapters "The Basement and Dalradian Rocks of the North Mayo Inlier" and "The Slishwood Division and Its Relationship with the Dalradian Rocks of the Ox Mountains") and lies along a major geophysical lineament, the Fair Head–Clew Bay line (FCL) that can be traced into the Highland Boundary lineament of Scotland and the Baie Verte–Brompton line of Newfoundland (Max and Riddihough 1975; Max et al. 1983). The FCL is generally regarded as the fault zone bounding the Laurentian margin (Chew 2003; Max et al. 1983; Ryan et al. 1995). It dips northwards (Ryan et al. 1983) and merges into a north-dipping lower crustal zone of reflectivity that can be continued to a keel in the Moho some 100 km farther north (Klemperer et al. 1991). The FCL records a long and complex movement history. It probably initiated at the ocean-continent transition during Neoproterozoic rifting (Kennedy 1980), formed a major décollement during the Grampian Orogeny (Chew 2003) and was subsequently reactivated during the late-Ordovician to Devonian (see for example Hutton and Dewey 1986; McCaffrey 1997).

The guide first examines the Deer Park and Clew Bay complexes in the north, believed to be components of an accretionary prism. The traverse then moves south to study the thick Ordovician deposits of the South Mayo Trough in the north limb of the Mweerea-Partry syncline, much of which are thought to have been deposited in a fore-arc basin. Finally, the more proximal sedimentary and volcanic rocks of the south limb of the fold which record the history of the arc and its subsequent stripping are studied.

2 The Clew Bay and Deer Park Complexes

The Clew Bay complex (CBC) principally comprises three highly-sheared greenschist-facies turbidite sequences: the South Achill Beg Formation (Chew 2003) on the north shores of Clew Bay; the Ballytoohy 'Series' of Clare Island; and the Killadangan 'formation' in the south (Fig. 1b). Some authors (e.g., Max 1989;

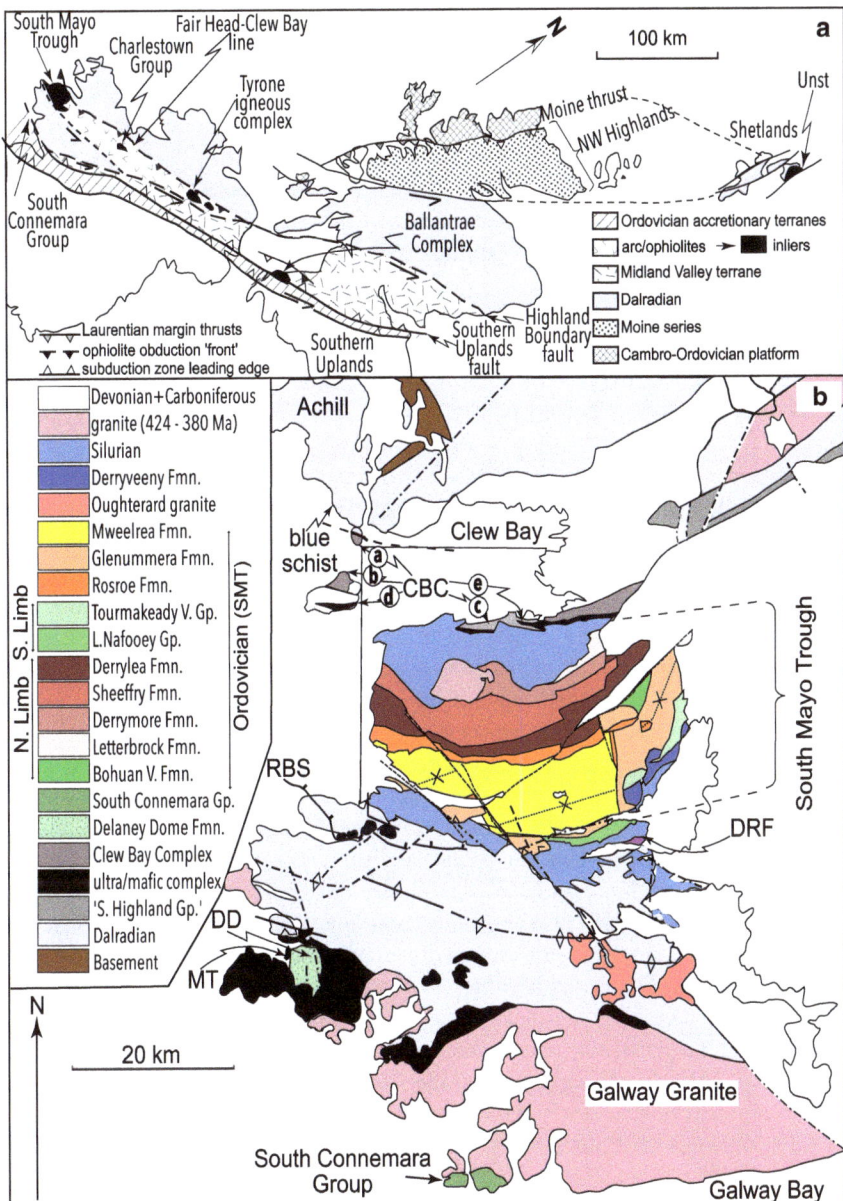

Fig. 1 a Summary tectonic map of the Irish and Scottish Caledonides redrawn from (Dewey 2005) **b** Geological map of western Ireland redrawn from (Ryan and Dewey 2011). Abbreviations: CBC = Clew Bay Complex; DD = Delaney Dome; DRF = Doon Rock Fault; MT = Mannin Thrust; RBS = Renvylle–Boffin Slide; SMT = South Mayo Trough. The letters in circles refer to the various components of the Clew Bay Complex: a = South Achill Beg Formation; b = Ballytoohy 'series'; c = Killadangan 'formation'; d and e = Deer Park Complex

Williams et al. 1997) have included the ophiolitic mélange of Deer Park (see below) within Clew Bay Complex. However, we here follow the usage of Chew et al. (2010) where these complexes are treated separately.

The Clew Bay complex can be traced east–west for 52 km along strike from west of Islandeady to the offshore Bills rocks, some 16 km west of Achill Beg (Graham et al. 1989; Max 1989). Lithologically similar metagreywacke and graphitic shale sequences (Westport Grit, Ardvarney and Cloonygowan formations) attributed to the Southern Highlands Group of the Dalradian continue eastwards to the Ox Mountains (Johnston and Philips 1995).

The Clew Bay complex underlies much of Clew Bay, giving an across-strike distance of 13 km. The South Achill Beg Formation comprises graphitic slate and limestones which are overlain by quartz psammites and micro-conglomerates (Chew 2003). The Ballytoohy 'Series' of Clare Island is a mélange with block-in-matrix structure (Max 1989; Williams et al. 1994). Typically, 10 m blocks of quartz-arenite are enclosed in an intensely sheared black shaley matrix that, locally, has manganese and iron carbonate concretions. Blocks commonly do not contain a shear fabric but show bedding disruption and shale injection features (Max 1989; Williams et al. 1994). There are also blocks of interbedded chert and siltstone, which still maintain bedding coherence, and of mafic volcanics interbedded with carbonates. Blocks are of variable shape and are at least 500 m along their longest axis parallel to the shear fabric, which is folded and strikes at ~100° east of north (Max 1989). Johnson and Philips (1995) argue that much of the bedding within the Ballytoohy 'series' is coherent. However, coherent terranes exist in, for example, the late-Cretaceous to early Tertiary Kodiak accretionary complex of Alaska (Fisher and Byrne 1987) that cover a much larger area than the total outcrop of the CBC. The Killadangan 'formation' on the south shore of Clew Bay comprises green, graded and poorly-sorted medium to fine-grained sandstones and sub-arkoses, pebbly sandstones with granules of strained, blue, rutilated vein quartz, siltstones, and graphitic and cherty argillites. This 'formation' also shows block-in-matrix texture with scaly phacoids of coarse-grained arenites with only sedimentary, often dewatering, fabrics lying in a matrix of highly sheared argillites. It is heavily quartz-veined. Veins are both deformed by and crosscut the shear fabric. The similarity in lithologies has led most authors to infer a general correlation between the main components of the CBC (see Max 1989). This view is supported by the similar Sm–Nd model T_{dm} ages for the three complexes (Chew 2003; Harkin et al. 1996) and the (~460 Ma) cooling age for S2 muscovites in all three blocks (Chew 2003). Heavy Mineral studies (Dewey and Mange 1999; Chew 2003), whole-rock geochemistry, petrographic analysis and Sm–Nd model ages (Harkin et al. 1996), Sm–Nd isotopes and detrital zircon profiles (Chew et al. 2010) all indicate that the CBC had similar provenance to the Dalradian.

The Deer Park Complex (Ryan et al. 1983) is a poorly-exposed ophiolitic mélange that can be traced for 18 km along the north slopes of Croagh Patrick and a further 12 km westwards to a fault bound inlier on the south shores of Clare Island (Fig. 1b, see section "Louisburgh–Clare Island Succession", chapter "The Silurian of North Galway and South Mayo"). It is in fault contact with the Killadangan 'formation' and this contact dips northwards (Ryan et al. 1983; Johnson and Philips 1995).

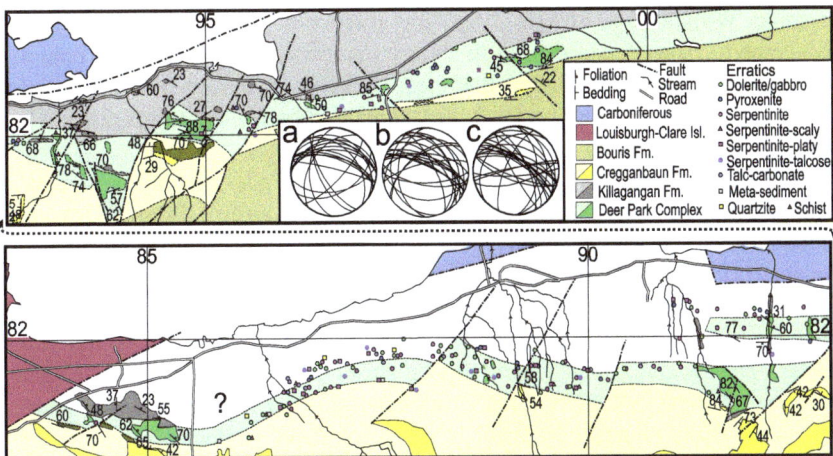

Fig. 2 Geological map of the Deer Park Complex on the south shore of Clew Bay after mapping by J. F. Dewey and P. D. Ryan. The Irish National Grid 1 km grid lines are shown. The three equal area stereonets refer to foliation in: **a** the Killadangan 'formation' **b** the Deer Park Complex and **c** the Silurian Cregganbaun Formation

It is generally poorly exposed with a thick glacial sediment cover. The discontinuous outcrop comprises small knolls commonly of low metamorphic grade mafic and ultramafic igneous rocks, interpreted as blocks, set in a highly sheared serpentinite or talc-carbonate matrix. Much of the map pattern is defined by concentrations of glacial erratics, which are often found to be close to the source bedrock (Fig. 2). The DPC is associated with a strong positive magnetic anomaly, which becomes weaker eastwards (Gibson et al. 1996) suggesting, as does the distribution of the erratics, that the matrix becomes more talcose in this direction. A deposit of some 3.5 million tonnes of talc-magnesite was recorded just south of Westport but never exploited. The following lithologies are recorded: meta-sediments; carbonate + calc + tremolite schists; amphibolites; serpentinites; metagabbros; metapyroxenites; metadolerites; pyroxenite breccia; and rodingites. Metasediments and amphibolites display a complex deformational history and contain assemblages typical of the epidote–amphibolite facies (Ryan et al. 1983). Blocks of schist are tectonically intercalated with the DPC and are referred to by some authors as the 'Deer Park schists' in this account they are assigned to the CBC, Killadangan 'formation' (Fig. 2). Ryan et al. (1983) and Chew et al. (2010) report a MORB geochemistry for the mafic igneous rocks. Pronounced Nb anomalies and initial Nd isotopic compositions ($\varepsilon Nd_{[490]} = +6$) suggest a juvenile, subduction related, origin for the metadolerites (Chew et al. 2007). The proportions of the various meta-igneous lithologies recorded in the erratics are: metadolerite (22.2%); metagabbro (2.7%); metapyroxenite (6.5%); talc-schist (21.1%); and serpentinite (47.6%). Assuming that the erratics provide a representative sample of the lithologies in the complex, the relative paucity of gabbros and layered gabbros (which, for example, comprise ~41% of the Bay of Islands ophiolite: Williams and Stevens 1974) and absence of volcanics suggests tectonic dismemberment as opposed to sedimentary mass wasting of a flat slab ophiolite. Dewey and Ryan (1990) suggest that these rocks were sourced from an ophiolitic backstop to an accretionary prism, a view supported by the Nd isotopic data (Chew et al. 2010).

The Ballytoohy Series in Clare Island have yielded a mid-Cambrian–Tremadoc (?) *Protospongia hicksi* (Rushton and Philips 1973) and an early- to mid-Ordovician conodont fauna of *Drepanoistodus* and *Panderodus* (Harper et al. 1989). Williams et al. (1994) report a microfauna that they interpret as Wenlock. Johnson and Philips (1995) state that a Wenlock age is incompatible with Llandovery–Wenlock sequences resting unconformably on both the Killadangan 'formation' and the Ballytoohy 'series'. Dewey and Mange (1999) argued strongly against a Silurian age because there are no fresh volcanic zircons or metamorphic detritus in the CBC; both are ubiquitous in the Silurian elsewhere in western Ireland. Graham (2001) provides a summary of this controversy. The S2 fabric which cuts the Clew Bay complex yields muscovite cooling ages of ~460 Ma (Chew 2003; Chew et al. 2010). These difficulties may be somewhat resolved by the recognition that mazuelloids (Burrett 1985) and trilete spores from Baltica (Rubinstein and Vadja 2019) extend at least to the Sandbian. The thermal alteration index for the spores from the Ballytoohy 'series' suggests these rocks have been heated to 250–350 °C. Studies on chlorite in the South Achill Beg Formation suggest P–T conditions of 10 kbar at 325–400 °C (Chew et al. 2003). An amphibolite in the Deer Park complex has yielded an ^{40}Ar–^{39}Ar plateau age of 514 ± 3 Ma whilst muscovite from an intercalated garnet-muscovite-albite schist gives a ^{40}Ar–^{39}Ar plateau age of 482 ± 2 Ma and a Rb–Sr age of 483 ± 7 Ma. THERMOCALC multi-equilibria suggest P–T values of ~3.3 kbar and 580 °C for this schist (Chew et al. 2010).

The CBC and DPC are interpreted as components of a northward-facing accretionary prism which lay outboard of continental blueschists but inboard of an oceanic forearc–arc complex (Chew et al. 2003; Dewey and Shackleton 1984; Ryan and Dewey 1991). Available dating evidence suggests that this accretionary prism was active for some 58 m.y. from the middle Cambrian (Series 2, ~514 Ma) to the late Ordovician (Sandbian, ~456 Ma). The slightly older cooling ages (Chew 2003) may reflect the complex relationship between sedimentation and uplift typical of accretionary complexes (e.g., Allen et al. 2008). Regionally, the geophysical lineament bounding this accretionary complex to the north, the Fair Head–Clew Bay line, can be traced eastwards to the Highland Border Complex of Scotland (e.g., Chew 2003; Max et al. 1983; Ryan et al. 1995; Tanner 2007), a distance of over 500 km. The trench associated with this 'Taconic' oceanic arc extended some 2500–3000 km westwards into the Appalachians (Dalziel and Dewey 2018). If approximately one third of this trench was accretionary as opposed to erosive (see Stern et al. 2012) the accretionary prism(s) may have extended some 1000 km, have been some 100–200 km in width (e.g., the Barbados accretionary prism, Maltman et al. 1997) and up to 30 km in thickness (P-T data from South Achill Beg Formation, Chew et al. 2003). The many tectonic processes which subsequently modified this prism and destroyed much of the evidence required to elucidate a full tectonic history are reviewed by Van Staal et al. (1998). However, the recorded metamorphic history may give some insights. The 514 ± 3 Ma age for an amphibolite in the DPC could either date the basal accretion of oceanic lithosphere as in the Kenting mélange of Taiwan (Zang et al. 2016) or the thrusting of the back-stop ophiolite over its accretionary prism (e.g., Stern et al. 2012; Topuz et al. 2013). The supra-subduction zone geochemistry of the associated gabbros (Chew et al. 2010) support the latter interpretation and suggest subduction initiated about 514 Ma, broadly coeval the with middle Cambrian initiation of the 'Taconic' arc (Dalziel and Dewey 2018). The relatively high heat flow 482 Ma event can be related to burial and exhumation paths within the accretionary prism (e.g., Batt et al. 2001). For example, garnet-biotite-albite schists occur in the Otago Schists (Mortimer 2000) a component of the Triassic–early Cretaceous accretionary prism of the Eastern Province of New Zealand. These higher-grade metamorphic rocks in the Otago Schists are associated with reactive fluid flow linked to dehydration (Breeding and Ague 2002) and their exhumation may have been related to subduction roll-back (Forster and Lister 2003). High heat flow is recorded in the Neogene–Quaternary Southern Apennines accretionary prism, possibly related to its being underlain by a hot mantle wedge related to roll-back (Doglioni et al. 1996). In summary, the complexity of stratigraphic, structural and metamorphic relationships preserved in the CBC and DPC are consistent with them being part of a Cambro–Ordovician accretionary complex.

3 Excursion to the Clew Bay and Deer Park Complexes

3.1 Aim

The aim of the excursion is to study the lithologies and fabrics present in an accretionary complex including off-scraped sediments and accreted ophiolite fragments.

3.2 Locality 6.1: Lecanvey Foreshore (53.774403°, −9.742747°)

Access and other information This short traverse of about 220 m is along the south shores of Clew Bay. Access is best gained by parking at the shore at Fallduff (53.774581°, −9.753202°) and walking 700 m east along the sandy shore (Fig. 3). This area is tidal and this traverse should only be undertaken at low water or an hour or so before low water.

From the Westport-Louisburgh road (R335) take the small road L18282 to its end, where there is parking for several cars. Walk down onto the beach and east along the shore. The section starts in the Knockmore Sandstone Formation of the

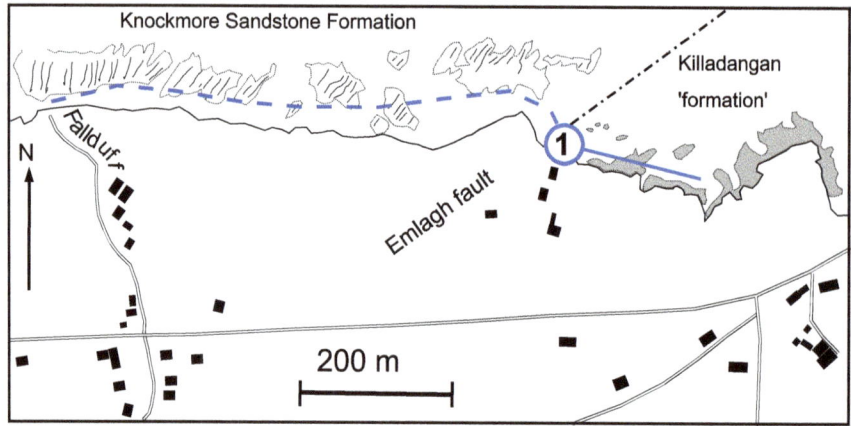

Fig. 3 Geological sketch map for locality 6.1. The traverse begins at the circled number '1'. See Fig. 6 for key

Silurian Louisburgh–Clare Island succession (see section "Louisburgh–Clare Island Succession", chapter "The Silurian of North Galway and South Mayo"). Bedding dips to the east but is overturned. Thin red mudrock horizons help define the bedding. Internal laminations are evident in some sandstones on wave polished sections, but many of the sandstones show little internal structure on normally weathered surfaces. Lamination is generally flat, which is largely a function of the fine grain sizes (the dune stability field is very limited at these grain sizes). Evidence for desiccation in the red mudrock layers supports the interpretation of a fluvial origin for the sandstones. Mudrocks are commonly reworked as clasts in overlying sandstones.

Small calcite veinlets are common as are fractures with brecciation. There are two small folds that plunge 15°–300° (Fig. 4a), but otherwise dips are consistently to the east and are progressively shallower (more overturned) approaching the Emlagh Fault. The fault runs through the bay at Sea View but is not exposed here.

After crossing a sandy bay where extensive quartz veining marks the trace of the Emlagh Fault (Fig. 4b), the section starts at a small private slip-way and continues east for 220 m over the rocky foreshore (Fig. 3). It shows the chaotic bedding and structural relationships typical of the Killadangan 'formation' part of the Clew Bay complex (Fig. 5a). Coarse, graded, commonly granule grade sandstones are interbedded with finer sands and mud rocks. The sandstones show little internal deformation, but they occur in phacoidal shaped bodies surrounded by highly sheared, mostly mud rocks with complex quartz veining. The sands show a variety of sedimentary structures associated with dewatering, which include: load and flame structures (Fig. 5b and c); ball and pillow structures (Fig. 5c); sheet structures; sandstone dykes; and dewatering veinlets. Bedding is broadly parallel with the foliation and dips shallowly north. Clear younging evidence commonly suggests inversion (Fig. 5c). The beds are highly-disrupted and cannot be traced along strike for any great distance. The main lithic components are chert and polycrystalline aggregates of slightly strained

The Ordovician Fore-Arc and Arc Complex

Fig. 4 **a** Open north–south oriented folds in the Knockmore Sandstone Formation **b** Quartz veining adjacent to the Emlagh Fault which runs through the beach. The near rocks are part of the Clew Bay Group and those on the far side of the beach belong to the Knockmore Sandstone Formation of the Louisburgh succession. Clew Bay and the Nephin Beg mountains form the background

Fig. 5 **a** Overview looking east of locality 6.1 showing irregularity of the bedding **b** Loose block showing vein quartz (white) and chert (black) pebble conglomerate loading into a granule grade sandstone. **c** Inverted granule grade sandstone whose base shows loading with ball and pillow structures. Flame structures are formed at the top of the stratigraphically underlying bed. **d** Complex cross-cutting patterns of quartz veining within the Killadangan 'formation' showing both folded and extensional veins

vein quartz some of which contain rutile needles (Phillips 1973) and can show a blueish hue. These are believed to be the rapidly recycled components of a deforming and dewatering accretionary prism (Dewey and Mange 1999). The quartz veins are concentrated in the sheared, commonly finer-grained, lithologies that mantle the more coherent sandstones. A complex sequence of injection and deformation is present and the state of strain can vary very locally (Fig. 5d). This suggests changing mechanical properties associated with dewatering, compaction and lithification of the sediments during progressive layer parallel shear (see, for example, Fisher and Byrne 1987).

3.3 Locality 6.2: Croagh Patrick, Reek Road (53.774384°, −9.640091°)

Access and other information Parking at the Croagh Patrick car park at Murrisk (53.779937°, −9.640084°) on the R335. Follow the signs for Croagh Patrick along the Reek Road. This is a pay car park which can become full and may be time limited in the tourist season. The carpark is built on Carboniferous limestones but neither these nor the major fault that separates them from the older rocks to the north are exposed in accessible localities here. Proceed up the small road to the start of the Reek Road or Pilgrims' Path. Near where the stairs commence, around the first seat, there is exposure of low-grade psammites of the Killadangan Formation. Estimated time for localities 6.2 and 6.3 is 2 h (Fig. 6). Locality 6.3 involves walking

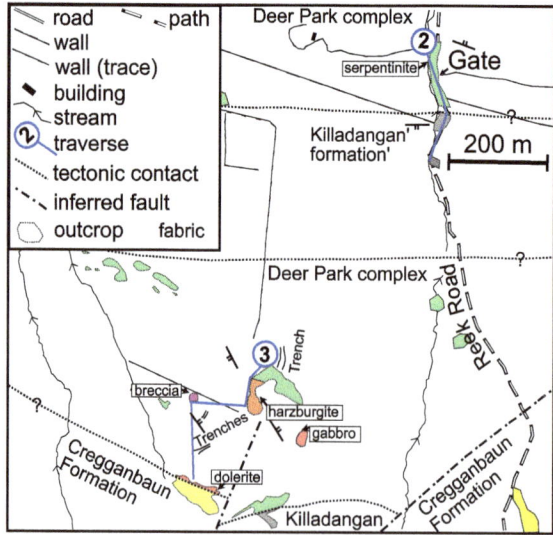

Fig. 6 Geological sketch map for localities 6.2 and 6.3 where the traverses begin at the circled numbers '2' and '3' respectively

Fig. 7 a Strongly sheared serpentinite with sinistral S–C fabrics **b** Strongly sheared serpentinite interleaved with more massive blocks that show much less internal deformation (to the right of the rucksack)

across an exposed mountain side with heather and boggy soil underfoot. Please take appropriate precautions and consult weather forecasts.

The section begins at the statue of St Patrick (53.776709°, −9.639690°). Walk uphill, following the Reek Road, noting that there are many erratics of meta-quartz sandstone of the Killadangan Formation (Fig. 6). Continue up the path to the next obvious exposures (53.774852°, −9.639991°), where strongly-deformed serpentinites, in part talcose schists, are exposed in foot-worn sections of the footpath. A sinistral penetrative fabric (S-fabric) that dips steeply north is cut by a closely spaced sub-vertical crenulation fabric (C-fabric) (Fig. 7a). Sinistrally-transpressive fabrics cut both Ordovician and Silurian rocks in this region (see section "Locality 8.14: Emlagh Point (53.744148°, −9.897374°)", Fig. 2a–c). Although much of the exposure contains these strong fabrics, there are parts that have escaped the strongest deformation and form resistant areas of serpentinite. Lenses of metasedimentary rocks are also included (Fig. 7b). These rocks form the Deer Park Complex, interpreted as representing the deformed margin of the ophiolite backstop where it is tectonically interleaved with the continental margin sediments of the Clew Bay Complex (Killadangan Formation—here interpreted to be part of the Dalradian succession). This large exposure continues, albeit with a small gap, as far as the gate (53.774384°, −9.640091°) where the path emerges onto open hillside. The mineralogy of the serpentinites on Croagh Patrick has not been studied in detail, but authors report that they are mainly antigorite (e.g., Phillips 1973; Ryan et al. 1983) which is formed by the hydration of peridotite at temperatures in excess of 250 °C (Evans et al. 2013).

Continue up the path to (53.772936°, −9.640123°) where the Killadangan Formation is exposed in the stream and in the path; the main fabric here has a steep southerly dip. After a further gap, more schistose serpentinite interleaved with Killadangan Formation with a strong fabric is exposed in the path adjacent to the stream at (53.772293°, −9.639789°).

3.4 Locality 6.3: Croagh Patrick, Lenacraigaboy (53.769148°, −9.645088°)

Access and other information Continue to climb on the Reek Road for about another 70 m passing around a small drumlin to the east. Leave the path where the road and steam diverge and walk 450 m on a bearing of 230° to the large outcrop at Lenacraigaboy (Figs. 7 and 8a) at the top of a small, steep mound of till (53.769148°, −9.645088°). The pale-weathering sandstones of the Silurian Cregganbaun Formation that forms the peak of the mountain are evident in the distant scree slopes and there are fine views of the corrie of Lugademnon on the north side of the mountain (Fig. 8a). This traverse is on an exposed hillside and should not be undertaken in adverse weather conditions.

The section starts at the northwest face of the exposure which is made up of phacoidal shaped blocks of massive dark green antigoritic serpentinite set in a highly sheared matrix with light green chrysotile fibres (Fig. 8b). The chrysotile overgrowths suggest retrogressive metamorphism associated with later stages of development of the shear fabric (Fig. 8c), that dips about 60° N. A large block of meta-harzburgite occurs on the west face of this outcrop. Relict orthopyroxene rhombs are set in an orange-weathering serpentinite matrix containing small black crystals of chromite (1–4 mm) which may be either dispersed (Fig. 8e) or layered. These are interpreted as representing depleted mantle.

Walk due west for 100 m, passing an exploration trench after about 80 m to a small, obvious knoll (53.768429°, −9.647338°). Here breccias contain angular blocks of mostly pyroxenite (1–30 cm) with marked reaction rims which are of oblate shape and are aligned in a weak planar fabric (Fig. 8d). Near the base of the north face of this outcrop a layer of serpentinite sand is parallel to this fabric (Fig. 8f). The breccias are interpreted (Dewey and Ryan 1990) as possibly having an origin in a serpentinite protrusion such as are found on oceanic transform zones and/or the serpentinite mélanges typical of supra-subduction-zone ophiolites (Bonatti 1976). The sedimentary reworking is consistent with this being an olistostrome sourced from an upward-migrating protrusion which penetrated the seafloor (see Lockwood 1971).

From the western end of this crag walk south for 150 m, towards a prominent rock face (53.767064°, −9.647470°). A steeply north-dipping fault juxtaposes part of the ophiolite to the north with quartzites of the Cregganbaun Formation to the south and west. The basal conglomerate is not seen at this locality. The ophiolitic mélange locally contains blocks of ophitic dolerites with asymmetric chilled margins that have MORB geochemistry (Ryan et al. 1983).

There are fine views from here of the impressive topographic feature formed by the Leck Fault on Clare Island. (Fig. 8g) and the partially-drowned drumlin field of Clew Bay (Fig. 8h). It is likely that Croagh Patrick is a palaeonunatak (Ballantyne et al. 2008). It is considered the holiest mountain in Ireland and is the site of an annual pilgrimage on Reek Sunday in July.

Fig. 8 a North face corrie of Croagh Patrick and the screes of Cregganbaun Formation on the side of the mountain **b** View (looking east) of the first outcrop in traverse 6.3. The meta-harzburgite with chromite layering shows forms the orange coloured north-dipping layer just left (north) of the centre of the outcrop. The field of view is approximately 70 m **c** Chrysotile fibres over-growing massive antigoritic serpentinite. This photograph was located at the extreme left (north) of the outcrop in Fig. 8b. **d** Pyroxenite breccia. Fragments show reaction rims and are set in a matrix of massive serpentinite with flecks of chromite (black). The tip of the walking stick is 2.5 cm across **e** Meta-harzburgite with dispersed chromite (black). This photograph was located at the approximate centre of the outcrop in Fig. 8b. The walking stick segment is 30 cm in length **f** ultramafic sand lying beneath the pyroxenite breccia. **g** the impressive topographic feature formed by the Leck Fault on Clare Island. (Fig. 8h) **h** Clew Bay drumlin field from the flanks of Croagh Patrick

Exercises

- Document the different lithologies in the Clew Bay and Deer Park complexes and discuss their likely provenance.
- Document the deformation styles and discuss in what tectonic environments such deformation might occur.
- What evidence is there that the Clew Bay and Deer Park complexes formed at a plate boundary?

4 The South Mayo Trough

The South Mayo Trough contains a thick sequence of Lower to Middle Ordovician sediments and intercalated volcanic rocks that are preserved in an approximately east–west trending mega-syncline, the Mweelrea–Partry Syncline (Fig. 9). The northern limb contains a mainly middle Ordovician sequence (Fig. 10) that exceeds 8 km in thickness in which early sediments show derivation from a primitive island arc complex, an ophiolitic backstop, and poly-metamorphic trench sediments. These are conformably overlain by marine clastic sequences derived from a more evolved arc complex and an emerging juvenile (Grampian) orogen (Dewey and Mange 1999). In the south limb, early Ordovician volcanic complexes are overlain by more proximal sediments which are correlated with the more distal deposits of

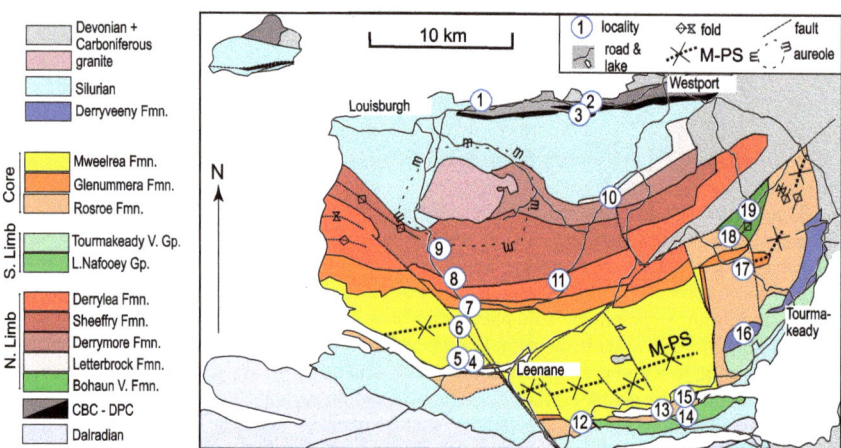

Fig. 9 Outline geological map of the Cambro-Ordovician of South Mayo based on Graham et al. (1989). Minor formations on the south limb and many minor roads are omitted. Abbreviations used MP-S = Mweelrea–Partry Syncline. The circled numbers '1–19' show the positions of localities 6.1–6.19 respectively

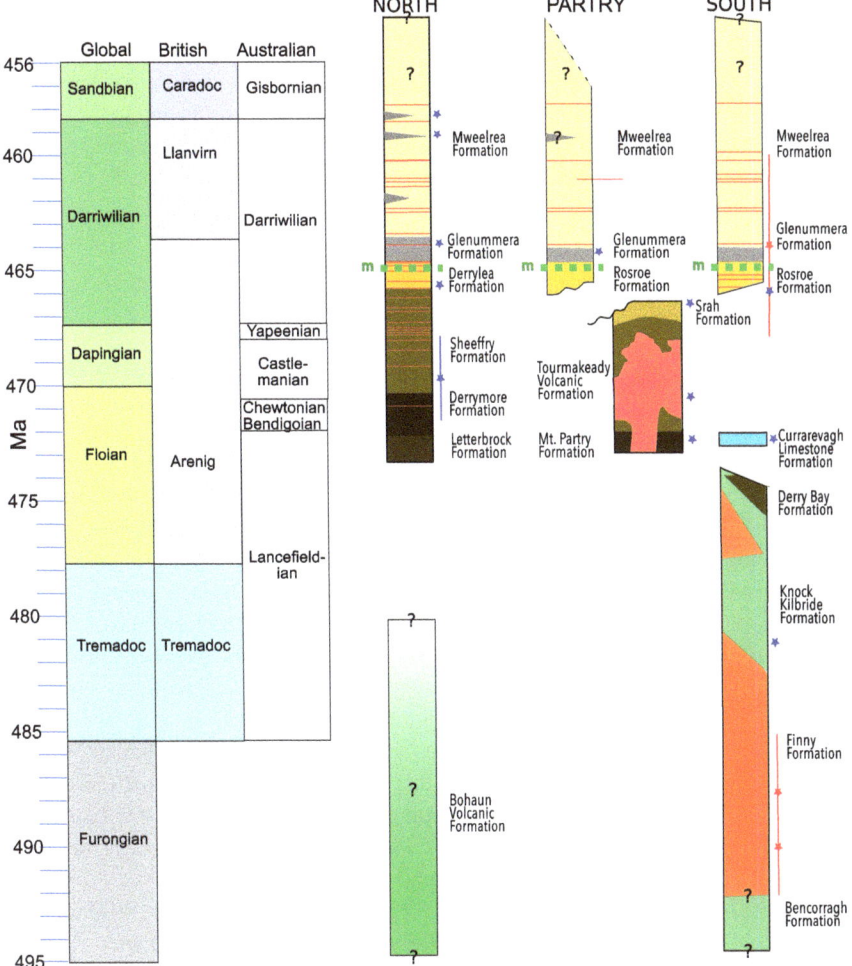

Fig. 10 Chronostratigraphic chart of the Ordovician deposits within the South Mayo Trough (modified after Ryan and Dewey 2011). The absolute ages are based on Cohen et al. (2013, updated) and refer to the Global stratotypes only. The correlations between the global and regional Ordovician stratigraphic units are after Bergstrom et al. (2009). Blue stars represent fossil horizons and the blue bar represents the uncertainty in the palaeontological age of the Sheeffry Formation. The red stars with bars represent absolute age dates and their associated error. The dotted green line labelled 'm' represents the stratigraphic level of the input of significant juvenile metamorphic detritus (Dewey and Mange 1999)

the north limb (see, for example, Graham et al. 1989). Ash layers occur throughout both limbs. Dewey and Ryan (1990) interpreted the South Mayo Trough as a fore-arc basin on the grounds that it was coeval with and lay between an arc (volcanic complexes of the south limb) and an accretionary prism (the Clew Bay and Deer Park Complexes). The Grampian Orogeny is attributed to the obduction of this arc complex over the Laurentian passive margin (Dewey and Shackleton 1984; Ryan and Dewey 1991). The South Mayo Trough remarkably provides a continuous record of sedimentation during Grampian deformation of the Laurentian margin (Dewey and Mange 1999). A possible analogy occurs in modern Taiwan (Ryan and Dewey 2011) where detritus from the oblique arc-continent collision is being transported towards the south-west and deposited in the Luzon fore-arc in the South China Sea.

4.1 Stratigraphy

The stratigraphy of the Ordovician of the South Mayo Trough (Fig. 10) has been studied since the 19th Century. Modern accounts are based on the work of Archer (1977), Dewey (1963), Dewey and Mange (1999), Graham (1987, 2009, 2019), Graham et al. (1989), McManus (1972), Pudsey (1984), Ryan et al. (1980) and Stanton (1960). The lithostratigraphy of the north limb is summarised in Table 1 and the south limb and the core of the fold in the east in Table 2. Graptolites provide palaeontological controls (Dewey et al. 1970; Skevington and Archer 1971; Williams and Harper 1994) as do shelly faunas commonly occurring within limestones associated with volcanic edifices of the south limb (Donovan 1986; Harper et al. 1988, 2010; Pudsey 1984; Stouge et al. 2016; Taylor and Curry 1985 and Williams and Curry 1985). On the north limb, the oldest graptolites occur some 1200 m below the top of the Sheeffry Formation (Stanton 1960) and are probably of Dapingian age though they may be slightly younger (Pudsey 1984 and Skevington and Archer 1971). The composition of the large number of tuff bands within the Sheeffry Formation support a general correlation with the Floian to Dapingian Tourmakeady Volcanic Formation of the south limb (Ryan and Dewey 2011). The youngest age, Sandbian, is recorded for a shelly fauna in the highest marine mud band within the Mweelrea Formation (Harper et al. 2010). In the south limb, late Tremadoc age graptolites (Williams and Harper 1994) recovered from the Knock Kilbride Formation approximately 1400 m above the base of the Lough Nafooey Group (Ryan et al. 1980) are the oldest fossils within the South Mayo Trough. The underlying volcaniclastic deposits and pillow basalts have little intercalated pelagic sediment suggesting that this volcanic complex was formed relatively quickly. Two plagiogranite clasts from the overlying Silurian conglomerates, believed to be derived from the Finny Formation, yield dates of 489.9 ± 3.1 and 487.8 ± 2.3 (Chew et al. 2007) suggesting that the Lough Nafooey Group formed over 12 m.y. to 18 m.y. period from the Furongian to the Floian. The youngest fauna are Darriwilian (Da 3) graptolites from the Rosroe Formation (Skevington 1971). A 464.4 ± 3.9 Ma U–Pb zircon age is reported by S. Noble in Dewey and Mange (1999) for the first ignimbrite band within the Mweelrea Formation. In the Partry region, the Lough Shee Mudrocks Formation, which correlate with the Glenummera Formation (Graham 1987), yield a shelly fauna of likely Darriwilian (Da 3/4) age Harper et al. (1988). The ignimbrite bands within the Mweelrea Formation crop out on both limbs of the syncline and, as the uppermost band is above the highest marine band, this formation must also extend into the Sandbian on the south

Table 1 Stratigraphy of the (? Cambrian) early to middle Ordovician rocks of the north limb of the Mweelrea–Partry syncline. Abbreviations Dw = Darriwilian; Dp = Dapingian; Fl = Floian; Fu = Furongian; Sa = Sandbian; Tr = Tremadoc

North Limb stratigraphy	
Formation, startigraphic age, thickness	Lithology
Mweelrea, Dw 3–Sa 1(?) > 3200 m	Alluvial fans, fan deltas and rivers flowing west. Coarse and thick fan conglomerates. 6 ignimbrite horizons and 3 marine slate bands thicken west near base. Abundant clasts of siliceous volcanics, high level granites and schist, some linked isotopically to Connemara. Derived from east and south-east
Glenummera, Dw 3 600 m	Green-grey slates and mudrocks. Common metamorphic and igneous detritus, rare chromite
Derrylea, Dw 2–Dw 3 1700 m	Greywacke turbidites with carbonate nodules. Chromite and chlorite rich turbidites from north and east. Volcaniclastic turbidites from south. Fresh metamorphic detritus becomes common near top. Several tuff bands
Sheeffry, Fl 3–Dw 2 2500 m	Volcaniclastic turbidites with andesitic-rhyolitic tuff bands. Significant ophiolitic detritus and recycled basement. Becomes more pelitic up-section. Provenance from north, east and south
Derrymore, Fl 3(?) 700 m	Fine-medium grained feldspathic turbidites and slates. Detritus: fresh volcanic; recycled basement; ophiolitic. Northerly to easterly derivation
Letterbrock, Fl 2–Fl 3 (?) > 250 m	Turbidites with conglomerates. Detritus: fresh volcanic; recycled basement; basic igneous at base passing up into ultramafic at top. Northerly derivation
Bohaun, Fu?–Tr? >1000 m	Pillow lavas (few boninites) and breccias, cherts and calcareous shales. Lavas have strongly positive ϵ_{Nd}

limb. There is a relative paucity of fossils in such a thick succession and correlation between regional and global stratotypes for both shelly and planktonic species is difficult, so there is some possibility of future minor modification of the scheme in Fig. 10. In summary, volcanism began in the Furongian and the sedimentary record spans Floian to Sandbian times.

4.2 Provenance

The tectonic setting of the volcanic complexes, based upon geochemistry, petrography and isotopic signature, is that of an oceanic island arc, which progressively shows a greater influence of continental melts up section (Fig. 11) (Chew et al. 2007; Clift and Ryan 1994; Draut and Clift (2001), Draut et al. 2004; Ryan and Dewey 2011 and Ryan et al. 1980). On the north limb the Bohaun Volcanic Formation is of pre-Darriwilian age as it is unconformably overlain by the Rosroe Formation (Graham

Table 2 Stratigraphy of the (? Cambrian) early to middle Ordovician rocks of the south limb and eastern fold closure of the Mweelrea–Partry syncline. Abbreviations Dw = Darriwilian; Dp = Dapingian; Fl = Floian; Fu = Furongian; Sa = Sandbian; Tr = Tremadoc

South Limb stratigraphy	
Formation, stratigraphic age, thickness	Lithology
Mweelrea, Dw 3–Sa 1(?) > 3200 m	Alluvial fans, fan deltas and rivers flowing west. Coarse and thick fan conglomerates. 6 ignimbrite horizons and 3 marine slate bands thicken west near base. Abundant clasts of siliceous volcanics, high level granites and schist, some linked isotopically to Connemara. Derived from east and south-east
Glenummera, Dw 3 300 m	Green-grey slates and mud rocks. Common metamorphic and igneous detritus, rare chromite
Rosroe Dw 2(?)–Dw 3 1200 m	Coarsening upwards alluvial and shallow marine conglomerates in the east. Deep-water proximal fans with mafic and silicic volcanic and granitoid clasts in the south. Metamorphic detritus occurs at higher levels. Coarser in east and derived from south to east
Srah, Dw 1 120 m	Granule grade sandstones and polymict conglomerates. Volcanic and sedimentary detritus
Tourmakeady, Fl 3–Dp 1 1000 m	Silicic tuffs, mud rocks, shales, limestones, which locally fringe volcanic plugs, and limestone breccias
Mt Partry, Fl 2–Fl 3 400 m	Chert, argillite and locally derived volcaniclastic sediments. Mainly volcanic detritus
Currarevagh Limestone Formation, Fl 2–Fl 3 Knock Kilbride Argillite Fl 2	Limestone pebble conglomerates and breccias (limited outcrop) Graptolitic mud rocks (limited outcrop)
Derry Bay, Fl 1(?) 130 m	Oldest silts, arkose and volcanic sandstones, probably a member near the top of the underlying formation
Knock Kilbride, Fl 1–Tr 3 390–600 m	Pillow lavas, pillow lava breccia, massive lavas, andesitic breccias intercalated with cherts and tuffs
Finny, Tr(?) 390–590 m	Andesite flows and thick units of breccia intercalated with tuffs, pillow lavas, and red and green cherts
Bencorragh Tr–Fu(?) >855 m	Highly vesicular pillow lavas. Large pillows form lava tubes. Little interstitial sediment. Lavas have positive ϵ_{Nd}

1987). It is generally correlated with the Bencorragh Formation of the Lough Nafooey Group based upon the primitive nature of the pillow basalts (Draut and Clift 2001), although this correlation is uncertain. In general these lavas are said to have boninitic affinities, but in fact some are boninites (Fig. 11a). On the south limb, early lavas of the Lough Nafooey Group have primitive arc affinities (Clift and Ryan 1994; Draut et al. 2004 and Ryan et al. 1980) and are of Cambrian (Furongian) to Tremadocian age. The relatively flat chondrite normalised rare earth element and rock/N-MORB profiles for the Bohaun and Bencorragh Formations (Fig. 11b and c) and the positive $\varepsilon Nd(t)$ values are similar to those found in modern oceanic island arcs (Clift

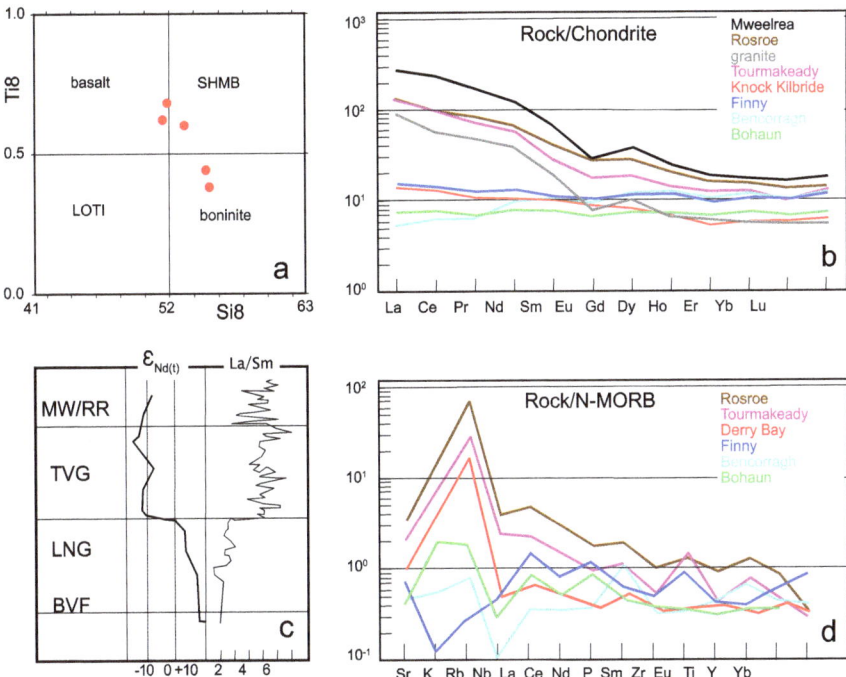

Fig. 11 Geochemistry of the Cambro–Ordovician volcanic complexes within the South Mayo Trough **a** Si8Ti8 diagram after Pearce and Reagan (2019) for the lavas in the Bohaun volcanic Group that have MgO values >8%, normalised to a 100% volatile-free basis, data from Clift and Ryan (1994). Two samples are boninites and one a siliceous high magnesium basalt (SHMB). LOTI = low titanium basalt **b** Selected chondrite normalised rare earth element plots after Clift and Ryan (1994), 'granite' refers to a granitic clast from the Rosroe Formation **c** εNd(t) values and La/Sm ratios for various lavas plotted in stratigraphic order. The marked decrease in εNd(t) and corresponding increase in La/Sm corresponds to greater continental input with decreasing age (after Draut and Clift 2001). BVF = Bohaun Volcanic Formation, LNG = Lough Nafooey Group, TVG = Tourmakeady Volcanic Group, RR = Rosroe Formation and MW = Mweelrea Formation **d** Multi-element 'spider diagrams' for selected volcanic horizons (replotted from Clift and Ryan 1994, and Draut and Clift 2001)

and Ryan 1994; Draut and Clift 2001). Younger, Floian to Dapingian, eruptives of the Tourmakeady Volcanic Group, the Rosroe Formation and the Mweelrea Formation are more fractionated and show an increasing proportion of Laurentian-derived melt up section as inferred from the decreasing εNd(t) values and increasing La/Sm ratios (Fig. 11c), increased enrichment in both light rare earth elements (Fig. 11b) and large ion lithophile elements (Fig. 11d) (Draut and Clift 2001; Draut et al. 2004) and detrital apatite analysis of the associated sediments (O'Sullivan and Chew 2020).

Sedimentary provenance data indicates four principal sources for the sediments of the South Mayo Trough: mature continental; ophiolitic; volcanic; and first cycle metamorphic, which vary both with stratigraphic height and across the limbs of the Mweelrea–Partry syncline (Dewey and Mange 1999). The variation and anomalously

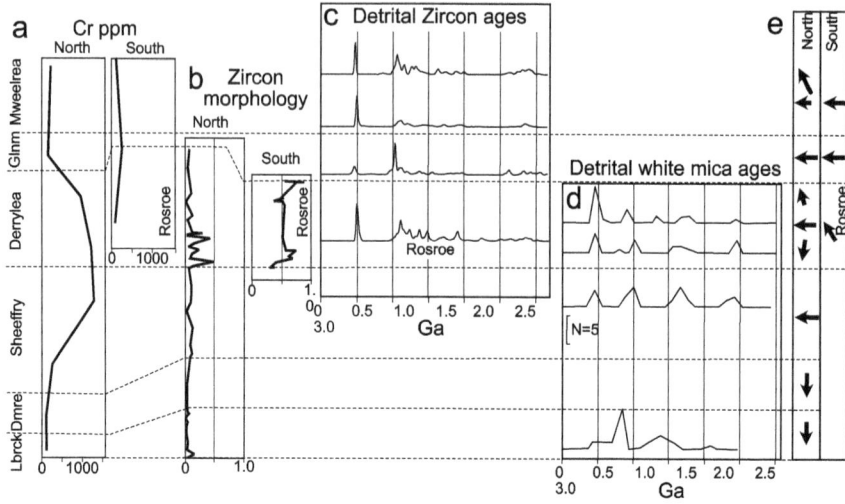

Fig. 12 Summary of provenance data for the Ordovician rocks of the South Mayo Trough **a** Stratigraphic variation in Cr (ppm) for sandstones of both the north and south limb of the Mweelrea–Partry syncline (Wrafter and Graham 1989) **b** Stratigraphic variation in zircon morphology, plotted as the combined frequency of combined euhedral and subhedral zircons as a proportion of euhedral, subhedral and rounded zircons north and south limb (data from Dewey and Mange 1999) **c** Kernel density estimate plots of detrital zircons from various formations. Data is from Clift et al. (2009, 2021); McConnell et al. (2009) **d** Detrital white mica ages from the north limb after Mange et al. 2010 **e** Palaeocurrent directions for each formation. Where there is more than one arrow, the larger indicates the dominant trend. Data is from Pudsey (1984) and Wrafter and Graham (1989). The formation names for the north limb are given on the left, abbreviations used are Lbrck = Letterbrock and Dmore = Derrymore. Where data refers specifically to the Rosroe Formation of the south limb, it is labelled as such

high Cr content within sandstones of the northern limb (Fig. 12a and e) with similar trends in MgO and Ni provides evidence for a northerly source of ultramafic ophiolitic detritus for the rocks of the Sheeffry and Derrylea Formations and gabbroic detritus for older formations (Wrafter and Graham 1989). Such anomalous values are not recorded in the equivalent Rosroe Formation (Fig. 12a) of the southern limb which were derived from the southeast (Wrafter and Graham 1989).

High resolution heavy mineral analysis (HRHMA) and petrographic analysis confirms this result (Dewey and Mange 1999) and shows that the sediments of the Letterbrock Formation on the northern limb were initially dominated by mature Laurentian continental detritus which became diluted by ophiolitic and fresh igneous material in the Sheeffry Formation followed by a sudden influx of first cycle metamorphic detritus in the upper part of the Derrylea Formation. The proportion of 'fresh igneous' zircons (see Dewey and Mange 1999) remains relatively low in the north limb sediments, only becoming significant in the lower part of the Derrylea Formation (Fig. 12b). The Rosroe Formation on the south limb is initially arc- derived with a high concentration of igneous zircons (Fig. 12b) but shows increasing juvenile metamorphic input. Similar patterns are recorded in the Rosroe Formation in the east, which in earlier literature is referred to as the 'Maumtrasna Formation', a term that Graham (2009) recommended be discontinued and these rocks assigned to the Rosroe Formation. No ophiolitic material is recorded in the Mweelrea Formation which evolves from arc to more quartz rich recycled orogen sources (Dewey and Mange 1999).

Detrital zircon analysis (Fig. 12c) for the upper part of the stratigraphy suggest derivation from Laurentia (Clift et al. 2009, 2020; McConnell et al. 2009) and two different arc complexes (McConnell et al. 2009). East-derived sandstones contain 487 Ma igneous zircons, whilst south-derived have 467–474 Ma igneous zircons, possibly sourced from Connemara. Detrital white mica populations for the sediments of the north limb show little sign of the 460–480 Ma peak recorded in the zircon populations until the Sheeffry Formation (Fig. 12d) but show several peaks in the 800–2500 Ma range demonstrating the importance of a mature continental source. These conclusions are confirmed by a recent study (O'Sullivan and Chew 2020) which shows that rift magmatism associated with the formation of the Iapetus Ocean was a source for early mafic volcanic detritus and that the apatites of Mweelrea Formation, derived from peraluminous igneous and metamorphic sources, are less than 700 Ma implying that the older zircons (Fig. 12c) were recycled. In summary, there is a marked change in provenance throughout time and space. In the north, early Laurentia and ophiolite sourced detritus becomes progressive diluted with arc products and, finally, material derived from the Grampian orogen. The proximal sediments of the south limb, which overly the main volcanic complexes, are dominated by arc detritus at lower levels and by both low- and high-grade metamorphic materials, derived from the Grampian orogen, at higher levels.

4.3 Tectonic History and Structure

The tectonic evolution of the South Mayo Trough will be discussed in chapter "The Geology of Western Ireland: A Record of the 'Birth' and 'Death' of the Iapetus Ocean", however, a brief synopsis of current tectonic models for its development follows. The South Mayo Trough originated in the late Cambrian-early Ordovician as the fore-arc to the Lough Nafooey-Tyrone oceanic arc (Dewey and Ryan 1990). It lies in an Ordovician volcanic arc belt that includes Tyrone and the Midland Valley of Scotland (Fig. 1b) and continues via Newfoundland into the Northern Appalachians. Mid-Ordovician arc-continent collision, involving obduction of this arc and fore-arc onto the hyper-extended Laurentian margin caused the Grampian Orogeny (e.g., Chew et al. 2010; Dewey 2005; Dewey and Ryan 2016; Dewey and Shackleton 1984; O'Sullivan and Chew 2020 and Ryan and Dewey 1991, 2011). The South Mayo Trough remained below sea level during its obduction (Ryan 2008) and continued, following subduction flip, as a successor basin above a north-dipping subduction zone beneath a continental arc at the edge of Laurentia (e.g., Clift et al. 2021; McConnell et al. 2009 and Ryan and Dewey 2011). Shortening of this arc coincided with deposition of the Mweelrea Formation and terminated with shortening leading to the initiation of the Mweelrea–Partry mega-syncline (the Mayoian event of Dalziel and Dewey 2018) and emergence above sea-level. Silurian marine sediments and calc-alkaline volcanics were subsequently deposited unconformably upon the Ordovician of the South Mayo Trough. Ordovician and Silurian deposits were then deformed during the Erian event, a "soft", late Silurian, strongly-oblique, sinistral collision that marked the closing of the Iapetus Ocean, followed by early Devonian transtension (Dewey and Strachan 2003). This event was responsible for cleavage development throughout this region (Dewey 1963, 1967; Dewey and Strachan 2003).

5 Excursion to the North Limb of the Mweelrea–Partry Syncline

5.1 Aim

The aim of this excursion (Fig. 13) is to show the stratigraphic variations within the Ordovician, the considerable thickness of this conformable succession, evidence for the provenance of the sediments at various stratigraphic levels, its relationship with the unconformably overlying Silurian strata and its structural and stratigraphic development during the mid-Ordovician Grampian, late-Ordovician Mayoian and late-Silurian Erian orogenies. It provides an unique opportunity to study the evolution of those parts of a fore-arc basin adjacent to a subduction-accretion complex.

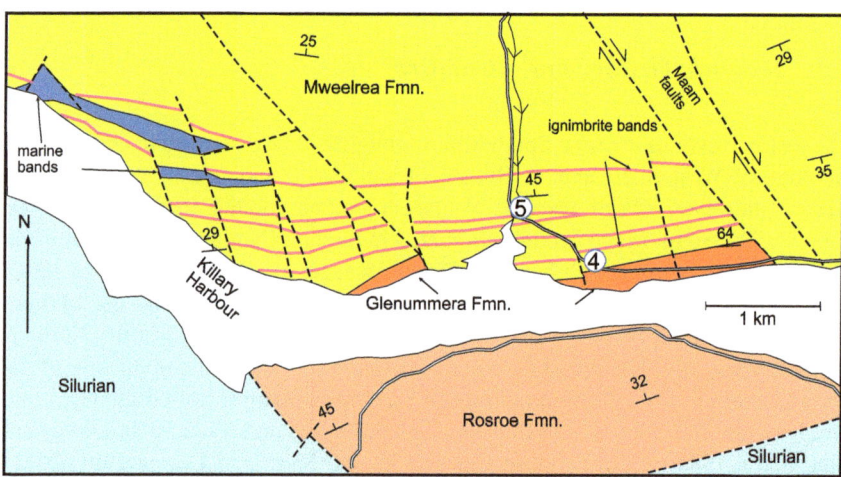

Fig. 13 Geological map covering localities 6.4 and 6.5. The circled numbers '4' and '5' refer to localities 6.4 and 6.5 respectively

Fig. 14 a Cross bedded sandstones near the base of the Mweelrea Formation, foresets dip shallowly to the west (left). The walking stick ferrule is 2.5 cm across **b** Cross bedded coarse sandstones of the Mweelrea Formation. The beds dip at a moderate angle to the north and the foreset dip about 20° west. The bed is the right way up

5.2 Locality 6.4: Tully (53.607459°, −9.750636°)

Access and other information The outcrop is on the north side of the R335. Parking for smaller vehicles is available on the soft curb or in gateways. There is parking for larger vehicles some 75 m west-northwest in a small lay-by near locality 6.5. This is a busy tourist route in summer. Care should be exercised.

This roadside exposure (Fig. 13) is of trough cross-bedded feldspathic sandstones near the base of the Mweelrea Formation. Cross-bedding sets are typically 10–30 cm (Fig. 14a) and are the dominant structure throughout this formation (Pudsey 1984). Clasts are dominated by sub-angular metamorphic quartz, with smaller amounts of oligoclase, orthoclase, detrital muscovite and jasper fragments (Dewey 1963).

The relative freshness of the fragments suggests rapid transport and a high rate of sedimentation. Statistically the fore-sets, which are variable in inclination and orientation, dip westwards suggesting a dominantly easterly source. The thickening of the marine bands to the west and their thinning and pinching out to the east and south imply deepening and more marine conditions to the west (Fig. 13). This locality is 600 m west of the Maam faults which run south-southeast through Leenane and displace the axis of the Mweelrea–Partry Syncline dextrally by 10 km (Figs. 6.2 and 6.13).

5.3 Locality 6.5: Bundorragha (53.607880°, −9.751577°)

Access and other information This locality is 75 m west-northwest of locality 6.4 (Fig. 13). Parking is available on the west side of the road (R335) in a small lay-by which is suitable for larger vehicles. The exposure is on the east side of the road. Care should be taken in crossing as it is close to a blind bend.

Fig. 15 a Orange fiamme of rhyolite in a dark welded tuff matrix containing clear quartz phenocrysts from tuff band 4 in the Mweelrea Formation. The photograph is taken approximately parallel to bedding in a fresh quarry face and the walking stick ferrule is 2.5 cm across. **b** Weathered portion of same quarry, the pumice lapillii weather out and exhibit a degree of flattening parallel to bedding. The photograph is taken looking along strike

The fourth of six ignimbrite bands within the Mweelrea Formation (Fig. 15) is exposed in a small quarry on the east side of the road. These were erupted in a short period in late Darriwilian to early Sandbian times. They are correlated around the Mweelrea–Partry Syncline and were one of the first welded sub-aqueous pyroclastic flows recorded (Dewey 1963). Orange pumice lapilli symmetrically flattened in the plane of bedding (Fig. 15b) and approximately circular when viewed parallel to bedding (~dipping 40° N) are set in a green glassy (welded) matrix containing clear quartz crystals, sericitized albite-oligoclase feldspar, magnetite and euhedral apatite. Generally, there is little variation throughout the flow which is approximately 10 m thick. It is possible that these eruptives are related to the flip in the subduction zone after the Grampian arc-continent collision event (Dewey 2005).

5.4 Locality 6.6: Lettereeragh (53.615318°, −9.751642°)

Access and other information A parking is available at a small lay-by on the east side of the road some 830 m north of locality 6.5. This is a roadside locality; care should be observed.

On a clear day, a view to the west (285°) shows rocks of the upper part of the Mweelrea Formation folded into the Mweelrea mega-syncline (Fig. 16) on the skyline some 5 km away. Oughty Craggy which occupies the northern part of the Mweelrea horseshoe occupies the middle ground to the northwest and the Sheeffry mountains lie due north. The axis of this fold is approximately horizontal and trends east–west. Although the main folding event in the region is late-Silurian, this syncline first formed during the mid-Ordovician causing angular unconformities between Ordovician and Silurian strata in the region. Mweelrea (the Grey King) at 817 m is the highest mountain in Connacht. Two corries separated by a truncated spur head the valley of Glenconnelly. Both have corrie lakes, or tarns, Lough Lugaloughan in

The Ordovician Fore-Arc and Arc Complex

Fig. 16 View looking west-northwest of the core of the Mweelrea–Party Syncline on the east face of Mweelrea mountain. The field of view is approximately 3 km

the south and Lough Bellawaum in the north. The fold closure is visible in the back wall of the southern corrie. This view also gives an impression of the thickness of the Mweelrea Formation (2500–3200 m).

5.5 Teevnabinnia: Locality 6.7 (53.642504°, −9.746739°)

Access and other information Parking is available with care at the road junction between R355 and L1824 (Drummin road). This is a viewing stop giving a good view of the south dipping Mweelrea Formation of the north limb of the Mweelrea–Partry Syncline. It is a roadside stop, care should be observed.

The Drummin road runs to the east up the Glenummera Valley which is eroded into the mud rocks and slates of the Glenummera Formation. This valley terminates at southern end of Doolough where the Glenummera Formation is displaced northwards by a minor splay of the dextral Maam Valley fault. To the west and southwest, on a clear day, there is a view of the lower part of the Mweelrea Formation (Fig. 17). Bedding progressively steepens looking north as the fold tightens. The Glenummera Formation occupies the grassy slopes on the west shores of Doolough where it is faulted, by a northwest trending splay of the Maam fault, against the Mweelrea Formation which comprises the higher ground to the west with rocky bluffs (Fig. 17). The fault trace runs parallel to the western shore of Doolough.

Fig. 17 View looking southwest down Doolough. The south dipping Mweelrea Formation forms the rocky bluff to the southwest of the valley which is separated by a splay of the dextral Maam Faults from the Glenummera Formation which occupies the grassy slopes by the lakeshore (bottom right)

5.6 Locality 6.8: Clashcame (53.660786°, −9.770517°)

Access and other information Parking is available on a track to the abandoned buildings of Clashcame on the east side of the R355. This is a roadside locality on a bend on a busy tourist route in the summer, care should be observed (Fig. 18).

Examine the outcrops along the lake shore and in the road cut about 180 m south-southwest of the parking at Clashcame. The Derrylea Formation, which here comprises axial turbidites (Dewey 1963), shows clear grading with bases of graded units having sharp discontinuities with the underlying well sorted fine sandstones or siltstones (Fig. 19a and b). This locality lies on the steep north limb of the Mweelrea–Partry syncline and bedding youngs to the south. Bedding, which is near vertical or overturned, dips more steeply than the clockwise transecting S2 cleavage and is consequently facing upwards on this cleavage (Fig. 19c). Cleavage refraction is shown in the finer tops of some of the grading beds. The S2 fabric is defined by sericite flakes and the long axes of limonite and siderite nodules lie within this fabric (Fig. 19d; Dewey 1967). The major folds in both Ordovician and Silurian strata were formed during D1. However, the associated S1 cleavage is only well developed north of the Corvock granite, about 5 km further north (Loc. 6.11). This results in variable

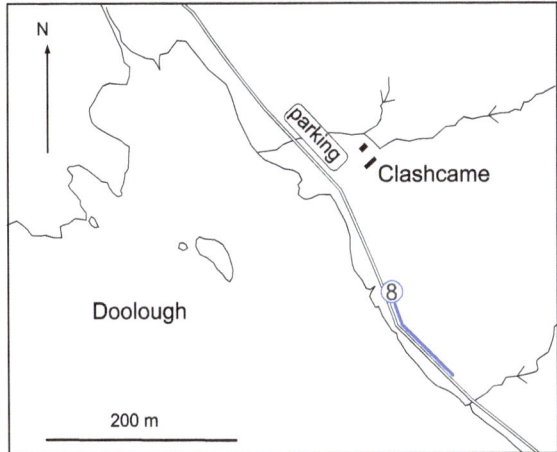

Fig. 18 Location map for locality 6.8. The traverse begins at the circled number '8'

Fig. 19 a The Derrylea Formation on the northeastern shores of Doolough, general view of outcrop looking south **b** Bedding (So) is inverted and dips steeply to the north (left) the cleavage (S2) dips northwards at a shallower angle **c** Graded bedding (S0) in the Derrylea Formation. A flame structure (arrow) and grading indicate younging to the south (right). The trace of the S2 cleavage is indicated **d** Siderite nodules flattened in the plane of the S2 cleavage

facing of bedding on the S2 cleavage (Dewey 1967) as it crosses the earlier structures. The cleavage development was caused by the Erian orogeny and will be discussed in a later traverse (Locality 6.11).

5.7 Locality 6.9: Cushinyen (53.674422°, −9.785051°)

Access and other information Before visiting this locality you may wish to first stop at the lay-by, some 500 m further south (53.670224°, −9.785416°), to observe the quartz veins that host a significant gold deposit (Fig. 20). Small rocky hummocks about 100 m west of this lay-by (53.670152°, −9.787240°) contain auriferous quartz veins that are part of an east–west shear zone at Cregganbaun within the Sheeffry Formation that contains 530,000 tonnes of gold ore at 6 gm/tonne that has not been exploited for environmental reasons. Gold concentrations are 6 ppm and it is generally not visible in hand specimens.

Parking for locality 6.9 is available on a lay-by on the east side of the R355 (53.674446°, −9.787432°). Access involves walking over wet boggy ground. Walk 85 m due east of the southern end of the lay-by to a small rocky bluff on the north wall of a minor valley (Fig. 20).

Green to grey-green turbidites of the Sheeffry Formation have a clear schistosity defined by chlorite and muscovite. They contain clasts of clear volcanic quartz, plagioclase, chromite and detrital muscovite. The S2 schistosity is well defined and plates of emerald green fuchsite mica (a chrome bearing muscovite, $K(Al,Cr)_2(AlSi_3O_{10})(OH)_2$) occur often with relict chromite clasts in their core

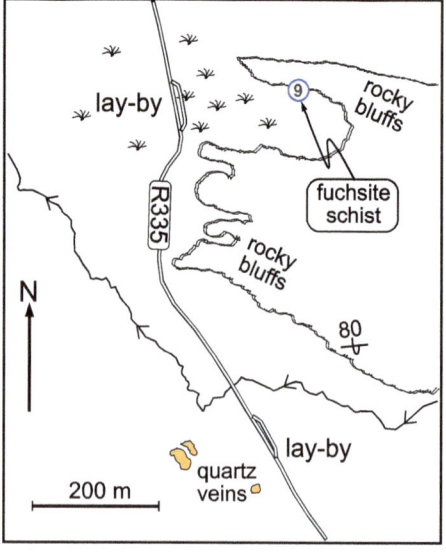

Fig. 20 Location map for Locality 6.9 which is marked by the circled number '9'

Fig. 21 Green fuchsite, mica locally cored by black chromite grains, defines the schistosity within the Sheeffry Formation. The field of view is approximately 5 cm

(Fig. 21). Loose oxidised slabs with fuchsite, parted along the foliation, often have distinctive orange-gold weathering colours. This locality provides clear evidence of ophiolitic detritus within the north limb succession (see Fig. 12a) and also highlights the increase of metamorphic grade northwards.

5.8 Additional Locality on This Route from Elsewhere in the Guide

Continue driving north on the R355 towards Louisburgh crossing the steeply dipping to vertical Sheeffry formation, which in the extreme north of its outcrop is folded into an isoclinal anticline (Stanton 1960), to the Silurian unconformity at Derryheagh (Locality 8.9). This locality is important as it demonstrates that the Ordovician was tilted and eroded during the Mayoian orogeny, during early stages of the development of the Mweelrea–Partry Syncline, prior to Silurian deposition. This relationship is also seen at Oughty (Dewey 1963; Dewey and McManus 1964) although this locality is not visited by this guide.

5.9 Locality 6.10: Letterbrock (53.722627°, −9.583785°)

Access and other information This location can be accessed from the L1824 a minor road that heads west–southwest from the crossroads at Liscarney on the Westport to Leenane road (N59). The fields can be entered via a large gate at the end of an old farm track (Fig. 22).

The first exposure just north of the track is of conglomerate typical of those in the Letterbrock Formation (Fig. 23), although they make up only a minor portion of it. Five polymict conglomerate horizons, up to 15 m in thickness, within the Letterbrock

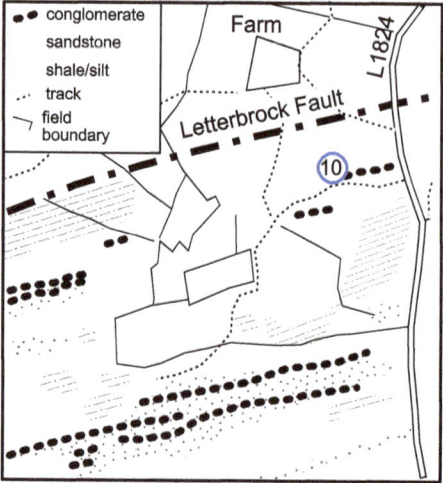

Fig. 22 Geological sketch map, after Pudsey (1984), of the Letterbrock region. The circled number 10 marks Locality 6.10

Fig. 23 Photograph of a conglomerate horizon with a sandy lens within the Letterbrock Formation

Formation are interbedded with blue-green slates (Fig. 22). The conglomerates are poorly graded and at their base large boulders have loaded into underlying sediments. Clast types consist of quartzite, greenschist, intra clasts, albite schists, metabasalt, metagabbro, and chert. They were probably emplaced by large submarine slides from the north (Dewey 1963) and have been interpreted as providing a sedimentary linkage between the Clew Bay Complex and the Ordovician of the South Mayo

Trough (Dewey and Ryan 1990; Fig. 9). Sandstones and mudrocks can be examined by walking a short distance to the SSW parallel to the road. A detailed map of a more extended area is given by Pudsey (1984)

5.10 Locality 6.11: Sheeffry Cliff (53.660219°, −9.640250°)

Access and other information Return to the L1284 and turn right. Continue westwards for 8.5 km passing through the village of Drummin to Owenmore Bridge. The L1284 turns south and climbs a steep hill to the east side of Sheeffry Cliff (Fig. 24). Park is available on the west side of the road at the base of the hill (53.662659° N, −9.642297° E). Additional information: This is a narrow road with a steep drop to the east protected by a wall. Care is needed to allow traffic to pass safely. There is no need to ascend the cliff section to the west; it is dangerous and all the features of the rocks in this section may be seen at roadside.

Traverse 450 m uphill (southwards) through the central third of the Derrylea Formation which is well exposed in the cliffs to the west. Graded beds of turbidites, typically 20–30 cm in thickness which may reach up to 1 m in thickness are interbedded with thinner bands of siltstone and mudstone (Fig. 25a). Bedding is steep to vertical, strikes east–west and youngs to the south. Bottom structures, mainly flute casts and groove casts which suggest flow from the east (Fig. 25b), occur on the base of some of the larger surfaces comprising the cliff face and are best viewed looking southwards. This current lineation is sub-parallel to the cleavage-bedding intersection lineation and care should be taken not to confuse the two. At the top of the

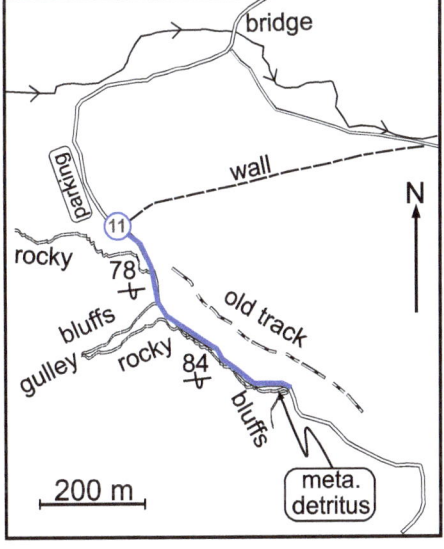

Fig. 24 Location map for Locality 6.11. The traverse begins at the circled number '11'

Fig. 25 a Vertical bedding in the Derrylea Formation at Sheeffry. The cleavage trace is marked by a thin yellow line. It is orthogonal to bedding in the thicker sandy layers but refracted in the silty layers. Lens cap is 5 cm in diameter **b** Flute casts (yellow arrow) on the base of a sandy layer within the Derrylea Formation. View is looking south. The cleavage bedding intersection lineation is marked with a short yellow line. Field of view approximately 1 m **c** John F. Dewey (approximtely 1.8 m tall) at the exposure at the top of the Sheeffry Cliff traverse where heavy minerals derived from juvenile metamorphic detritus first flood into the north limb succession **d** Sandy layer 30 cm thick in the Derrylea Formation which shows late open buckle folds with a sub-horizontal axial surface. Extensional quartz views, orthogonal to bedding and utilising partings parallel to the S2 cleavage are marked with yellow arrows

section (53.657710° N, −9.635464° E, Fig. 25c), the last substantial outcrop on the west side of the road marks the stratigraphic level at which the first influx of juvenile metamorphic detritus is recorded in high resolution heavy mineral studies (Dewey and Mange 1999, see Fig. 10 and Table 1). This level represents the exhumation of middle crustal Barrovian metamorphic rocks of the Grampian orogeny during the Darriwilian, which resulted in an influx of metamorphic detritus as the trough was still below sea-level.

The section illustrates the cleavage development on the north limb of the Mweelrea–Partry Syncline. Cleavage in the coarser sandy beds is almost orthogonal to bedding, implying layer parallel shortening during early fold development (Fig. 25a). This folding began during D1, which is more intense in the Silurian to the north, and no clear D1 structures are visible in this traverse. The cleavage visible here, regionally assigned as S2 (Dewey 1967), was refracted by shear in the finer layers which accommodated bedding slip in the more competent bands (Fig. 25a). The beds rotated towards the vertical as the fold tightened and the refracted cleavage became approximately parallel to the axial surface of the syncline. Sub-horizontal quartz veins cut the cleavage in some coarser bands (Fig. 25d). Their emplacement was associated with vertical extension of the bedding pile in the north limb during late stages of D2 (Dewey 1967). The youngest structures seen in this section are gentle buckles (Fig. 25d) with sub-horizontal to northward-dipping axial surfaces in some of the sandy horizons. These are associated with the vertical lock-up of the pile followed by the onset of vertical shortening associated with horizontal collapse.

Exercises

- The South Mayo Trough remained below sea-level for the whole arc-continent collision process. Discuss mechanisms that might account for this.
- The thickness of the north limb deposits is estimated at 9 km. In what tectonic environments are such thicknesses recorded, for example, would you expect such a thickness in a rift basin?
- Prepare a stereographic net showing both bedding from Locs 6.4, 6.8, 6.9, 6.10, and 6.11 and plot a Beta diagram to determine the statistical fold axis for the Mweelrea-Partry syncline. Plot cleavage readings from Locs 6.8 and 6.11 on the same net. Is the cleavage axial planar to this fold? This exercise can be undertaken using digital mapping techniques.

6 Excursion to the South Limb of the Mweelrea–Partry Syncline and the Volcanic Complexes

6.1 Aim

The aim of this excursion is to show the provenance and more proximal nature of the south limb sedimentary succession, the progressive evolution of an intra-oceanic arc as more continental material becomes entrained up section and evidence for the stripping of the arc and the unroofing of the Grampian orogen. The traverse illustrates the progressive petrological evolution of an island arc before, during and after arc-continent collision.

6.2 Locality 6.12: Currarevagh (53.572419°, −9.600713°)

Access and other information Parking is available with care on the south side of L1301 some 90 m west of the outcrop (53.572355°, −9.602093°). This is a roadside locality and appropriate care should be observed.

The Bunnacunneen Member conglomerates of the lower part of the Mweelrea Formation are exposed on the south side of L1301 (Figs. 9 and 26a). In this section the Mweelrea Formation is conglomerate dominated (Williams 1984) and is interbedded with the 5 ignimbrite bands. This cobble conglomerate is clast supported and, therefore, deposited by traction currents (Pudsey 1984). Clasts, up to 90 cm across, include

Fig. 26 **a** General view of the outcrop of the Bunnacunneen Member at Currarevagh. The ignimbrite bands capped by conglomerates form the marked features on and below the skyline **b** Rounded cobbles of metamorphic rock (below lens cap) and granite (left of lens cap) within the Bunnacunneen Member of the Mweelrea Formation. Lens cap is 5 cm in diameter

granite, quartz-porphyry, psammite and vein quartz (McConnell et al. 2009). This locality is stratigraphically above ignimbrite band 3. Dip slopes of bands 2 and 3 each overlain by conglomerates (dip is approximately 40° towards the north), form prominent ridges on the hillside due south, with band 2 occupying the skyline. Williams (1984) interpreted the conglomerates as deposits of an alluvial fan. McConnell et al. (2009) report that the granitoid clasts are similar to the Connemara orthogneiss suite in both age and composition, suggesting derivation from Connemara to the south.

6.3 Locality 6.13: Cummer (53.576627°, −9.511426°)

Access and other information Parking is available, with care, on the south side of R300 (53.578427°, 9.510211°) at a T junction with a 'bog road' than runs south (Fig. 27). Access involves walking over wet boggy ground. The 'bog road' should be used for access to the outcrops in Loc 6.15 on the south side of the Finny River.

Examine the two large outcrops of the Rosroe Formation (Fig. 27). The first is 150 m away from this T junction on a bearing of 190° and the second a further 40 m southwest. Here, conglomerate units between 1 and 10 m in thickness with a channelled base are interbedded with poorly sorted arenites containing mainly rhyolitic debris (Fig. 28a and b). Sand-grade material is usually angular to subangular. Rounding increases with grain size and pebbles are often rounded. Clasts in

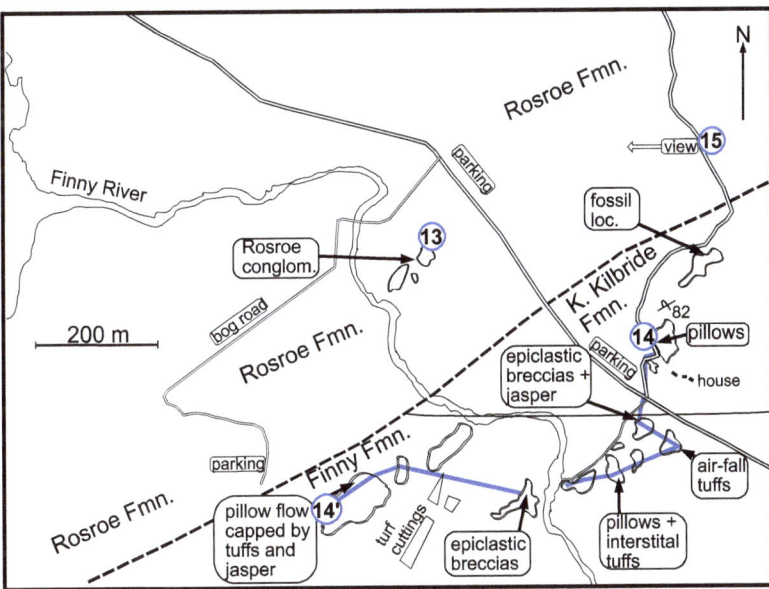

Fig. 27 Location map for Locs 6.13, 6.14, 6.14′ and 6.15, whose positions are marked by the appropriate circled number

Fig. 28 **a** Conglomerate layers in the Rosroe Formation at Cummer, Lough Nafooey, which channel into underlying sands. Field of view is approximately 3 m **b** Cobbles within the Rosroe Formation conglomerates mostly consist of high level (~30% modal quartz) granites. A mafic igneous clast of volcanic origin is marked with a yellow arrow. A metamorphic clast is marked with a white arrow. Lens cap is 5 cm in diameter

the conglomerate consist of pink high-level granites, rhyolites and psammitic metamorphic debris with minor components of metabasalt and chert. These are interpreted by Archer (1984) as being deposited in the upper (conglomerates) and middle (sandstone) zones of a deep-water submarine fan which accumulated on a north-facing slope and whose main channel lay some 2.5 km to the east. The principle palaeocurrent indicators in the sandstones imply northeast–southwest flow that is interpreted as axial on a regional scale (Archer 1984).

6.4 Locality 6.14: Lough Nafooey, East End (53.576627°, − 9.511426°)

6.4.1 North of the Finny River (Loc 6.14)

Access and other information Drive 500 m southeast along R300 where parking is available on the roadside at a minor crossroads where a small track heads south to the Finny River (Fig. 27). Access involves walking over wet boggy ground.

Walk 75 m northwards up a bridle path to a large outcrop of pillow basalts of the Knock Kilbride Formation (Fig. 29a) by the gateway to a house at a sharp bend. Hyaloclastites to the extreme south of the outcrop are overlain by pillows which are flattened at the base of the flow but become more equant in shape higher up (Fig. 29b). Bun shaped tops and lobate bases of overlying pillows provide good way-up indicators. Bedding is slightly overturned and younging is to the north. The pillow unit is then overlain by a massive lava flow. Return southwards 35 m to the next bend in the track. Pillows are exposed in a small quarry to the east of the track, these display red and green chert infilling in the interstices, small drain outs filled with carbonates or chalcedony and are vesicular.

Fig. 29 a Pillow lava flow within the Knock Kilbride Formation. View is looking east. Pillow lava breccia at the base of the flow (on the right) are overlain by flattened pillows (white arrow marks the contact). Well-formed bun shaped pillows with lobate bases (yellow arrow) indicate younging to the north. Field of view is approximately 4 m. **b** Epiclastic breccias of the Finny Formation. Blocks, mainly andesitic, are typically 15 cm on the longest axis. Field of view is approximately 1.5 m **c** Scoriaceous block within the Finny Formation **d** Bedded red (radiolarian?) cherts lie on top of a pillow lava flow (to the right) within the Finny Formation. The view is looking eastwards and the field of view is approximately 70 cm **e** Volcanic (andesite) bomb penetrates layering in tuffs (yellow arrow) which lie above bedded red cherts or 'jaspers'. The view is looking east and the pen is 14 cm in length **f** Disrupted red chert band lying within the epiclastic breccias of the Finny Formation eastwards along strike from photos Fig. 6.29d and e. Head of the walking stick is 9 cm in length

Continue southwards and cross the R300. The outcrops to the east of the track comprise the top of the Finny Formation. A red chert, or jasper, band lies in coarse epiclastic volcanic breccias with matrix supported angular blocks of andesite and pumice (Fig. 29c). Careful examination of finer facies shows thin bands of extremely well sorted air-fall tuffs. These continue down to the river and are intercalated with a small flow with large porphyritic pillows with interstitial tuff. The vesicularity of the pillows suggest a high volatile content as the background sedimentation of pelagic chert indicates deep water conditions. More felsic lavas are locally scoriaceous (Fig. 29c) and may have been erupted subaerially (see Loc 6.14' below).

6.4.2 South of the Finny River (Loc 6.14')

Access and other information Access for the part of the section to the south of the Finny River is provided by the 'bog road' at Cummer. Return to the parking spot for Locality 6.14. Turn left and proceed southwest. Drive 900 m following the bog road, which is not suitable for large vehicles, to a parking spot in a small quarry. The outcrops are 135 m due east (Loc 6.14', Fig. 27).

The sequence of outcrops to the south of the river comprises, from west to east (Fig. 27), a pillow flow which is overlain by finer epiclastic breccias and tuffs interbedded with a slumped bright red chert (jasper) band (Fig. 29d) and a volcanic bomb (Fig. 29e). The bomb indicates subaerial eruption for some of the andesites. If this band is followed along strike eastwards to the river it becomes progressively disrupted in a coarse epiclastic flow (Fig. 29f) containing blocks of andesite up to 1 m across. This relationship was interpreted by Ryan et al. (1980) to be the result of accumulation of pelagic deposits during a period of quiescence after the eruption of the pillow flow which were then disrupted by a submarine flow of andesitic debris which was channeled between the pillow flows. The overall composition of the Lough Nafooey Group is consistent with it being erupted in an inter-oceanic arc setting (Ryan et al. 1980; Fig. 11b–d). The basaltic pillow lavas within the Finny Formation (Fig. 29d) have MORB composition (Fig. 11b and d) and probably represent rifting in the arc similar to that which occurred in the Izu-Bonin arc during the mid-Oligocene (Fujioka et al. 1989).

This location is a good place to examine the tradition of turf cutting by hand. This practice is being discontinued on environmental grounds.

6.4.3 Additional Locality from Elsewhere in the Guide

The Bencorragh Formation, the oldest unit of the Lough Nafooey Group, underlies the Finny Formation. A traverse to the summit of Bencorragh mountain to examine the spectacular pillow lavas near the exposed base of this formation is included in chapter "The Silurian of North Galway and South Mayo" (Loc 8.3) for logistical reasons.

6.5 Locality 6.15: Lough Nafooey Viewing Platform (53.578585°, −9.503884°)

Access and other information Proceed northwards up the minor road for 400 m from locality 6.14. Parking is available at a lay-by with a spectacular view to the west. This view is weather dependent.

The view looks westwards down the axis of Lough Nafooey (Fig. 30). Bencorragh forms the skyline to the southwest and is made up of a thick sequence of tholeiitic pillow basalts that underlie the Finny Formation (Loc 8.3). The ridge to the immediate south of Lough Nafooey is occupied by the Rosroe Formation that rests unconformably on the Lough Nafooey Group (Graham 2019). The colour contrast picks out the trace of this contact. A major fault (Derry Bay Fault) with a northerly downthrow runs through Lough Nafooey and then along the valley to the west of Bunnacunneen that is visible at the western end of the lough. The Maumtrasna hills forming the skyline to the north of the lough comprise northward dipping Mweelrea Formation and the axis of the Mweelrea Partry Syncline runs just to the south of the plateau at their summit. A small Carboniferous outlier of basal Viséan sandstones, occurs on the beveled summit whilst the main Carboniferous outcrop is to the east on the shores of Lough Mask, some 600 m lower (Dewey and McKerrow 1963). This differential uplift has been attributed to Eocene–Oligocene block rotation during normal faulting and Miocene to present-day compression (Dewey 2000).

Fig. 30 View down Lough Nafooey looking west

6.6 Locality 6.16: Glensaul School (53.634384°, −9.428533°)

Access and other information Limited parking is available on the roadside by the now abandoned school buildings, more and better is available at 53.635622°, −9.425787°). This is a narrow, single-track road. Access involves walking over wet boggy ground. The field can become overgrown and may have stock in it. In this case it might be best to just visit the rhyolite breccia locality by the school.

Walk through the schoolyard and cross the wall to the north (Fig. 31). Examine the west face of the outcrops 30 m northeast of the northeast corner of the school yard of auto-brecciated rhyolites that underlie the Tourmakeady Volcanic Formation (Fig. 32a and b). These are believed to be part of the carapace of a submarine rhyolite tholoid (Graham et al. 1989). The main body of the lava dome comprises the hill to the

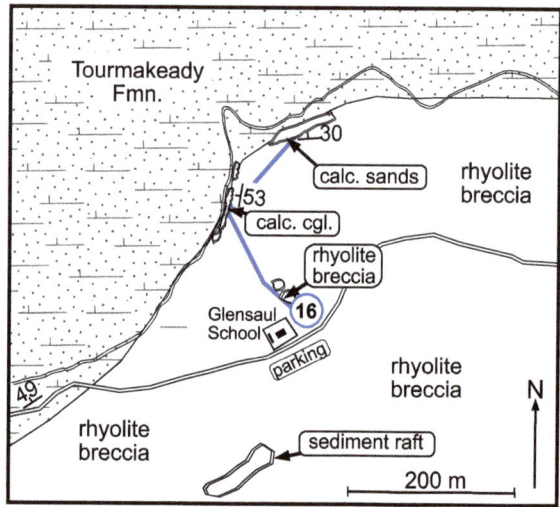

Fig. 31 Sketch geological map for Locality 6.16. The traverse begins at the circled number '16'

Fig. 32 **a** Auto-brecciated rhyolites of the Tourmakeady Volcanic Formation near Glensaul School. Field of view is 1 m **b** A typical outcrop of rhyolite breccia to the south of Glensaul School

south (about 400 m southwest), which also incorporates a large >50 m across slab of volcanic sediments (Fig. 31). The Glensaul River wraps around the field containing this outcrop. Careful examination of the stream banks and bed 150 m northwest of the first outcrop show Tourmakeady Volcanic Formation conglomerates with a calcareous matrix whose strike trends 005°–185°, parallel to the stream. At the sharp bend in the river well bedded calcareous tuffs dip at 38° and strike 080°. This swing in strike is believed to reflect the geometry of the carapace of the tholoid which may have produced 500 m of positive relief on the seabed allowing colonization by shelly faunas of the overlying Tourmakeady Volcanic Formation. The higher silica concentrations of these eruptives are related to melts which contain significant continental material (Draut et al. 2004) and are interpreted as marking the onset of arc-continent collision as the leading edge of Laurentia was subducted beneath the 'Lough Nafooey arc'.

6.7 Locality 6.17: Drumcoggy

Access and other information Take the L16022 heading northwest from Tourmakeady village and turn left after 1 km. Looking east-southeast there are fine views on a clear day of Lough Mask as the road crests Ballybanaun Mountain (Fig. 34a, 53.668558°, −9.405647°). Locality 6.17a is an additional locality visiting a Tertiary dolerite sill. Parking is available on the roadside (Fig. 33) on the western face of this hill. This is a narrow, single-track road.

6.7.1 Locality 6.17a: Drumcoggy Sill (53.670881°, −9.411125°)

The Palaeocene Drumcoggy dolerite sill whose emplacement was associated with North Atlantic opening at 55 Ma (Mohr et al. 1984) can be accessed in a small abandoned roadside quarry on the north side of the road. The sill is deeply weathered but contains less weathered 'core stones' and has zeolite mineralogy. The weathering probably took place under humid conditions not present today. It provides evidence that less than 150 m has been stripped off the landscape during the Cenozoic (Dewey 2000).

6.7.2 Locality 6.17: Drumcoggy (53.678127°, −9.425031°)

Cobble conglomerates of the Rosroe Formation (previously termed the 'Maumtrasna Formation' in this area, see Graham 2009), interbedded with rare boulder and pebble conglomerates and sandstones, are well-exposed on the east side of the road. Mudrocks are absent either as clasts or interbeds. The conglomerates are mainly clast supported and lack sedimentary structures (Fig. 34b and c) and it is difficult to

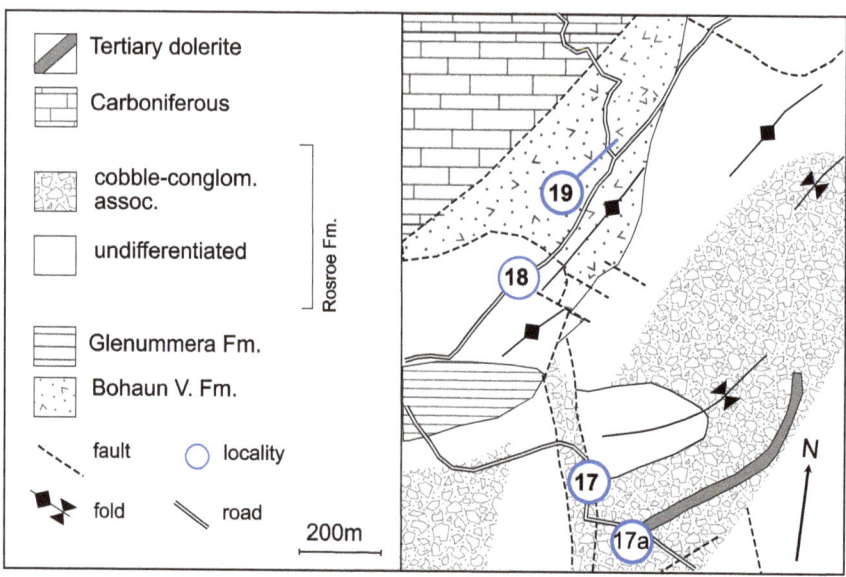

Fig. 33 Sketch geological map covering Locs 6.17, 6.18 and 6.19. Abbreviations used: Fm. = Formation; conglom. = conglomerate; V. = Volcanic. Based on Graham 1987, 2009) and Graham et al. (1989). The numbers refer to the following localities: 17 = Loc 6.17, 17a = Loc 6.17a, 18 = Loc 6.18, 19 = Loc 6.19

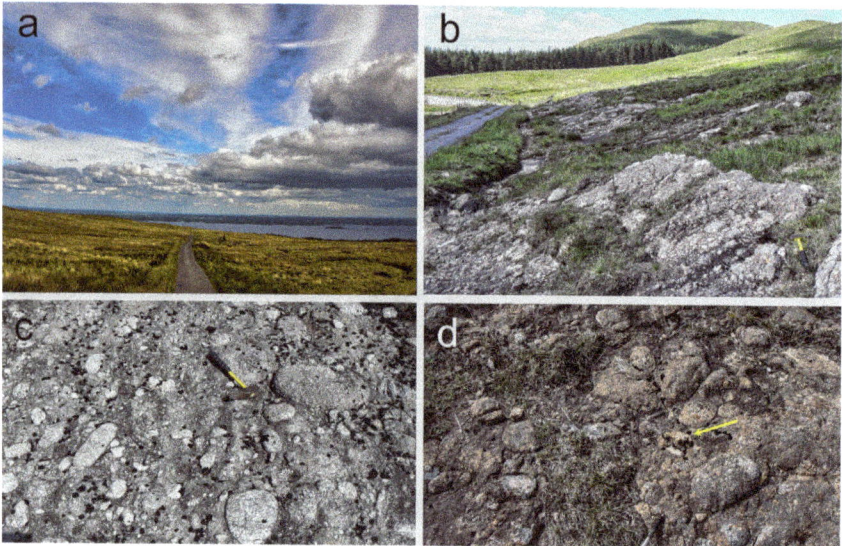

Fig. 34 a View over Lough Mask from the summit of Ballybanaun Mountain **b** General view of the Rosroe Formation cobble-conglomerate facies **c** large granitoid boulder (at head of hammer) within the Rosroe Formation **d** Conglomerates of the Rosroe Formation. Cobbles are mostly of high-level granite (~30% modal quartz). A rhyolite cobble is marked by the yellow arrow. Head of the walking stick is 9 cm in length

determine bed thickness. Rare imbrication suggests currents from the east or southeast. Clasts are typically less than 40 cm across. Detritus in the Rosroe Formation is composed of high-level, quartz-rich granite or of quartz-feldspar porphyry (Fig. 34d). The association of high-level granites and petrographical similar porphyries suggests stripping of the upper levels of an evolved arc and deposition in coarse grained alluvial fans (Graham 1987). This cobble conglomerate facies occurs at high levels within the Rosroe Formation (Fig. 33). In the synclinal core at Lough Shee, these conglomerates are conformably overlain by fossiliferous mudrocks of the Glenummera Formation (Fig. 33; Harper et al. 1988). The underlying Rosroe Formation is, therefore, a correlative of the Derrylea Formation of the north limb.

6.8 Locality 6.18: Croaghcrom (53.700554°, −9.435422°)

Access and other information Continue northwards following the road for 3.5 km then turn right and drive 2.4 km (Fig. 33). Parking is available on the roadside. This is a narrow, single-track road.

The Rosroe Formation is best seen on the south side of the road and on the northwestern slopes of Croaghrimcarra hill. Bedding dips steeply to the south-southeast and is right way-up (Fig. 35a). McManus (1972) reports sandstones that he interpreted as lateral turbidites. They contain angular quartz grains and partly-rounded feldspars with bottom structures suggesting transport from the east-southeast. Silicic porphyries are the commonest clast type in these conglomerates. Locally, McManus (1972) reported down-cutting channels that contain clasts of granite, quartzite, slate, sandstone, jasper, black chert and spilite. The rocks seen here, despite their strong deformation (Fig. 35b), have all the characteristics of the Rosroe Formation (Graham 2009) rather than the Derrylea Formation as assigned by McManus (1972).

Fig. 35 Rosroe Formation at Croaghchrom **a** General view, looking north, of interbedded coarse sandstones and conglomerates typical of the Rosroe Formation (see Loc 6.13) **b** Close up of a conglomerate layer showing the main cleavage wrapping around a granitoid clast

Fig. 36 a General view looking north of pillow lavas at Bohaun **b** Stretched pillows elongated within the main fabric **c** Small pillow with lobate base and bun shaped top showing way-up, younging is to the north

6.9 Locality 6.19: Bohaun (53.710772°, −9.423749°)

Access and other information Continue northwards for 1.4 km. This is a narrow, single-track road. The section traverses the outcrops on the northwest side of the road and runs for 230 m to the outcrops to the north of the 'T' junction in Bohaun where parking is available (53.712474°, −9.421747°) on the roadside.

The Bohaun Volcanic Formation comprises pillow lavas, pillow lava breccias, calcareous tuffs, and rare chert lenses (Fig. 36a). Pillows are deformed, lie in a steep north-northeast striking foliation and are stretched (Fig. 36b) defining a linear fabric that has a variable plunge (20°–60°) to the south-southwest. Sections orthogonal to this fabric show the pillows often to be small (25–50 cm across) and younging to the north (Fig. 36c). In sections parallel with this lineation, pillows may exceed 1 m in length and show extension joints perpendicular to the stretching direction (Fig. 36b). Ratios of long to intermediate axes are up to 9:1. Larger pillows occur, but the predominance of smaller forms is consistent with their high magnesium values (Fig. 11a). The pillows have vesicles elongated in the stretching direction. There is little interstitial sediment. Pillows have a chlorite-albite-epidote greenschist-facies assemblage and selvages are commonly epidotic. Pillow lava breccias that contain blocks of pillows up to 50 cm across become more important up-section to the north. The post-tectonic thickness of this unit has been estimated at 1000 m (McManus 1972) to 1700 m (Graham et al. 1989), but a lower value is more likely. The occurrence of boninites (Fig. 11a) within this formation are consistent with their eruption in a fore-arc environment, possibly during subduction initiation (see Dewey and Casey, 2011; Pearce and Reagan 2019).

The state of strain of these pillows is perhaps greater than that of the Rosroe Formation at locality 6.18. The contact between the Bohaun Volcanic and the Rosroe Formations is a high strain zone, which may represent a tectonised unconformity as spilite clasts are recorded locally in the base of the Rosroe Formation (McManus 1972; Graham 1987, 2021). This contact, where the Bohaun Formation structurally overlies the Rosroe Formation is folded and may represent a pre-Erian structure.

Exercises

- Do we see similar evidence for the increasing involvement of continental material in arc magmas during arc-continent collision elsewhere in the geological record? Does this date the onset of arc-continent collision?
- What mechanisms might have controlled the rapid unroofing of the Grampian Orogen, do such mechanisms operate in Taiwan today?
- Identify and log the compositions of cobbles or boulders at Locs 6.10, 6.12, 6.13 and 6.17. Prepare bar charts, grouped by size, of clast lithologies and place in stratigraphic order. What does this tell you about the nature of the various source areas and their variation throughout time?

References

Allen R, Carter A, Najman Y, Bandopadhyay PC, Chapman HJ, Bickle MJ, Garzanti E, Vezzoli G, Andò S, Foster GL, Gerring C (2008) New constraints on the sedimentation and uplift history of the Andaman-Nicobar accretionary prism, South Andaman Island. Spec Pap Geol Soc Am 436:223–256

Archer J (1977) Llanvirn stratigraphy of the Galway-Mayo border area, western Ireland. Geol J 12:77–98

Archer JB (1984) Clastic intrusions in deep-sea fan deposits of the Rosroe Formation, Lower Ordovician, western Ireland. J Sediment Petrol 54:1197–1205

Ballantyne CK, Stone JO, McCarroll D (2008) Dimensions and chronology of the last ice sheet in Western Ireland. Q Sci Rev 27:185–200

Batt GE, Brandon MT, Farley KA, Roden-Tice M (2001) Tectonic synthesis of the Olympic Mountains segment of the Cascadia wedge, using two-dimensional thermal and kinematic modeling of thermochronological ages. J Geophys Res-Sol Ea 106:26731–26746

Bergstroem SM, Chen X, Gutiérrez-Marco JC, Dronov A (2009) The new chronostratigraphic classification of the Ordovician System and its relations to major regional series and stages and to $\delta 13C$ chemostratigraphy. Lethaia 42:97–107

Bonatti E (1976) Serpentinite protrusions in the oceanic crust. Earth Planet Sc Lett 32:107–113

Breeding CM, Ague JJ (2002) Slab-derived fluids and quartz-vein formation in an accretionary prism, Otago Schist, New Zealand. Geology 30:499–502

Burrett C (1985) Problematic, phosphatic microspheres (mazuelloids) from the Ordovician of Tasmania, Australia. Alcheringa 9:158–158

Chew D (2003) Structural and stratigraphic relationships across the continuation of the Highland Boundary Fault in western Ireland. Geol Mag 140:73–85

Chew DM, Daly JS, Magna T, Page LM, Kirkland CL, Whitehouse MJ, Lam R (2010) Timing of ophiolite obduction in the Grampian orogen. Geol Soc Am Bull 122:1787–1799

Chew DM, Daly JS, Page L, Kennedy M (2003) Grampian orogenesis and the development of blueschist-facies metamorphism in western Ireland. J Geol Soc Lond 160:911–924

Chew DM, Graham JR, Whitehouse MJ (2007) U-Pb zircon geochronology of plagiogranites from the Lough Nafooey (= Midland Valley) arc in western Ireland: constraints on the onset of the Grampian orogeny. J Geol Soc Lond 164:747–750

Clift PD, Carter A, Draut AE, Van Long H, Chew DM, Schouten HA (2009) Detrital U-Pb zircon dating of lower Ordovician syn-arc-continent collision conglomerates in the Irish Caledonides. Tectonophysics 479:165–174

Clift PD, Luther AL, Avery ME, O'Sullivan PB (2021) U–Pb detrital zircon constraints on active margin magmatism and sedimentation after the Grampian Orogeny in western Ireland. J Geol Soc Lond 178, jgs2020-093. https://doi.org/10.1144/jgs2020-093

Clift PD, Ryan PD (1994) Geochemical evolution of an Ordovician island arc, South Mayo, Ireland. J Geol Soc Lond 151:329–342

Cohen KM, Finney SC, Gibbard PL, Fan J-X (2013) The ICS international chronostratigraphic chart (2013: updated v2018/08). Episodes 36:199–204

Dalziel IW, Dewey JF (2018) The classic Wilson cycle revisited. Geol Soc Lond Spec Pub 470, SP470-1

Dewey JF (1963) The stratigraphy of the Lower Palaeozoic rocks of central Murrisk, Co., Mayo, Eire, and the evolution of the South Mayo Trough. J Geol Soc Lond 119:313–344

Dewey JF (1967) The structural and metamorphic history of the Lower Palaeozoic rocks of central Murrisk, County Mayo, Eire. Q J Geol Soc Lond 123:125–155

Dewey JF (2000) Cenozoic tectonics of western Ireland. P Geologist Assoc 111:291–306. https://doi.org/10.1016/S0016-7878(00)80086-3

Dewey JF (2005) Orogeny can be very short. P Natl Acad Sci USA 102:15286–15293

Dewey JF, Casey J (2011) The origin of obducted large-slab ophiolite complexes. In: Arc-continent collision. Springer, pp 431–444

Dewey JF, Mange M (1999) Petrography of Ordovician and Silurian sediments in the western Irish Caledonides: tracers of a short-lived Ordovician continent-arc collision orogeny and the evolution of the Laurentian Appalachian-Caledonian margin. Geol Soc Lond Spec Pub 164:55–107

Dewey JF, McKerrow W (1963) An outline of the geomorphology of Murrisk and north-west Galway. Geol Mag 100:260–275

Dewey JF, McKerrow W, Moorbath S (1970) The relationship between isotopic ages, uplift and sedimentation during Ordovician times in western Ireland. Scot J Geol 6:133–145

Dewey JF, McManus J (1964) Superposed folding in the Silurian rocks of Co., Mayo, Eire. Geol J 4:61–76

Dewey JF, Ryan PD (2016) Connemara: its position and role in the Grampian Orogeny. Can J Earth Sci 53:1246–1257

Dewey JF, Ryan PD (1990) The Ordovician evolution of the South Mayo Trough, western Ireland. Tectonics 9:887–901

Dewey JF, Shackleton RM (1984) A model for the evolution of the Grampian tract in the early Caledonides and Appalachians. Nature 312:115–121

Dewey JF, Strachan R (2003) Changing Silurian-Devonian relative plate motion in the Caledonides: sinistral transpression to sinistral transtension. J Geol Soc Lond 160:219–229

Doglioni C, Harabaglia P, Martinelli G, Mongelli F, Zito G (1996) A geodynamic model of the Southern Apennines accretionary prism. Terra Nova 8:540–547

Donovan S (1986) Pentameric pelmatozoan columns from the Arenig Tourmakeady Limestone, county Mayo. Irish J Earth Sci 7:93–97

Draut AE, Clift PD (2001) Geochemical evolution of arc magmatism during arc-continent collision, South Mayo, Ireland. Geology 29:543–546

Draut AE, Clift PD, Chew DM, Cooper MJ, Taylor RN, Hannigan RE (2004) Laurentian crustal recycling in the Ordovician Grampian Orogeny: Nd isotopic evidence from western Ireland. Geol Mag 141:195–207

Evans BW, Hattori K, Baronnet A (2013) Serpentinite: what, why, where? Elements 9:99–106

Fisher D, Byrne T (1987) Structural evolution of underthrusted sediments, Kodiak Islands, Alaska. Tectonics 6:775–793

Forster M, Lister G (2003) Cretaceous metamorphic core complexes in the Otago Schist, New Zealand. Aust J Earth Sci 50:181–198

Fujioka K et al (1989) Arc volcanism and rifting. Nature 342:18–20

Gibson P, Lyle P, George D (1996) Ground-based magnetic patterns across selected Irish rocks and structures. Irish J Earth Sci 15:129–143

Graham JR (1987) The nature and field relations of the Ordovician Maumtrasna Formation, County Mayo, Ireland. Geol J 22:347–369

Graham JR (2001) The geology of Clare Island: perspective and problems. In: New survey of Clare Island, Vol 2: Geology, Royal Irish Academy, pp 1–17

Graham JR (2009) Ordovician of the North. In: Holland C, Saunders A (eds) The geology of Ireland. Dunedin Academic Press Edinburgh, pp 43–67

Graham JR (2019) The structure and stratigraphical relations of the Lough Nafooey Group, South Mayo. Irish J Earth Sci 37:1–18

Graham JR (2021) Geology of South Mayo: a field guide. Geological Survey Ireland, Dublin. 158pp

Graham JR, Leake BE, Ryan PD (1989) The geology of South Mayo. Scottish Academic Press

Harkin J, Williams D, Menuge J, Daly J (1996) Turbidites from the Clew Bay Complex, Ireland: provenance based on petrography, geochemistry and crustal residence values. Geol J 31:379–388

Harper D, Graham J, Owen A, Donovan S (1988) An Ordovician fauna from Lough Shee, Partry Mountains, Co., Mayo, Ireland. Geol J 23:293–310

Harper DA, Parkes MA, McConnell BJ (2010) Late Ordovician (Sandbian) brachiopods from the Mweelrea Formation, South Mayo, western Ireland: stratigraphic and tectonic implications. Geol J 45:445–450

Harper DA, Williams D, Armstrong HA (1989) Stratigraphical correlations adjacent to the Highland Boundary Fault in the west of Ireland. J Geol Soc Lond 146:381–384

Hutton DH, Dewey JF (1986) Palaeozoic terrane accretion in the Western Irish Caledonides. Tectonics 5:1115–1124

Johnston JD, Phillips WE (1995) Terrane amalgamation in the Clew Bay region, west of Ireland. Geol Mag 132:485–501. https://doi.org/10.1017/S0016756800021154

Kennedy MJ (1980) Serpentinite-bearing melange in the Dalradian of County Mayo and its significance in the development of the Dalradian basin. J Earth Sci-Dublin 3:117–126

Klemperer S, Ryan PD, Snyder DB (1991) A deep seismic reflection transect across the Irish Caledonides. J Geol Soc Lond 148:149–164

Lockwood J (1971) Sedimentary and gravity-slide emplacement of serpentinite. Geol Soc Am Bull 82:919–936

Maltman A, Labaume P, Housen BA (1997) Structural geology of the décollement at the toe of the Barbados accretionary prism. Proc ODP Sci Res 156:279–292

Mange M, Idleman B, Yin Q-Z, Hidaka H, Dewey JF (2010) Detrital heavy minerals, white mica and zircon geochronology in the Ordovician South Mayo Trough, western Ireland: signatures of the Laurentian basement and the Grampian orogeny. J Geol Soc Lond 167:1147–1160

Max MD (1989) The Clew Bay Group: a displaced terrane of Highland Border Group rocks (Cambro-Ordovician) in Northwest Ireland. Geol J 24:1–17

Max MD, Riddihough RP (1975) The continuation of the Highland Boundary Fault in Ireland. Geology 3:206–210

Max MD, Ryan PD, Inamdar D (1983) A magnetic deep structural geology interpretation of Ireland. Tectonics 2:431–451

McCaffrey K (1997) Controls on reactivation of a major fault zone: the Fair Head–Clew Bay line in Ireland. J Geol Soc Lond 154:129–133

McConnell B, Riggs N, Crowley QG (2009) Detrital zircon provenance and Ordovician terrane amalgamation, western Ireland. J Geol Soc Lond 166:473–484

McManus J (1972) The stratigraphy and structure of the Lower Palaeozoic rocks of eastern Murrisk, Co. Mayo. P Roy Irish Acad B 72:307–333

Mohr P, Mussett AE, Kennan PS (1984) The Droimchogaidh Sill, Connacht, Ireland. Geol J 19:1–21

Mortimer N (2000) Metamorphic discontinuities in orogenic belts: example of the garnet–biotite–albite zone in the Otago Schist, New Zealand. Int J Earth Sci 89:295–306

O'Sullivan GJ, Chew DM (2020) The clastic record of a Wilson Cycle: evidence from detrital apatite petrochronology of the Grampian-Taconic fore-arc. Earth Planet Sc Lett 552:116588. https://doi.org/10.1016/j.epsl.2020.116588

Pearce JA, Reagan MK (2019) Identification, classification, and interpretation of boninites from Anthropocene to Eoarchean using Si-Mg-Ti systematics. Geosphere 15:1008–1037

Phillips WE (1973) The pre-Silurian rocks of Clare Island, Co., Mayo, Ireland, and the age of metamorphism of the Dalradian in Ireland. J Geol Soc Lond 129:585–606

Pudsey CJ (1984) Ordovician stratigraphy and sedimentology of the South Mayo inlier. Irish J Earth Sci 6:15–45

Rubinstein CV, Vajda V (2019) Baltica cradle of early land plants? Oldest record of trilete spores and diverse cryptospore assemblages; evidence from Ordovician successions of Sweden. Geol Foren Stock for 14:181–190

Rushton AW, Phillips WE (1973) A Protospongia from the Dalradian of Clare Island, Co Mayo, Ireland. Palaeontology 16:231–237

Ryan PD (2008) Preservation of forearc basins during island arc-continent collision: some insights from the Ordovician of western Ireland. Spec Pap Geol Soc Am 436:1–9

Ryan PD, Dewey JF (1991) A geological and tectonic cross-section of the Caledonides of western Ireland. J Geol Soc Lond 148:173–180

Ryan PD, Dewey JF (2011) Arc–continent collision in the Ordovician of western Ireland: stratigraphic, structural and metamorphic evolution. In: Arc-continent collision. Springer, pp 373–401

Ryan PD, Floyd PA, Archer JB (1980) The stratigraphy and petrochemistry of the Lough Nafooey Group (Tremadocian), western Ireland. Q J Geol Soc Lond 137:443–458

Ryan PD, Sawal VK, Rolands AS (1983) Ophiolitic melange separates ortho-and para-tectonic Caledonides in western Ireland. Nature 301:50–52

Ryan PD, Soper NJ, Snyder DB, England RW, Hutton DH (1995) The Antrim-Galway Line; a resolution of the Highland Border Fault enigma of the Caledonides of Britain and Ireland. Geol Mag, Mag 132:171–184

Skevington D (1971) Palaeontological evidence bearing on the age of Dalradian deformation and metamorphism in Ireland and Scotland. Scot J Geol 7:285–288

Skevington D, Archer JB (1971) A review of the Ordovician graptolite faunas of the west of Ireland. Irish Nat J 17:70–78

Stanton WI (1960) The lower Palaeozoic rocks of south-west Murrisk, Ireland. Q J Geol Soc Lond 116:269–290

Stern RJ, Reagan M, Ishizuka O, Ohara Y, Whattam S (2012) To understand subduction initiation, study forearc crust: to understand forearc crust, study ophiolites. Lithosphere 4:469–483

Stouge S, Harper DA, Sevastopulo GD, O'Mahony D, Murray J (2016) Lower and Middle Ordovician conodonts of Laurentian affinity from blocks of limestone in the Rosroe Formation, South Mayo Trough, western Ireland and their palaeogeographic implication. Geol J 51:584–599

Tanner P (2007) The role of the Highland Border ophiolite in the 470 Ma Grampian event, Scotland. Geol Mag 144:597–602

Taylor PD, Curry GB (1985) The earliest known fenestrate bryozoan, with a short review of Lower Ordovician Bryozoa. Palaeontology 28:147–158

Topuz G, Çelık ÖF, Şengör AC, Altıntaş İE, Zack T, Rolland Y, Barth M (2013) Jurassic ophiolite formation and emplacement as backstop to a subduction-accretion complex in northeast Turkey, the Refahıye ophiolite, and relation to the Balkan ophiolites. Am J Sci 313:1054–1087

Van Staal C, Dewey JF, Mac Niocaill C, McKerrow W (1998) The Cambrian-Silurian tectonic evolution of the northern Appalachians and British Caledonides: history of a complex, west and southwest Pacific-type segment of Iapetus. Geol Soc Lond Spec Pub 143:197–242

Williams A, Curry GB (1985) Lower Ordovician Brachiopoda from the Tourmakeady Limestone, Co., Mayo, Ireland. Bull Br Mus Nat His Geol 38:183–269

Williams D (1984) The stratigraphy and sedimentology of the Ordovician Partry Group, southeastern Murrisk, Ireland. Geol J 19:173–186

Williams D, Harkin J, Armstrong H, Higgs K (1994) A late Caledonian melange in Ireland: implications for tectonic models. J Geol Soc Lond 151:307–314

Williams D, Harkin J, Rice A (1997) Umbers, ocean crust and the Irish Caledonides: terrane transpression and the morphology of the Laurentian margin. J Geol Soc Lond 154:829–838

Williams H, Stevens R (1974) The ancient continental margin of eastern North America. In: The geology of Continental Margins. Springer, pp 781–796

Williams S, Harper D (1994) Late Tremadoc Graptolites from the Lough Nafooey Group, South Mayo, Western Ireland. Irish J Earth Sci 13:107–111

Wrafter JP, Graham JR (1989) Ophiolitic detritus in the Ordovician sediments of South Mayo, Ireland. J Geol Soc Lond 146:213–215

Zhang X, Cawood PA, Huang C-Y, Wang Y, Yan Y, Santosh M, Chen W, Yu M (2016) From convergent plate margin to arc–continent collision: Formation of the Kenting Mélange, Southern Taiwan. Gondwana Res 38:171–182

The South Connemara Group

Paul D. Ryan and John F. Dewey

Abstract The South Connemara Group is a middle Ordovician (Dapingian(?) to Darriwilian) subduction-accretion complex that is preserved onshore as a roof pendant in the Devonian Galway granite and crops out on the islands of Lettermullen and Gorumna on the north shores of Galway Bay. It is in structural continuity with similar lithologies of the offshore Skird Rocks, some 16 km west, and is believed to trace the westwards continuation of the Southern Uplands fault in Ireland. It comprises a sequence of MORB volcanics and shallow intrusives of the Gorumna Formation which are overlain by manganiferous deep sea cherts of the Golam Formation that pass gradationally upwards into sandstones and conglomerates of the Lettermullen Formation. Tectonism during accretion produced mélange zones and structural repetition. The succession represents the oldest part of the Longford–Down–Southern Uplands accretionary prism and provides evidence for subduction reversal with very rapid uplift at late stages of the short-lived Grampian arc-continent collision orogeny. This excursion provides an opportunity to study ocean floor lithologies that were overlain by continental detritus as they approached a trench.

1 Introduction

The Ordovician South Connemara Group (SCG) (McKie and Burke 1955; Ffrench and Williams 1984; Ryan and Dewey 2004) crops out on the islands of Lettermullen, Gorumna and Crappagh on the north-western shores of Galway Bay (Fig. 1). The SCG occurs as a roof pendant in the Devonian Galway Granite but is essentially in situ as it can be traced 16 km offshore to the west of the batholith where it is in fault contact with the Connemara Dalradian (Leake 1963; Max and Ryan 1975). Fabrics

---

P. D. Ryan (✉)
School of Natural Sciences, Earth and Ocean Sciences, National University of Ireland Galway, University Road, Galway, Ireland
e-mail: paul.ryan@nuigalway.ie

J. F. Dewey
University College, High Street, Oxford OX1 4BH, UK

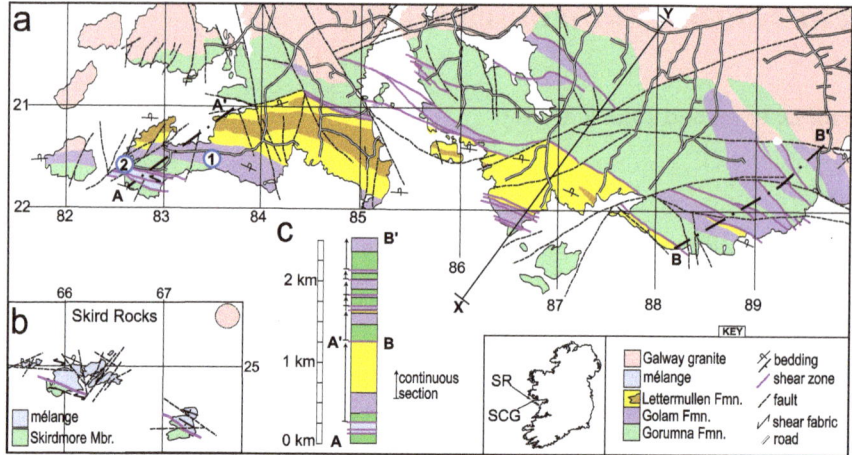

Fig. 1 **a** Geological map of the South Connemara Group (SCG) after Ryan and Dewey (2004). The numbers in circles refer to the two traverses shown in Fig. 8 **b** Geology of the Skird Rocks (SR) after Ryan and Max (1975, modified). The pink circle represents the nearest submarine outcrop of the Galway granite. The Irish National Grid 1 km squares are shown **c** A schematic stratigraphical column constructed along lines A–A' and B–B' showing the degree of structural repetition. Continuous sections in which the original bedding contacts are either observed or inferred are show by an arrow (after Ryan and Dewey 2004, modified). Localities are marked, see Fig. 8 for details

in the offshore outcrops of the Skird Rocks Formation on the small rocky islands of Skirdmore and Doonguddle have similar orientations to those onshore (Ryan and Max 1975). Marine surveys indicate that the SCG is overlain unconformably by Carboniferous limestones some 2.5 km southwards in Galway Bay (Max and Ryan 1975; Kenirons et al. 2009). Steep, inverted bedding implies a structural thickness of up to 7 km (Ryan and Max 1975). An onshore gravity survey (Fairhead and Walker 1977) suggests that it is a wedge-shaped body about 400 m thick in the south (Fig. 2) fading to a feather edge in the north at the contact with the granite. Onshore

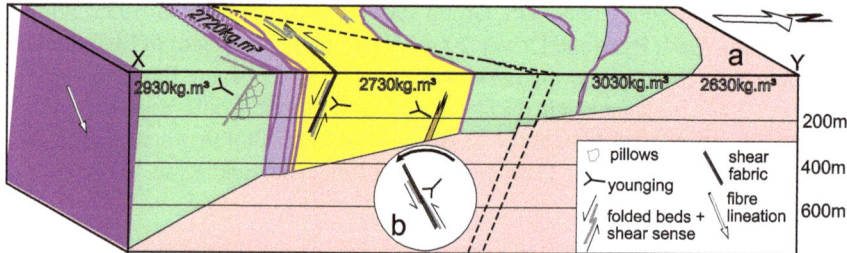

Fig. 2 **a** True scale block diagram looking west along line of section X–Y (Fig. 1). The densities and the granite contact at depth are after Fairhead and Walker (1977) **b** Shear sense becomes north side up after rotating bedding through the vertical to become right way-up. The map ornament is the same as in Fig. 1

the rocks of the SCG, which were initially of low metamorphic sub-greenschist (?) grade, have been subjected to hornblende hornfels facies contact metamorphism but original fabrics are locally preserved and exposure is excellent, although commonly lichen covered. Relationships are best seen on weathered surfaces in the wave zone.

2 Stratigraphy

The SCG is interpreted as an accretionary complex (Ryan and Dewey 2004) that contains three stratigraphic units, which are, in order of decreasing age: the Gorumna Formation; the Golam Formation and the Lettermullen Formation. Bedding uniformly dips steeply to the south but youngs to the north (Fig. 1a). In spite of these units being structurally repeated, sufficient primary contacts are observed to establish the stratigraphy described below. This account relies mainly on those of Ryan and Dewey (2004) and Ffrench and Williams (1984).

2.1 Gorumna Formation

The Gorumna Formation (Ryan and Dewey 2004) consists of mafic volcanic rocks and shallow intrusives which probably correlate with the Skirdmore member of the Skird Rocks Formation offshore (Ryan and Max 1975; Fig. 1b). It is repeated in seven separate tectonic slices (Fig. 1a, c) but consistently youngs to the north. Primary fabrics are relatively rare with outcrops dominated by fine-grained meta-basalts with the shear fabric partly annealed by hornblende hornfels contact metamorphism (Ryan and Dewey 2004). Where primary textures are preserved, the volcanic component consists of massive to pillowed basalts inter-layered with pillow lava breccias, scoriaceous and hyaloclastite units. Vesicular pillows up to 1 m in diameter with bun-shaped tops (Fig. 3a) have interstitial limestone and red chert on Lettermullen, Gorumna and Golam Head. Pillows are commonly isolated in form with rare elongate lava lobes. The intrusive component consists of ophitic dolerites (Fig. 3b), which occur either as sills or in tectonic slices where their texture is that of deeper levels in sheeted dike complexes, although no chilled margins are preserved. On the south shores of Gorumna Island ($53.225178°$, $-9.667372°$), pillow lavas cut by fine grained ophitic sills young into and are overlain directly by manganiferous cherts of the Golam Formation.

The Gorumna Formation exhibits MORB (mid-ocean ridge basalt) geochemistry (Ryan et al. 1983). In the MORB normalization diagram (Pearce et al. 1981) the high field strength elements (HFSE) show a relatively flat, uniform profile (Fig. 4a). Large ion lithophile elements, particularly Rb, are enriched with respect to MORB. However, these element profiles may have been altered during contact metamorphism. These lavas fall in MORB fields on discriminant diagrams which rely on HFSE (Ryan et al. 1983; Fig. 4b–d).

Fig. 3 **a** Pillow lavas of the Gorumna Formation on the cliff face south of Lough Faoileán. Field of view is approximately 4 m **b** Contact metamorphosed ophitic dolerite within the Gorumna Formation on the north side of the car park. The tip of the walking stick is 25 mm in width

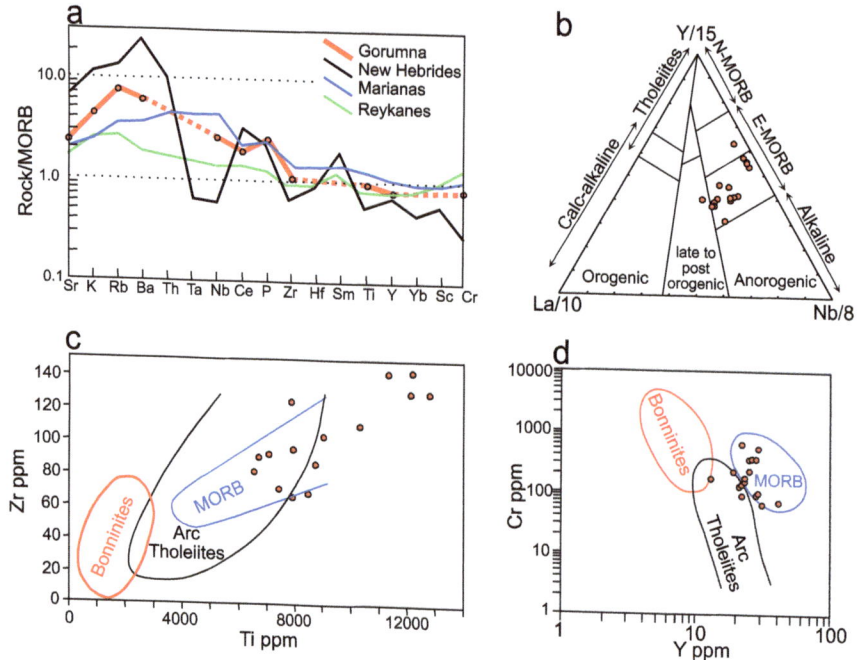

Fig. 4 Geochemical discrimination plots for the Gorumna Formation meta-basalts, data is from Ryan et al (1983). **a** Rock/MORB normalization diagram after Pearce et al (1981). Reykanes = MORB from the Reykanes Ridge; Marianas = marginal basin ocean floor basalts from the Marianas Trough; New Hebrides = arc tholeiites from the New Hebrides arc **b** La/10-Y/15-Nb/8 diagram after Cabinis and Lecolle (1989) **c** Ti/Zr plot after Dilek and Furnes (2009) **d** Y/Cr plot after Dilek et al. (2007)

2.2 Golam Formation

The Golam Formation is dominated by white-weathering grey cherts with bed thicknesses varying from a few centimetres to a few millimetres. These are interbedded with maroon, green and grey argillites (Fig. 5a). These cherts are commonly highly manganiferous near their base at the contact with the Gorumna formation. Locally the cherts have specific gravities of up to 3150 kg.m^3 and contain up to 19% pyrolusite, although thin seams (<2 mm) contain up to 90% pyrolusite (Ryan and Dewey 2004; Fig. 5b). The MnO$_2$ content systematically decreases up section indicating dilution of the manganiferous cherts by more siliceous components associated with a colour change from deep red or deep green at the base to lighter reds and whites near the top. The post tectonic thickness of this formation on Lettermullen is 324 m. It is impossible to assess the original thickness because of bedding- parallel shearing and small-scale structural repetition. However, the gradational change up section indicates that there is no large-scale folding and that this formation contains a considerable thickness of abyssal sediment deposited below the carbonate compensation depth (Ffrench and Williams 1984; Ryan and Dewey 2004). This formation is interpreted as a deep-sea deposit which became progressively diluted up-section with distal continentally derived detritus, similar to that comprising the overlying Lettermullen Formation (Ffrench and Williams 1984). Ryan and Dewey (2004) suggest that the cherts lay on oceanic crust that was transported towards a continental margin and became intercalated with distal turbidite flows and thickened tectonically as the trench was approached similar to the situations in the Nankai Trough (Taira et al. 1992) and the Kuril Trench (Schnürle et al. 1995). The highest pyrolusite concentrations occur in thin layers (<2 mm) in the lower part of the formation. They are bedded Fig. 5b), not nodular, in form and are hence interpreted as active ridge sediments (see Böstrom 1975).

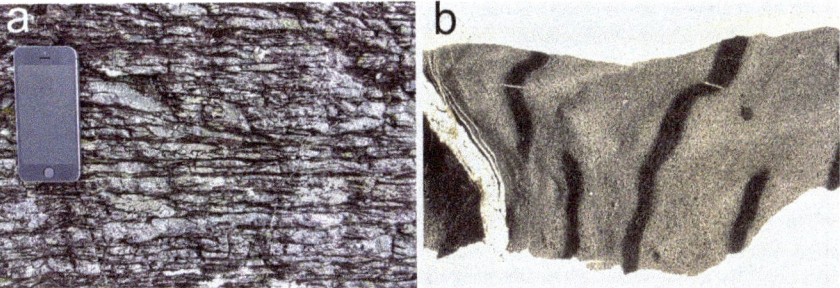

Fig. 5 a Sheared interbedded cherts and siliceous shales of the Golam Formation. A bedding cut-off is visible to the right of the mobile phone (58 mm by 123 mm). The S1 fabric,sub-parallel to bedding, is only visible in weathered outcrops **b** Photomicrograph (plane light) of disrupted manganiferous beds, now replaced by pyrolusite, of the Golam Formation. Field of view is 20 mm

2.3 Lettermullen Formation

The Lettermullen Formation comprises turbiditic sandstones, with slump folds, which coarsen upwards to thicker granule-grade beds of over 1 m containing conglomeratic layers and then subsequently fine upwards. The thickness on Lettermullen Island, where a continuous section is apparently preserved, is 618 m (Ryan and Dewey 2004). Ffrench and Williams (1984) divided the formation into 5 facies (Table 1), the last of which (facies 5) is now interpreted as a tectonic repetition of the Golam Formation (Ryan and Dewey 2004).

The lower part of this formation coarsens upwards. Facies 4 or 3 (Fig. 5a) occur at the base, lying conformably upon the Golam Formation. These are, in turn, overlain by facies 3 (Fig. 6a) or 2 (Fig. 6b) and 1 (Fig. 6c, d). In its upper part, on Lettermullen, the sequence fines upwards with facies 1 passing up into facies 3 and then facies 4 (Ffrench and Williams 1984: Fig. 2). Facies 1 has attracted the most attention because the metamorphic and meta-igneous clast lithologies within these conglomerates (Table 2, Fig. 6c, d) contain lithologies characteristic of the adjacent Grampian metamorphic tract of Connemara. Palaeocurrent analysis (Ffrench and Williams 1984), after allowing for deformation, indicates a dominant, strike-parallel, easterly source (105°) and a minor component from the north (25°). Ffrench and Williams (1984) argue that these immature lithic arenites show some of the characteristics of trench fill deposits.

Table 1 Facies types within the Lettermullen Formation (after Ffrench and Williams 1984; Mckie and Burke 1955, modified)

Facies	Lithology
1	Boulder, cobble and pebble polymict conglomerates with subordinate coarse sandstones involving transport by both sheet-flow and in a channel system. Conglomerates may be diffuse or framework supported. Imbrication is rare and grading, where recorded, is normal. Maximum recorded boulder size (long axis) is 1.8 m. Metamorphic clasts (Table 2; Fig. 6c, d) increase up section
2	Thick sandstones (0.2–1.1 m) with thin mudstone inter-beds. Beds are normally graded and often laterally continuous with angular quartz and microcline in a groundmass of quartz, plagioclase, biotite and muscovite. Bouma sequences T_{a-e}, T_{b-e} and $T_{a,b,e}$ (Fig. 6b)
3	Medium-grained thin sandstones (0.05 m) with mudstone inter-beds (0.01 m). Beds exhibit parallel or cross-laminations with minor basal scours. Bouma sequences T_{b-e} with T_b dominant (Fig. 6a)
4	Fine grained sandstones, siltstones and mudstones (0.02 m). Beds are laterally continuous. Bouma sequences $T_{d,e}$

Fig. 6 a Graded meta-sandstones of facies 3 of the Lettermullen Formation. The beds are inverted with the sharp basal contacts of the sandstones being on top. Ripples occur at the bottom of the photograph and slight channelling (arrowed) near the centre. Pseudomorphs after cordierite occur in the mudstones giving an impression of reversed grading **b** 1.2 m thick sandstone at the base of facies 2 of the Lettermullen Formation which contains quartz granules at the base (arrowed). The stick is 90 cm in length and south is to the right **c** and **d** Conglomerates of facies 1 of the Lettermullen Formation. Note the wide variety of lithologies including: various granitoids (g); meta-gabbro (m); foliated quartzites (q); silicic volcanic rock (r); schists (s); and ultramafic rock (u). The tip of the walking stick is 25 mm in diameter

Table 2 Clast types in the conglomeratic facies of the Lettermullen Formation (Ffrench and Williams 1984; Ryan and Dewey 2004)

Type	Lithology	Bulk (%)
Igneous	High level quartz-rich granites both foliated and infoliated Foliated tonalites Migmatitic grey andesine/oligoclase gneisses Metagabbros (rare) Granodiorite Quartz- diorite Quartz-plagioclase porphyries	37–69
Metamorphic	Pebbly quartzites Quartzites showing refolds Marble (rare) Biotite, muscovite, and amphibole semipelitic schists Greenschist phyllites and slate	31–50
Sedimentary	Intraformational sedimentary clasts	1–13

3 Age and Regional Correlations

The Golam Formation is assigned an age of Arenig (Floian ~ 478 Ma) to lower Llanvirn (Darriwilian: Da1 ~ 466 Ma) based on a single chitnozoan attributed to *Lagenochitina esthonica* recorded some 70 m below the top of the formation (Williams et al. 1988). Isotopic ages here cited are for the bases of the various stages using the timescale of Cohen et al. (2013). *Lagenochitina esthonica* is also reported in the Tremadoc (Tr3 ~ 480 Ma) of Sweden (Grahn and Nolvak 2007) allowing a possibly slightly older age for this formation. Williams et al. (1988) suggest a deposition rate of 6 m/m.yr., using the calculated sedimentation rates for Ordovician and Silurian shales in the Southern Uplands of Scotland (Churkin et al. 1977), a possible along-strike correlative. This implies that deposition of the Golam formation may have spanned more than 50 m.yr. However, greater sedimentation rates are reported for active ridge sediments (e.g. Böstrom et al. 1974; Metz et al.1988) and the formation is tectonised, meaning that such an analysis should be treated with caution. The increasing continental component up section is consistent with deposition having spanned a period that exceeded the time needed to drift away from the active volcanic zone of the ridge. In the fast-spreading East Pacific Rise (~160 mm/yr) off-axis vulcanism is recorded between 4 and 20 km from the axial zone (Vithana et al. 2016) suggesting a minimum period of deposition of a few hundred thousand years. The much higher sedimentation rates of ~100 m/m.yr. reported for some active ridge sediments (Metz et al. 1988) and some hemipelagic sediments adjacent to continents (Wetzel and Uchman 2012) are consistent with deposition within significantly less than 5 m.yr.

Although there is no paleontological evidence for the age of the Lettermullen Formation, the facies 1 conglomerates (Table 2, Fig. 6c, d) represent the rapid stripping of an emerging metamorphic terrane, most probably Connemara or its along strike, now buried, equivalent (Ryan and Dewey 2004; Dewey and Ryan 2016). The South Mayo Trough fore-arc basin (Dewey and Ryan 1990) to the north remained below sea-level during early Grampian arc-continent collision (Dewey and Mange 1999; Ryan 2007) but records a sudden influx of metamorphic detritus during the Darriwilian (Da2, Dewey and Mange 1999) associated with over an order of magnitude increase in sedimentation rate (Ryan and Dewey 2011). Any such unroofing would have shed orogenic detritus in all directions during the short lived (Dewey 2005) Grampian Orogeny. It is, therefore, reasonable to suggest that facies 1 of the Lettermullen Formation has a maximum age of Darriwilian (Da2). This makes it likely that the upper Golam Formation is Darriwilian (Da1) at the top of the range of *Lagenochitina esthonica* (Paris 1996). If the lower part of the Golam Formation contains active ridge sediments, then these must effectively date the Gorumna Formation. If the Golam Formation was deposited in less than 5 m.yr., the Gorumna Formation would span the interval lower Darriwilian (Da1) to Dapingian. If deposition took more than 13 m.yr., the Gorunma Formation would be Tremadoc in age.

The South Connemara Group has been correlated eastwards along strike with the northern belts of the Longford-Down massif and the Southern Uplands of Scotland

(Philips et al. 1976; Morris 1983; Ffrench and Williams 1984; Williams et al. 1988). Williams et al. (1988) provide a full discussion and suggest correlation of the SCG with Tract 2 (after Leggett et al. 1979) of the Northern Belt of the Southern Uplands. Comparison of the revised stratigraphies of the SCG (Ryan and Dewey 2004) and the Southern Uplands of Scotland (Floyd 2000) suggests that the Gorumna and Golam Formations may correlate with the Raven Gill Formation of the Crawford Group. Metamorphic detritus occurs in late Ordovician (Sandbian–Katian) sandstones of the Northern Belts of the Longford-Down inlier (Morris 1987) and the Southern Uplands (e.g. Floyd 2000), which are probably too young to allow correlation with the Lettermullen Formation. Westwards, in Newfoundland, the Annieopsquatch mafic complex is in tectonic contact with the southern margin of the post-flip Notre Dame arc (Van Staal et al. 1998) and is in the same tectonic position as the SCG.

4 Structure and Tectonic Significance

The bedding both onshore and offshore dips steeply to the south but youngs to the north (Figs. 1 and 2). Local small to medium scale folds, some of which occur in single horizons and may be slump related, are most common in the lower portion of the Lettermullen Formation. Bedding faces upwards on the clockwise transecting, sub-parallel cleavage which dips to the south slightly less steeply (Ryan and Dewey 2004; Fig. 2). Cleavage development is zonal. A strong fabric is evident in the Golam Formation (Fig. 5a) but is often absent in the Lettermullen Formation (Fig. 6a) except within shear zones.

A mélange zone on the south side of Lettermullen separates the Golam and Gorumna formations (Fig. 7a, b). Within this zone, the matrix varies in well-defined layers on a scale of 0.1 m to several meters from muddy, volcanic, sandy to calcareous

Fig. 7 a Phacoidal cherts and disrupted sands within the mélange zone south of Lough Faoileán. The walking stick is 90 cm long. **b** Large sub-angular block of manganiferous chert in the mélange zone south of Lough Faoileán adjacent to the tectonised contact with the pillow lavas of the Gorumna Formation (immediately above the head of the hammer)

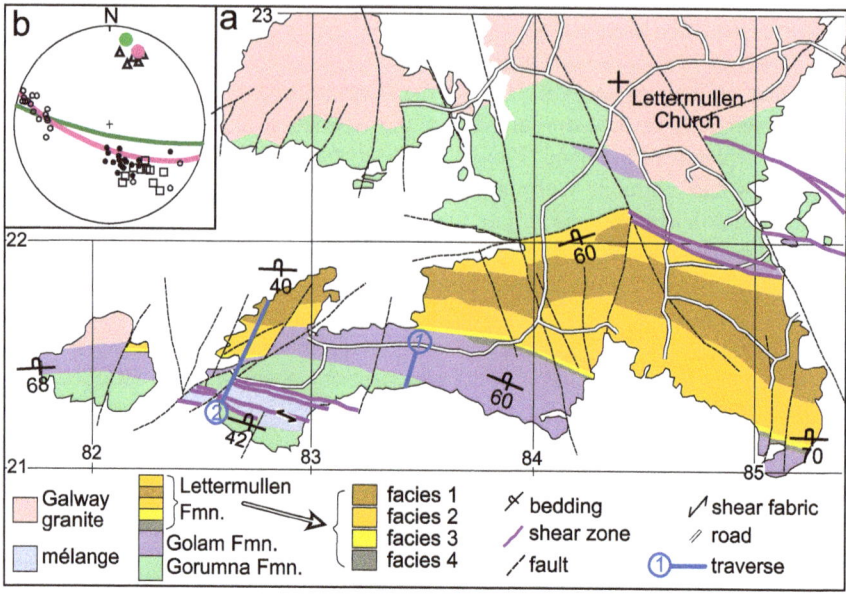

Fig. 8 a Geological map showing the traverses at localities 7.1 and 7.2 which begin at the encircled numbers '1' and '2' respectively. The geology is after Ryan and Dewey (2004), the individual facies (1–4) within the Lettermullen Formation are after Ffrench and Williams (1984, modified) **b** Structural data from Ryan and Dewey (2004). The following colour codes or ornaments are used, green = bedding, mauve = foliation, open circles = fold axes, closed circle = fibre lineations, open squares = mullions, open triangles = poles to shears with down to south-southeast sense of displacement

material. The sub-rounded to sub-angular blocks (0.1–1.0 m) comprise white, red (Fig. 7b), and epidotic green chert, meta-basalt, sandstone and limestone. They are elongated in the foliation and show considerable variations in their state of strain (Fig. 7a, b). Fibre lineation plunges steeply SSE (Figs. 2a and 8b). Fold geometry, ramp cut-offs, asymmetric clast tails in the mélange and cleavage orientation are consistent with north-side up sinistrally oblique transport when bedding is rotated through the vertical to become right-way up (Ryan and Dewey 2004, Fig. 2b). Ryan and Max (1975) correlated this zone, which they termed the 'Mixed member' with the Moyrus member of the Skird Rocks (Fig. 1b).

The correlation of the SCG with similar lithologies (meta-basalt, cherts, quartz sandstones) in the Northern Belts of the Longford-Down Massif of north-eastern Ireland and the Southern Uplands of Scotland has led all authors since Leake (1963) to propose that the Southern Uplands Fault can be traced westwards to the north shores of Galway Bay (Fig. 1 in chapter "The Late Silurian to Upper Devonian Galway Granite Complex (GGC)"). This fault is the northern boundary of a southward-accreting prism at the Iapetus margin of the Laurentian continent which operated from mid-Ordovician to Silurian (Wenlockian) times (Leggett et al. 1979). It formed as a result of a subduction flip following the short-lived early to middle Ordovician arc-continent collision Grampian orogeny (Dewey and Shackleton 1984; Ryan and Dewey 1991; Dewey 2005). The SCG represents one of the oldest, if not the oldest tectonic slice within this complex. It certainly is the oldest such tract where normal stratigraphic contacts occur between oceanic basalts, deep sea sediments and post-orogenic sandstones and is, therefore, likely to be closest to the continental backstop. This has considerable tectonic significance for the location of the Galway Granite batholith (see Feely et al. chapter "The Late Silurian to Upper Devonian Galway Granite Complex (GGC)", this volume).

5 Excursion

5.1 Aim

The aim of the excursion is twofold. First, to demonstrate that the SCG records accretion of deep-sea sediments at a subduction interface and second to examine the various lithologies occurring within it and their stratigraphic relationships.

5.2 Locality 7.1 Lettermullen (53.231890°, −9.746154°)

Access and additional information: This short traverse of about 120 m is across a field to the south of L77. This is a very narrow, single track road only suitable for cars or small minibuses. There is no space for off-road parking. Vehicles can be parked at the'T' junction 410 m farther west. This traverse is on privately owned land. Estimated time 1 h.

Enter the field opposite a passing place on the north side of the road and walk southwards towards the skyline. The section is working down stratigraphy through the lower part of the Golam formation towards its contact with the Gorumna Formation (Fig. 8a). Thinly bedded (millimetres to centimetres), white weathering cherts are interbedded with siliceous argillites and green or grey argillites (Fig. 5a, b). Increasing MnO_2 concentration in the cherts at lower stratigraphic levels (further south) in this formation results in a deeper red or green colour in some bands. There is considerable bedding parallel shearing. Locally, the cleavage is parallel with the axial surfaces of small-scale asymmetric folds. Small thrusts occur with a cut-off angle of up to 20°. Deformation is zonal.

5.3 Locality 7.2 Lettermullen Western Fore-Shore (53.230572°, −9.757983°)

Access and additional information: Continue westwards for 400 m and turn left at'T' junction. Drive 600 m along narrow road to the parking space by the sea. Access to the shore requires crossing privately owned land. The locality is best visited at low tide. The southern part of the section is on the exposed foreshore where rocks may be slippery. On windy days breaking waves may make the southernmost part of this section inaccessible. Estimated time 2 h.

This section works from the Gorumna Formation in the south, through a mélange zone into the contact of the Golam and Lettermullen Formations (Fig. 8a). The section continues northwards through a coarsening upwards sequence (facies 3 then facies 2) to the conglomerates of facies 1 of the Lettermullen Formation on the north shores of the island. It is best to go to the base of the section in the south and then work northwards up section. Take the gate through the stone wall to the south of the car park and skirt the western side of the small brackish Lough Faoileán (Seagull Lake). The Gorumna Formation crops out on the exposed rocky headland 35 m south of the narrow channel draining the lough at its southern extremity.

The Gorumna Formation, in the extreme south of this traverse occupying the headland, consists of pillow lavas in which the bun-shaped crusts are facing downwards indicating inversion (Fig. 3a). The interstices are filled with meta-limestone, chert and pillow breccia. The cherts imply deposition below carbonate compensation depth (CCD); therefore, the limestones must have formed on volcanoes with significant topography. Ryan and Dewey (2004) interpret this to mean that they formed as a fringing reef on an accreted seamount.

The mélange, bounding the Gorumna Formation to the north (approximately 80 m north of the headland), has a variable matrix with muddy, mafic volcanic, sandy and calcareous material occurring in well-defined layers on a scale of 0.1 m to several metres. Blocks (0.1–1.0 m longest dimension) or 'knockers' of chert, metabasalt, sandstone and limestone show variable states of strain but have a strong layer-parallel shape fabric (Fig. 7a, b). Angular blocks generally lack pressure shadows and phacoidal blocks may have either symmetric or sinistrally asymmetric tails. There are fine grained zones, where the penetrative mylonitic foliation which has been hornfelsed is obvious on weathered surfaces. Zones with more calcareous material, such as on the small hillock to the immediate south of Lough Faoileán, have negative weathering patterns. The structural relationships are summarized in Figs. 2 and 8b.

The mélange, which is about 225 m in thickness, is in fault contact to the north with a meta-dolerite (Fig. 3b) which extends from the northern shores of Lough Faoileán under the car park and is then in faulted contact with the Golam Formation.

Cross the car park, take the gate to the north to access the foreshore. Cherts, locally red, pass upwards into finely bedded graded sandstones of facies 3 of the Lettermullen Formation (Fig. 6a) on the north side of the small bay immediately north of the car park. Grain size and bed thickness quickly increase northwards, up section, to facies 2 sandstones (Fig. 6b). Graded bedding, ripples and bottom structures provide way-up evidence. The muddy lithologies in the upper part of the beds may have contact metamorphic 'spotting' with sericitized pseudomorphs after cordierite a few mm across giving the impression the beds are coarsening upwards. Sandstones coarsen with granule grade quartz becoming a common component (Fig. 6b). The first conglomerate beds of facies 1 appear after approximately 150 m. The clasts of facies 1 (see Table 2; Fig. 6c, d) provide the first record of post-orogenic (Grampian) sedimentation within the Southern Uplands accretionary prism (see above).

Exercises

(1) Document the different lithologies in this complex and discuss what tectonic environment is likely to produce this, albeit disrupted, stratigraphy.
(2) Document the different deformation styles and discuss in what tectonic environment such discontinuous deformation might occur.
(3) Study the regional and contact metamorphic fabrics and discuss their likely grades and relative timings.
(4) Discuss the likely depositional environments of the Lettermullen Formation. Sedimentary logs may be prepared throughout this formation on the foreshore north of the car park.
(5) The mélange zone separating the Gorumna and Golam Formations south of Lough Faoileán represents a short-lived plate boundary: discuss.

References

Boström K, Joensuu O, Kraemer T, Rydell H, Valdés S, Gartner S, Taylor S (1974) New finds of exhalative deposits on the East Pacific Rise. Geol Fören Stock För 96(1):53–60. https://doi.org/10.1080/11035897409454258

Boström K (1975) Origin and fate of ferromanganoan active ridge sediments. In: Hsü KJ, Jenkyns HC (ed) Pelagic sediments: on land and under the sea. The International Association of Sedimentologists, pp 401–401

Cabanis B, Lecolle M (1989) Le diagramme La/10-Y/15-Nb/8: un outil pour la discrimination des séries volcaniques et la mise en évidence des processus de mélange et/ou de contamination crustale. CR Acad Sci II 313:2023–2029

Churkin Jr M, Carter MC, Johnson BR (1977) Subdivision of Ordovician and Silurian Time Scale Using Accumulation Rates of Graptolitic Shale. Geology 5(8):452–456

Cohen K, Stanley M, Finney C, Gibbard PL, Fan J-X (2013) The ICS International Chronostratigraphic Chart. Episodes 36(3):199–204

Dewey JF (2005) Orogeny can be very short. P Natl Acad Sci USA 102(43):15286–15293

Dewey JF, Mange M (1999) Petrology of Ordovician and Silurian Sediments in the Western Irish Caledonides: tracers of Short-Lived Ordovician Continent-Arc Collision Orogeny and the Evolution of the Laurentian Appalachian-Caledonian Margin. In MacNiocaill C, Ryan PD (ed) Continental Tectonics Spec Publ Geol Soc London 164:55–108

Dewey JF, Ryan PD (1990) The Ordovician evolution of the South Mayo Trough Western Ireland. Tectonics 9:887–901

Dewey JF, Shackleton RM (1984) A model for the evolution of the grampian tract in the early caledonides and appalachians. Nature 312(5990):115–121

Dewey JF, Ryan PD (2016) Connemara: its position and role in the Grampian Orogeny. Can J Earth Sci 53:1246–1257

Dilek Y, Furnes H (2009) Structure and geochemistry of Tethyan ophiolites and their petrogenesis in subduction rollback systems. Lithos 113:1–20

Dilek Y, Furnes H, Shallo M (2007) Suprasubduction zone ophiolite formation along the periphery of Mesozoic Gondwana. Gondwana Res 11:453–475

Fairhead JD, Walker P (1977) The geological interpretation of gravity and magnetic surveys over the exposed Southern Margin of the Galway Granite Ireland. Geol J 12(1):17–24

Ffrench GD, Williams DM (1984) Sedimentology of the South Connemara Group, Western Ireland-a possible Ordovician trench-fill sequence. Geol Mag 121:505–514

Floyd D (2000) Southern Uplands Terrane: a stratigraphical review. T Roy Soc Edin-Earth 91(3–4):349–362

Grahn Y, Nolvak J (2007) Chitinozoa and Biostratigraphy from Skåne and Bornholm Southernmost Scandinavia—an overview and update. B Geosci 82(1):11–26

Kenirons M, Curran O, Cunniffe J, Ryan JL, Ryan PD, Shearer A (2009) The MarineGrid Project in Ireland with Webcom. Comput Geosci 35(2):205–213

Leake BE (1963) The Location of the Southern Uplands Fault in Central Ireland. Geol Mag 100(5):420–423

Leggett JK, McKerrow WS, Eales MH (1979) The Southern Uplands of Scotland: A Lower Palaeozoic Accretionary Prism. J Geol Soc London 136(6):755–770

Max MD, Ryan PD (1975) The Southern Uplands Fault and its relation to the metamorphic rocks of Connemara. Geol Mag 112(6):610–612

Mckie D, Burke KC (1955) The geology of the islands of South Connemara. Geol Mag 92:487–498

Metz S, Trefry JH, Nelsen TA (1988) History and geochemistry of a metalliferous sediment core from the mid-atlantic ridge at 26°N. Geochim Cosmochim Ac 52(10):2369–78 https://doi.org/10.1016/0016-7037(88)90294-3

Morris JH (1983) The stratigraphy of the lower palaeozoic rocks in the Western End of the Longford-Down Inlier. Irish J Earth Sci 5(2):201–218

Morris JH (1987) The Northern belt of the Longford-Down Inlier Ireland and Southern Uplands Scotland: an ordovician back-arc basin. J Geol Soc London 144(5):773–786

Paris F (1996) Chitinozoan Biostratigraphy and Palaeoecology. In Jasonius J, McGregor DC (ed) Palynology: Principles and Applications, AASP 2: 531–52

Pearce JA, Alabaster T, Shelton AW, Searle MP (1981) Oman ophiolite as a cretaceous arc-basin complex: evidence and implications. Philos T Roy Soc A 300(1454):299–317

Philips WEA, Stillman CJ, Murphy TA (1976) A caledonian plate tectonic model. J Geol Soc London 132(6):579–605

Ryan PD (2007) Preservation of fore-arc basins during island arc-continent collision: some insights from the ordovician of Western Ireland. Spec Pap Geol Soc Am 436:1–9

Ryan PD, Dewey JF (1991) A geological and tectonic cross-section of the caledonides of Western Ireland. J Geol Soc London 148(1):173–180

Ryan PD, Dewey JF (2004) The South connemara group reinterpreted: a subduction-accretion complex in the caledonides of galway bay Western Ireland. J Geodyn 37(3–5):513–529

Ryan PD, Dewey JF (2011) Arc-continent collision in the ordovician of Western Ireland: stratigraphic structural and metamorphic evolution. In: Ryan PD (ed) Brown D. Arc-Continent Collision Springer, Heidelberg, pp 373–401

Ryan PD, Max MD (1975) The South Connemara Group Report of the Geological Survey of Lreland 75(3):24–34

Ryan PD, Max MD, Kelly, (1983) The petrochemistry of the basic volcanic rocks of the South connemara group (Ordovician) Western Ireland. Geol Mag 120(2):141–152

Schnürle PS, Lallemand E, von Huene R, Klaeschen D (1995) Tectonic regime of the Southern Kurile trench as revealed by multichannel seismic lines. Tectonophysics 241(3–4):259–277

Taira A et al (1992) Sediment deformation and hydrogeology of the Nankai Trough accretionary prism: synthesis of shipboard results of ODP Leg 131. Earth Planet Sci Lett 109(3–4):431–450

Van Staal CR, Dewey JF, MacNiocaill C, McKerrow WS (1998) The Cambrian-Silurian tectonic evolution of the Northern Appalachians and British caledonides: history of a complex West and Southwest pacific-type segment of iapetus. Geol Soc London Spec Pub 143:197–242

Vithana MVP, Xu M, Zhao X, Zhang M, Luo Y (2019) Geological and geophysical signatures of the east pacific rise 8°–10°N. J Solid Earth Sci 4(2):66–83. https://doi.org/10.1016/jsesci201904001

Wetzel A, Uchman A (2012) Hemipelagic and pelagic basin plains. In Knaust D, Bromley RG (ed) Developments in sedimentology, vol 64, pp 673–701 Elsevier, Amsterdam. https://doi.org/10.1016/B978-0-444-53813-0.00022-8

Williams DM, Armstrong HA, Harper DAT (1988) The age of the south connemara group Ireland and its relationship to the Southern uplands zone of Scotland and Ireland. Scot J Geol 24(3):279–287

The Silurian of North Galway and South Mayo

John R. Graham, John F. Dewey, and Paul D. Ryan

Abstract Silurian rocks are present in three successor basins that have no spatial and probably no temporal overlap. The southernmost Killary Harbour-Joyce country succession everywhere masks the contact between the Ordovician rocks of the South Mayo Trough and the Connemara Dalradian. It was deposited in c 5 m.y. from mid-Telychian (Late Llandovery) to Sheinwoodian (mid-Wenlock). The succession shows rapid deepening from terrestrial through shallow marine to a deeper water turbidite dominated package and then a subsequent shallowing. There are significant provenance changes throughout the succession and detrita that relate to source areas that are no longer extant. The Croagh Patrick succession was deposited on Ordovician rocks of the South Mayo Trough that had already been folded to form the open Mweelrea-Partry Syncline. This quartzose succession of Laurentian provenance is poorly dated but is probably Llandovery. It is of shallow marine origin with some indications of more offshore conditions to the west and is strongly deformed. The northernmost Louisburgh-Clare Island succession is non-marine and rests unconformably on the Clew Bay Complex and is in faulted contact with the Croagh Patrick succession. The lower part is of Ludlow or Pridoli age. It is not currently possible to delimit the geometry or extent of any of the depositional basins in which these three disparate successions formed.

J. R. Graham (✉)
Department of Geology, Trinity College Dublin, Museum Building, Dublin Dublin 2, Ireland
e-mail: jrgraham@tcd.ie

J. F. Dewey
University College, High Street, Oxford OX1 4BH, UK

P. D. Ryan
School of Natural Sciences, Earth and Ocean Sciences, National University of Ireland Galway, University Road, Galway, Ireland

© The Author(s), under exclusive license to Springer Nature Switzerland AG 2022
P. D. Ryan (ed.), *A Field Guide to the Geology of Western Ireland*, Springer Geology Field Guides https://doi.org/10.1007/978-3-030-97479-4_8

1 Silurian

South Mayo has outcrops of three Silurian successions that differ markedly from each other. They are either physically or structurally separated and there is currently no clear evidence that there is any age overlap although this remains possible (Fig. 1). From south to north these are:

- Killary Harbour–Joyce Country succession
- Croagh Patrick succession
- Louisburgh–Clare Island succession.

The essential features of these three were summarised in Graham et al. (1989) which contains references to the early and pioneering work.

1.1 Killary Harbour–Joyce Country Succession (KHJC)

This is the best known and biostratigraphically best constrained of the three Silurian successions. It crops out in an east–west belt from the Atlantic coast to Loughs Mask and Corrib in the east everywhere masking the contact between the Connemara Dalradian to the south and the Ordovician arc and South Mayo Trough sequences to the north. The outline stratigraphy is shown in Fig. 2 The outcrop is bisected by the

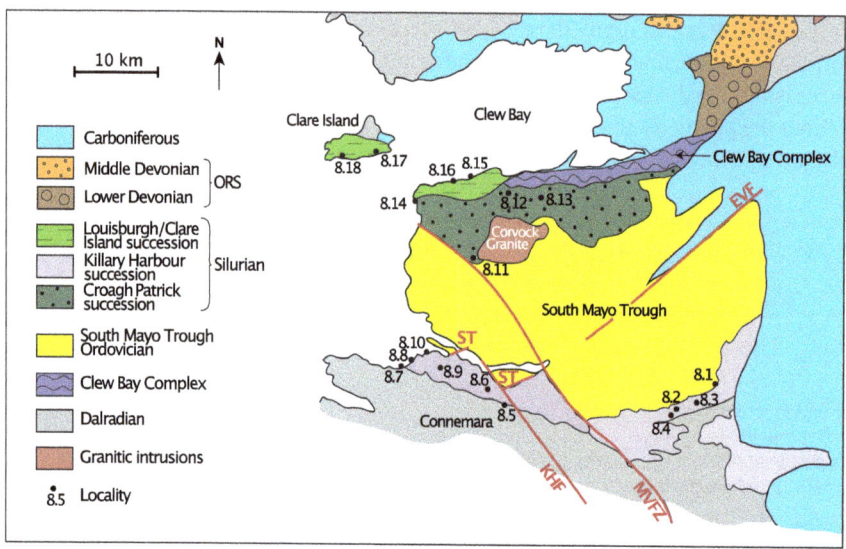

Fig. 1 General distribution of Silurian successions with field localities 8.1 to 8.18 shown. Abbreviations: EVF = Erriff Valley Fault; KHF = Killary Harbour Fault; MVFZ = Maam Valley Fault Zone; ST = Salrock Thrust

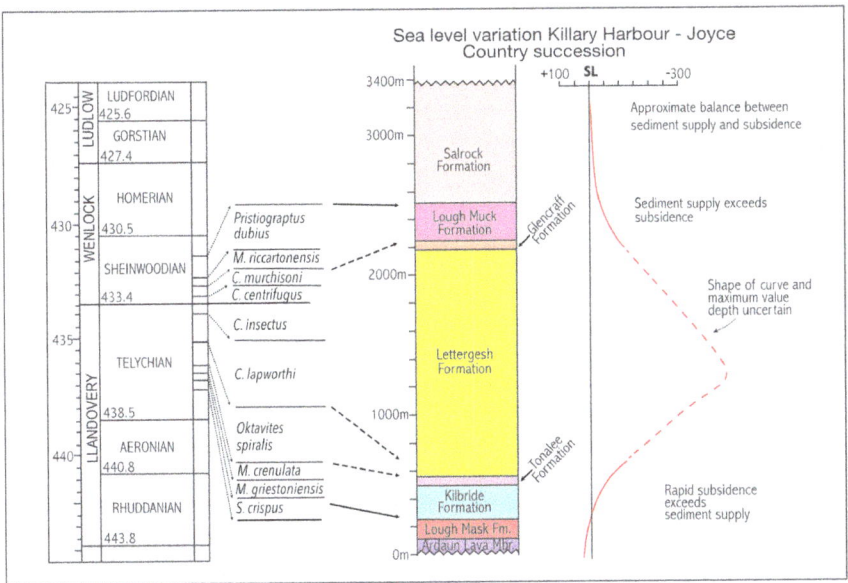

Fig. 2 Outline stratigraphy and interpreted sea level curve for the KHJC succession. The Silurian time scale and biostratigraphical control for the KHJC succession are shown on the left. Biostratigraphic data from Melchin et al. (2020) and Boydell (2012)

complex Maam Valley Fault Zone that is known to have significant post-Wenlock dextral displacement. However, there are also significant stratigraphic differences that occur from east to west where the changes approximate this fault zone suggesting that it had an earlier history.

The lower part of the succession (Lough Mask to Tonalee formations) is best developed to the east of the Maam Valley Fault Zone (Joyce Country) where it is much thicker than in areas west of the Maam Valley Fault Zone where it is still present but either thinner or truncated by contacts of younger age (McKerrow and Campbell 1960; Piper 1972). This sequence goes from terrestrial, in places with a basal conglomerate or breccia, through marginal and shallow marine to deep shelf and possibly beyond. The lowermost Lough Mask Formation can be seen to rest unconformably on the Dalradian to the south (Locs. 8.5 and 8.6) and on the Ordovician arc rocks to the north (Loc. 8.2). It also rests unconformably on the enigmatic Derryveeny Formation (Loc. 8.1) which may be of either Late Ordovician or early Silurian age (Graham and Badley 2021). It has been noted by Laird and McKerrow (1970) and Leake (2014) that the Dalradian rocks beneath the unconformity are stained red for several metres at many localities demonstrating that the basal Lough Mask Formation was deposited on a surface that had been exposed to erosion for a considerable time period. The presence of a distinctive and laterally extensive lava (Ardaun Lava Member) in the eastern part of the area and the laterally consistent nature of the Lough Mask Formation suggest that this surface had limited vertical relief. The bulk of the Lough Mask Formation, which reaches up to 200 m in this eastern area, is formed by moderately well sorted feldspathic red sandstones that display both cross-bedding and flat bedding. Thin intervening mudrock horizons are red and show desiccation cracks and red mud clasts are present in many of the sandstones. These sandstones were interpreted as the deposits of bedload dominant rivers by Piper (1972). Piper (ibid. Fig. 5) provides the only substantial palaeocurrent data that suggest south-flowing rivers in this eastern area. The well sorted sediment in the Lough Mask Formation suggests a drainage basin of significant scale that would have extended well beyond the currently outcropping pre-Llandovery geology.

Recent detrital zircon data from Finny and Dooros demonstrate a significant young input that is only marginally older than the interpreted age of the Lough Mask Formation (Riggs et al. press). In thin section there is the common presence of fine-grained volcanic rock fragments that do not resemble anything seen in the older parts of the local stratigraphy. These are thought to be the likely carriers of the younger part of the detrital zircon spectrum which otherwise is consistent with a Laurentian origin.

The Ardaun Lava Member is a fine grained albite trachyte lava (keratophyre in many earlier publications) that reaches 115 m in thickness at Ardaun and on Dooros (Leake 2014). It can be traced northwards to the northern outcrop limits on the west side of Lough Mask but is seen to thin out and disappear westwards. Leake (2014) suggested that the lava was confined to valleys in the sub-Silurian land surface. The lateral extent with little obvious variation suggests that any such valleys must have been broad and of limited relief and also that the lava was of low viscosity. In the field the lava is distinctive but offers little information as in most places it is difficult to distinguish even the feldspar crystals. There are notable exceptions where coarse feldspar-phyric facies exist, but these are not common (Gardiner and Reynolds 1912; Graham 2021 Fig. SE12.6). The lack of internal variation makes it difficult to assess whether there are numerous or a very few flows, but evidence of inter-lava horizons has never been reported. It remains a puzzle as to why there is this one significant volcanic horizon in a succession which otherwise only contains thin tuffs and bentonites.

Piper (1972) demonstrated lateral variation in the thickness of the Lough Mask Formation and suggested a topography that had relief of hundreds of metres. It is difficult to reconcile this interpretation with the evidence from the eastern area that suggests deposition on a relatively flat braid plain.

West of the Maam Valley Fault Zone the Lough Mask Formation rests unconformably on the Dalradian commonly with a basal schist breccia (Locs. 8.5 and 8.6). Above this the feldspathic sandstones appear lithologically similar to the Lough Mask Formation seen in the east. However, their detrital zircon signature is markedly different with a predominance of Ordovician gains and a lack of the younger Silurian component (Riggs et al. in press). Moreover samples from the westernmost exposures (Loc. 8.6) contain some 8% of detrital zircons that have a typical peri-Gondwanan signature. There are no reliable published palaeocurrent data for the Lough Mask Formation west of the Maam Valley Fault Zone where it appears only to reach a few tens of metres in thickness but retains the characteristics of a fluvial succession.

Thus one must question the physical connectivity of the fluvial distribution systems between these two areas. Clearly more detailed work is needed here.

The contact with the overlying Kilbride Formation is gradational. It is defined by a change from feldspathic to more quartzose sandstones that is generally accompanied by a change from red to grey or green-grey colours. The Kilbride Formation represents the first marine strata in the succession and the lower part is characterised by the common presence of vertical trace fossils (*Skolithos*) indicative of deposition in high energy shallow marine environments (Loc. 8.2). Piper (1972) suggested a fourfold subdivision of this formation on the Kilbride peninsula where it reaches 320 m in thickness whilst subsequent work has shown that a threefold subdivision is mappable in this area (Graham and Badley 2021). Whilst there are few fossils in the lowest Finny Coarse Arenite Member, shelly faunas are common in the overlying Finny School Laminated Sandstone Member and the uppermost Drin Siltstone Member (Gardiner and Reynolds 1912; Donovan et al. 1992; Doyle 1994; Harper et al. 1995; Donovan and Harper 2003). Faunas described by Harper et al. (1995) indicate the upper part of the *griestoniensis* biozone of mid Telychian age. The faunas belong to the *Eocelia* and *Clorinda* depth communities of Ziegler et al (1968) that indicate shallow to mid-shelf conditions. A progressive deepening through the Kilbride Formation into the gradationally overlying Tonalee Formation is indicated both by the faunas and by the sedimentology. Sparse faunas from the Tonalee Formation (Doyle et al. 1990, 1991) indicate the deep outer shelf. The age of the Tonalee Formation is based on the stratigraphically overlying graptolite and sparse shelly fauna of the *crenulata* biozone (mid to late Telychian) described by Rickards (1973). The Tonalee Formation, which is up to 75 m thick on the Kilbride peninsula, is characterised by red mudrocks despite its outer shelf environment. Red marine horizons are known from elsewhere in Llandovery strata of similar age (Ziegler and McKerrow 1975) and may reflect a widespread oceanic event that allowed oxidation of these sediments.

The contact with the overlying Lettergesh Formation is gradational but laterally variable and represents a significant change in provenance and in depositional processes. Piper (1967, 1970, 1972) regarded the lowest unit of the Lettergesh Formation as essentially a mudrock unit (Benbeg Member) with thick lenses of conglomerate that he termed the Gowlaun Member. Subsequent workers have retained the name Gowlaun Member for this lowest unit and have concentrated on the conglomeratic element (Locs. 8.5 and 8.6). It is evident on the geological map (Graham et al. 1989) that the base of the Gowlaun Member cuts across the lower formations in the stratigraphy such that it rests directly on the Connemara schists SW of Lough Fee at Altnagaighera. Laird and McKerrow (1970) give a thickness of 370 m here whereas Williams and O'Connor (1987) suggest a maximum thickness of 200 m south of Lough Fee.

There remains uncertainty both concerning the petrography of the clasts and the transport directions in what is essentially a single two dimensional section through the stratigraphy. McKerrow and Campbell (1960) state that the conglomerates are 80% quartzite clasts and 20% vein quartz and red chert, whereas Laird and McKerrow (1970), based on outcrops on the flanks of Lettershanbally recognise red and white quartzite (40–50%), sandstone and siltstone (30%), basic or intermediate volcanics

(15%) with much of the rest being jasper. Williams and O'Connor (1987) report pink and white psammite (55%), sedimentary clasts, mainly red sandstone, (25%) and fine grained basic to intermediate volcanics (20%) whereas Bluck and Leake (1986) note that the commonest clasts are pale pink sandstones rich in K-feldspar and suggest many of the clasts may be recycled from earlier conglomerates. Some of this disparity may reflect real lateral variation but some may also reflect the difficulty of distinguishing truly metamorphic quartzites from well cemented sandstones. Some more careful petrographic work is clearly needed here. The two reports of transport directions are by Piper (1970) based both on imbrication of conglomerate clasts and on cross-bedding in associated sandstones and Williams and O 'Connor (1987) based on clast imbrication in the conglomerates. Both indicate flows from the north-east. Piper (1970) interpreted the conglomerates as filling deep water channels whereas O'Connor and Williams (1999) suggest a shallow water origin.

The nature of the relationships of the Gowlaun Member to the underlying strata has also been the subject of considerable discussion. Whilst all workers have suggested the presence of major channels in the Gowlaun Member, there is some uncertainty regarding the relative importance of erosive and laterally interfingering contacts. A diachronous base with persistent location of channels has been suggested by O'Connor and Williams (1987, 1999), Williams and O'Connor (1987) and Piper (1970). Based on exposures south of Lough Fee interfingering with Lough Mask and Kilbride Formation has been reported (Piper 1970; Williams and O'Connor 1987) and at Lettershanbally Gowlaun conglomerates overlie the Kilbride Fm but are divided in two by 8 m of red mudrock of the Tonalee Formation (Williams and O'Connor 1987; O'Connor and Williams 1999). Sandstones associated with the Gowlaun conglomerates are usually massive or slightly graded. These persistent interfingering contacts in this area just west of the Maam Fault zone are perhaps suggestive of some structural control on their location.

The overlying bulk of the Lettergesh Formation is more consistent in nature although it still poses many intriguing questions. Only the lower 800 m is seen east of the Maam Valley Fault Zone but over 1500 m are reported on the slopes of Benchoona (Laird and McKerrow 1970) where both the base and top of the formation can be seen. The formation is dominated by the deposits of sediment gravity flows, mainly high density turbidity currents (Loc. 8.4). Despite the extensive outcrop there are very few indications of flow directions. Piper (1972) reports flows to the west or south-west based on exposures south of Lough Nafooey and Laird and McKerrow (1970) report flows to the south-east, south-west and south although no palaeocurrent data are presented. Graptolites from the Lettergesh Formation near Clonbur (Rickards and Smyth 1968) indicate a Wenlock age (*murchisoni* and *riccartonensis* biozones) (Sheinwoodian).

Petrographically the Lettergesh Formation is markedly different to anything seen lower in the succession. The commonest clasts are of intermediate to acidic volcanic clasts that can be seen in the coarser hand samples as distinctive red, pink, or purple weathering grains and are striking in thin section. The appearance of the volcanic clasts in these lithic arenites is very fresh and they do not resemble anything known from the local subjacent stratigraphy. These have been described by Williams et al.

(1992) who also note some clasts of volcanic glass and welded tuffs. They also note the presence of granitic clasts, serpentinites and detrital micas. They used published discriminant plots to suggest an arc source. The age and location of this arc source is unknown. Some detrital zircon data would clearly be useful here to investigate whether there is a very young component to the source area. Menuge et al. (1995) used a variety of reasoned assumptions along with Nd isotopic data to suggest that the arc was situated on Mesoproterozoic crust (1440–1600 Ma).

The upper contact of the Lettergesh Formation is gradational into the Glencraff Formation (60 m) seen west of the Maam Valley Fault Zone. This formation is characterised by finer-grained, thin-bedded sandstones and mudrocks that have been interpreted as outer shelf storm deposits (Nealon 1989; Nealon and Williams 1988) such that the shallowing of the environment must commence here or somewhere lower in the Lettergesh Formation. It is not possible to determine the maximum depth that this area reached during Lettergesh times.

The overlying Lough Muck Formation (340 m) indicates further shallowing and contains a shallow water fauna and evidence of bipolar currents (Williams and Nealon 1987) (Loc. 8.7). The faunas belong to the *Eocelia* community and indicate a mid-Wenlock age (Rickards and Smyth 1968). The overlying Salrock Formation (800 + m) (Loc. 8.8) is largely red coloured but has yielded some *Lingula* and poorly preserved gastropods (Laird and McKerrow 1970) indicating a continued marine influence. The depositional environment has been interpreted as a coastal plain but few details have been published. The faunas are not age diagnostic.

Thus using the current geological time scale (Melchin et al. 2020) the whole of the Killary – Joyce country succession may represent only 5 Ma for a post-compactional thickness of more than 2500 m. The pattern of apparently rapid deepening where subsidence exceeds sediment supply followed by shallowing where the reverse is true is seen in several basin types. Whilst this is a typical pattern seen in many foreland basins, there are no indications in the regional geology of the Laurentian margin to suggest this as an explanation. The pattern is also common in many pull-apart basins seen in transcurrent margins and given the indications for oblique collision in Silurian times (Dewey and Strachan 2003; Waldron et al. 2014) this may be a more likely explanation. However, it is difficult to delimit either the basin shape or extent on the basis of what is currently known. There are no indications of typical basin margin facies such that the geographical extent of the original basin cannot be determined. Although subsequent outcrop belts commonly match original basin geometries this cannot be assumed here. Such palaeocurrent data as do exist are predominantly orthogonal to the current strike of the outcrop. There are therefore many unanswered questions concerning this succession. There are some suggestions that there were some important intrabasinal faults such as the Maam Valley Fault Zone (Mike Badley pers. comm.) and the Doon Rock Fault, but delimitation of their importance awaits more detailed investigation.

Two suites of intrusive rocks occur within this succession. There is a concentration of sills and dykes of microgranitic nature between Lough Fee and the coast which are interpreted as being intruded into sediments that were not fully lithified (Mohr 1998). In the eastern part of the outcrop a series of doleritic sills appear to be associated with larger structures such as the Doon Rock Fault (Mohr 1990).

1.2 Croagh Patrick Succession

The rocks now assigned to the Croagh Patrick succession were originally considered to be part of the metamorphic basement (Dalradian) (Griffith 1838). The earliest workers were, not surprisingly, greatly influenced by the strong deformation displayed by these rocks and it was only clear that they were Silurian following discovery of faunas by the early Geological Survey geologists (Kinahan et al. 1876; Kilroe 1907a, b). Although the subject of ongoing research, little of biostratigraphic significance has been discovered since these early workers with the exception of Anderson (1960). A reassessment of the faunal collections by Williams and Harper (1991) has suggested that a Llandovery age is most likely rather than the original designation of Wenlock, but additional data are needed here. The succession rests unconformably on the Clew Bay Complex in the north and on Ordovician rocks of the South Mayo Trough in the south.

A stratigraphy of these rocks was presented by Anderson (1960) and Dewey (1963), the latter based on the eastern part of the outcrop, and subsequently supported and modified by Bickle et al. (1972) from the area around Croagh Patrick. Most subsequent workers have agreed that the detailed subdivisions evident in the area studied by Dewey (1963) are difficult to recognise throughout the whole outcrop. However, there is general agreement that the succession comprises a lower quartzose sandstone unit, in many places with a basal conglomerate, a middle more variable unit that contains significant amounts of mudrock, and an upper sandstone unit that is less pure than the lower one and locally slightly calcareous. This threefold subdivision presented in Graham et al. (1989) is shown in Fig. 3. Kelly and Max (1979) suggested that there was a unit of calcareous mudrocks and marbles present beneath the quartzose conglomerate of the Cregganbaun Formation at Derryheagh that they termed the Derryheagh Formation but subsequent work has shown that this cannot be substantiated.

The high level of deformation (Dewey 1967) and the possibility of structural excision or repetition of parts of the stratigraphy (Bickle et al. 1972) mean that there is the possibility of circular argument when looking at lateral variations in the succession. This probably also explains the large variation in thicknesses that have been reported for each of the formations and those shown in Fig. 3 should be treated as approximate. However, it does seem that the middle unit, the Bouris Formation, shows more sandy intercalations in the north-east and a more consistent muddy aspect in the western coastal sections (Loc. 8.12). The outcrop in the south-west near Cregganbaun also appears to show a somewhat sandier aspect at this level than the western coastal outcrop. However, exposure is very limited in the area west of the Corvock granite and the revised map pattern (Graham 2021) is only

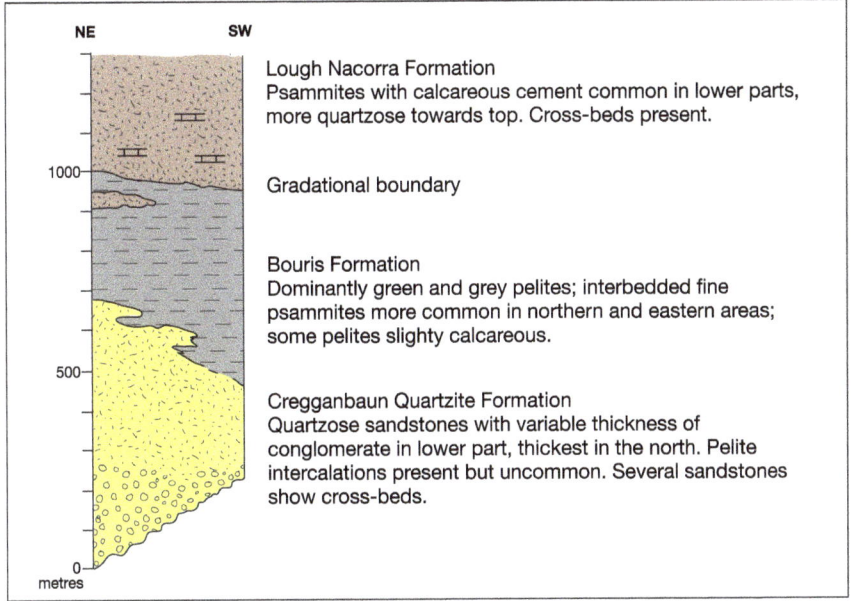

Fig. 3 Outline stratigraphy of the Croagh Patrick succession

one of several permissible representations. The tentative suggestion is that the area was sandier in the east and muddier, and presumably deeper water, in the west. This would be consistent with the few available palaeocurrent data given by Dewey (1963) suggesting flows from the east and north-east. The highly quartzose nature of the succession suggests derivation from the Laurentian margin of North Mayo. The detrital zircon data are dominated by Grenville and Laurentian grains but also contain a small percentage of Tonian grains (Riggs et al. in press). The succession is unusual in the context of South Mayo in that there is no clear evidence for contemporaneous volcanic activity. The Croagh Patrick succession is more strongly deformed than many of the Ordovician strata of the South Mayo Trough, particularly the southern limb of the Mweelrea–Partry syncline. It seems that the deformation is spatially controlled and it is thought most likely to relate to the position of major structures such as the Clew Bay Fault Zone (Highland Boundary Fault correlative). It has recently been shown that this deformation is of Lower Devonian age and the obvious questions relating to this succession concern its age and why there is such a high level of deformation. Work is in progress to delimit how the timing of this deformation which can be linked to the oblique collision history of this part of the Laurentian margin (Graham and Riggs 2021).

1.3 Louisburgh–Clare Island Succession

These rocks have variously been assigned to the Silurian and to the Old Red Sandstone (Bishopp 1950). They were investigated in detail by Phillips (1966, 1974) and Phillips et al. (1970) who established a stratigraphy that could be correlated between Clare Island and the mainland. This stratigraphy is shown in Fig. 4. The succession rests unconformably on rocks of the Clew Bay Complex (Loc. 8.15) on Clare Island and is in faulted contact with the Croagh Patrick succession to the south. The basal unit seen on Clare Island is the Kill Sandstone Formation with a local basal conglomerate dominated by Dalradian clasts. This has been interpreted as a fluvial unit by Maguire and Graham (1996). The overlying Strake Banded Formation is a dominantly red unit of mudrocks and sandstones that contains numerous volcaniclastic and volcanic horizons on Clare Island that pinch out eastwards such that volcanic levels are uncommon on the mainland. The bipolar current pattern noted by Phillips (1974) and interpreted as representing a tidal current regime was shown to be due to two different fluvial distribution systems by Maguire and Graham (1996). The volcanic source lay to the west of the current outcrop. The contact with the overlying Knockmore Sandstone Formation is best seen on the mainland and may represent a

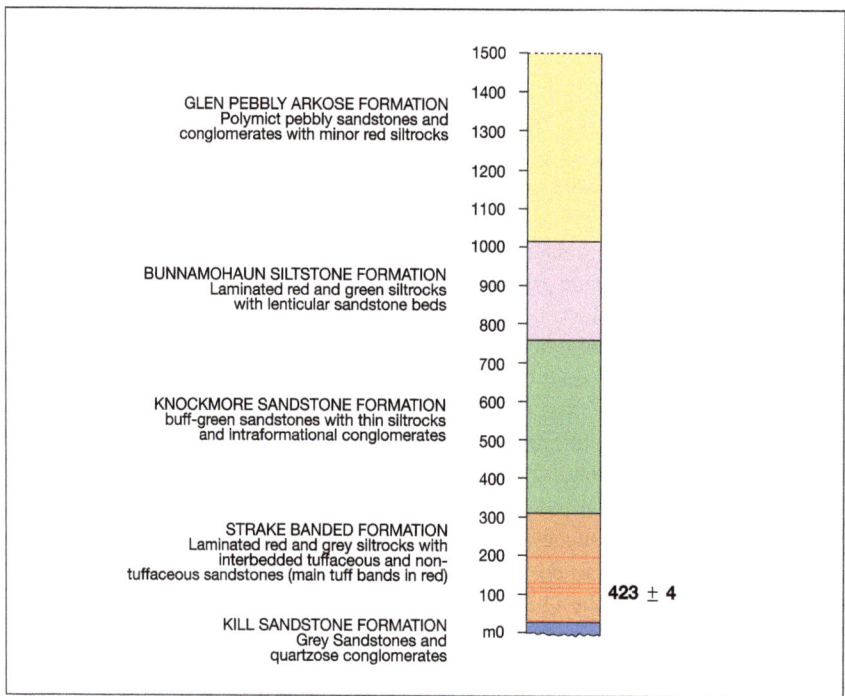

Fig. 4 Outline stratigraphy of the Louisburgh–Clare Island succession

time break within the succession although there is no indication of structural discordance (Loc. 8.13). The sandstones of the Knockmore Sandstone Formation are relatively fine-grained and well sorted implying that the fluvial system that delivered the sediment was of considerable length relative to the present outcrop. The overlying Bunnamohaun Siltstone Formation is interpreted as a flood plain deposit but with no significant difference in provenance.

There is a gradational contact up into the Glen Pebbly Arkose Formation which represents a significant coarsening of the sediment load and possibly a different provenance (Loc. 8.15). Although the petrography has been described to some extent by Phillips (1966; 1974) and Maguire (1989) there is much more information to be gleaned from a study of this formation. It has been suggested that most of the clasts can be matched with parts of the local Clew Bay geology.

The succession was interpreted as a marginal marine succession by Phillips (1974) based on the identification of bipolar current patterns, *Lingula* and some sea pens. The Silurian age was based on regional considerations and also on the presence of derived Silurian fossils in the uppermost Glen Pebbly Arkose Formation. This interpretation was rejected by Maguire and Graham (1996) who showed that the *Lingula* were spiral coprolites produced by fish (Gilmore 1992) and that the 'sea pens' were an enigmatic fossil initially described by Peach et al. (1899) from the Midland Valley of Scotland and termed *Glauconome*. Re-assessment of these fossils by Graham and Evans (2001), who renamed them as *Peltoclados clarus,* has shown that they are of uncertain biological significance but are possibly related to charophytic algae. Unfortunately they provide no biostratigraphical information. Maguire and Graham (1996) reinterpreted the succession as being entirely terrestrial in origin.

A single fossil fish sample, no longer extant, from the Bunnamohaun Siltstone Formation was discovered by Dr D Palmer (Palmer et al. 1989) and identified as *Birkenia* cf. *elegans.* The general uncertainty of age ranges of early fish along with the cf specific assignation means that the suggested Wenlock age cannot be considered accurate. A first reliable age for the lower part of the succession (Strake Banded Formation) was recently presented by Graham et al. (2020) who demonstrated a Ludlow to Pridoli age (423 ± 4 Ma) for a tuff in the Strake Banded Formation. This indicates that there is no age overlap with the Killary Harbour – Joyce Country succession and there is most unlikely to be any overlap in age with the Croagh Patrick succession.

In the case of this basin the palaeocurrents are subparallel to the strike of the outcrop. However, there are no indications of basin margin facies such that the original scale of the depositional basin cannot be determined. The presence of a totally non-marine succession is part of a progression that continued into the Devonian as the Laurentian margin gradually stabilised and elevated during continued oblique collision (chapter "Devonian and Carboniferous").

The location of a series of ultramafic minor intrusions into this succession appears to be spatially and genetically related to major faults (Upton et al. 2001).

2 Killary Harbour–Joyce Country Succession

2.1 Aim of Excursion

The aims of this excursion are to examine the evidence for the proposed deepening—shallowing basin fill; to examine provenance changes throughout the succession and to compare and contrast the successions developed in the eastern and western parts of the outcrop. This evidence can then be used to constrain models of basin development.

2.2 Locality 8.1: Derry Bay North (53.597583°, −9.434674°)

Access and other Information From Tourmakeady, take the R300 south towards Derry Bay, then take the small unnamed road that leads to the shore of Lough Mask at 53.597583°, −9.434674°; the road is gated part-way - make sure to close the gate after passing through. There is some roadside parking just through the gate near a set of obvious exposures to the west of the road. These are typical, poorly sorted pebbly sandstones of the Ordovician Rosroe Formation that are dominated by volcanic clasts.

Proceed to the end of the road where there is a picnic site and parking (Fig. 5). Walk east to the lake shore. The first few outcrops are of the conglomeratic facies of the Derryveeny Formation. Sandstone layers demonstrate a northerly dip (Fig. 6a) and also provide reference horizons for examining clast orientations. Imbrication is visible from clast long axes in cross-section that consistently dip to the east more steeply than bedding indicating palaeoflow from roughly east to west here. Limited three dimensional exposure suggests that the longer dimensions seen in cross section are the b-axes of the clasts.

The Derryveeny Formation contains a large range of clast types (Fig. 6a, b), many of which have strong metamorphic fabrics. There are also granitic clasts without obvious fabrics and undeformed acidic porphyry clasts. Some of the metamorphic clasts show migmatitic textures (Fig. 6b, c) and contain sillimanite. Clasts are very large with average maximum clast size in some coarser layers exceeding 80 cm and some individual clasts reaching 1.2 m (Fig. 6c). This clast assemblage is interpreted as being locally derived from the Connemara massif. The metamorphic rock types are either present or likely to have been present locally in the Kilbride Formation that presently crops out just 2–3 km away on the southern side of the Kilbride Peninsula (Graham et al. 1991). The igneous clasts have been interpreted as parts of the Connemara magmatic arc, from undeformed near-surface volcanic rocks to deeper plutons.

At 53.597478°, −9.428097°, just north of a small stone pen, is a critical exposure that shows red feldspathic sandstone typical of the Silurian Lough Mask Formation resting unconformably on conglomerate of the Derryveeny Formation (Fig. 6d). The strikes of the two units are almost perpendicular. The precise age of the Derryveeny

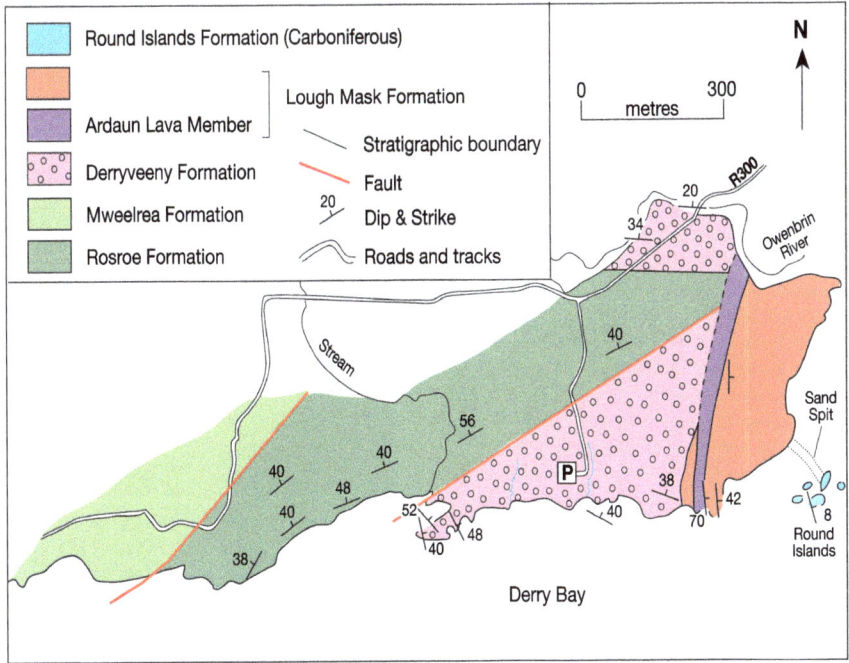

Fig. 5 Geological map of the area north of Derry Bay

Formation is not known, but the unconformity shows that it must be older than the late Telychian Lough Mask Formation, while clasts dated to c. 462 ± 7 Ma (Middle Ordovician) and detrital zircons of c. 450 Ma (Late Ordovician) (Riggs et al. in press) indicate its maximum age. Thus, either a Late Ordovician or early Silurian age is possible.

There is a gap in exposure before the next outcrop to the east, which is of the trachytic Ardaun Lava Member of the Lough Mask Formation. This widespread member is fine-grained and typically weathers with a pink or purple colour (Fig. 6e). Although there is a gap at the eastern end of this exposure, the lava must be at least 10 m thick here, assuming a constant dip in the Silurian strata.

The next exposure to the east is of red feldspathic sandstone of the Lough Mask Formation. The sandstone is cross-bedded (Fig. 6f) with apparent northerly derivation. The strata strike almost north–south, a c. 70° rotation from the western end of the Kilbride Peninsula. Palaeomagnetic studies have suggested that this swing in strike is a regional feature that is interpreted to have occurred in late Silurian times (Smethurst et al. 1994).

Return by the same route to the car park on the shores of Derry Bay.

Fig. 6 **a** Imbricated cobble conglomerate beneath sandstone lens **b** Migmatite clasts in Derryveeny Formation conglomerate **c** Poorly sorted Derryveeny conglomerate with outsize porphyry clast **d** Derryveeny Formation conglomerates, here dipping away from the camera, unconformably overlain by red sandstones of the Lough Mask Formation. The marked difference in strike is evident **e** Field appearance of the Ardaun Lava Member on a glaciated lakeshore exposure **f** Cross-bedded feldspathic sandstones of the Lough Mask Formation (foresets are visible just left of the hammer)

2.3 Locality 8.2: Finny Church (53.567923°, −9.487486°)

Access and other Information Park at Finny Church (53.570798°, −9.492675°) on the R300. Stops at this locality are adjacent to a moderately busy road so take care of traffic and wear high visibility clothing. Most stops involve entering private land through gates which should be carefully shut.

Walk east until the remnants of a small quarry are seen in a reddish-weathering sandstone, which provides fresh samples (53.567923°, −9.487486°). This rock is typical of the majority of the Lough Mask Formation and represents deposition in a fluvial environment. A set of small low-lying crags just to the west of the loose, fresh debris display the rocks in section (Fig. 7a) and cross bedding is visible, along with some thin, muddy partings and reworked red mud clasts. It is clear that this was one of the localities sampled for palaeomagnetic studies. In thin section these sandstones

Fig. 7 a Red, feldspathic sandstones of the Lough Mask Formation **b** Thin section of red Lough Mask sandstone: note the iron enriched volcanic clasts and the iron oxide grain coatings that give rise to the red colour

display numerous fine-grained volcanic rock fragments (Fig. 7b). These volcanic rock fragments are thought to be the carriers of the young (Silurian) component of the detrital zircon profile (Riggs et al. in press).

Return to the road and round a small bend and then cross the road when it is safe to do so. Enter the field on the right through a metal gate and examine the low lying rocks that dip towards the river to the south. These rocks belong to the lower part of the Kilbride Formation and are pale grey rather than the red of the Lough Mask Formation. They represent the first marine strata in this transgressive sequence and are characterised by mainly horizontal or low-angle stratification that is interpreted as being wave generated (Fig. 8a). The beds also contain numerous, closely spaced vertical burrows (Fig. 8b). These have been named *Skolithos* and they represent dwelling burrows of filter feeders that lived in the high energy shallow marine conditions. These types of trace fossils occur in rocks from the Cambrian to the Holocene, but it is by no means certain that they were produced by the same types of organisms through geological time.

Fig. 8 a Thin bedded, grey, quartzose sandstones in the lower part of the Kilbride Formation **b** *Skolithos* bearing grey sandstones seen in profile, lower part of the Kilbride Formation

2.4 Locality 8.3: Kilbride (53.570084°, −9.451363°)

Access and other Information Park at the road junction on the R300 where a minor, unnamed road runs northeast (53,570,084°, −9.451363°). Walk to the northeast along the minor road (Fig. 9).

Greywacke sandstones of the Lettergesh Formation are well exposed in the fields on the western side of the road where they form a series of low ridges. Continue to a large gateway at (53.572893°, −9.445347°) and then follow the track to the left passing through a series of farm buildings, beneath which are small exposures of the Tonalee Formation, and head towards the open hillside where there are visibly quarried areas.

On commencing the upslope track towards the quarries the ditch section on the right (E) side is through the middle and upper parts of the Kilbride Formation and also shows a thin, pale-weathering bentonite horizon (Fig. 10a). Just above the gate through the main boundary wall at 53.5747°, −9.4469° there are extensive and spectacular exposures through the fossiliferous sandstones and siltrocks of the Kilbride Formation (Fig. 10b, c). Some large bedding plane exposures are rich in fossil corals.

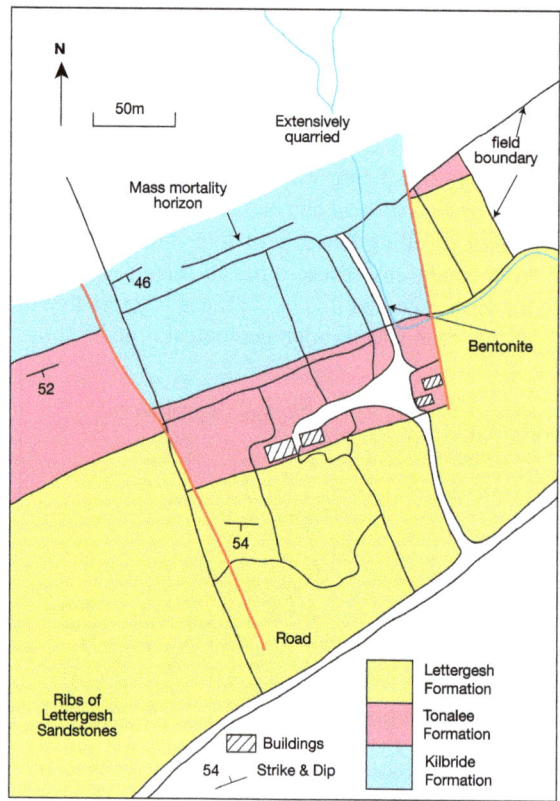

Fig. 9 Map of Kilbride locality

Fig. 10 **a** Bentonite in upper part of the Kilbride Formation **b** Quarried section in the middle part of the Kilbride Formation **c** Siltstones of Kilbride Formation with weathered out coral concentrations forming prominent horizons **d** Mass mortality horizon where corals are buried beneath thin, pale weathering ash layer rich in muscovite flakes

The farthest downhill of these has a thin volcanic ash layer (bentonite) covering it that is likely to have been responsible for mass mortality of the corals (Fig. 10d; Harper et al. 1995).

2.5 Locality 8.4: Drin and Bencorragh (53.564764°, −9.493817°)

Access and other Information Proceed down an unmarked minor road opposite Finny Church. After crossing a bridge over the Finny River continue around a bend to the left, then right for c. 800 m and park in a small layby (Fig. 11) at 53.564764°, −9.493817°.

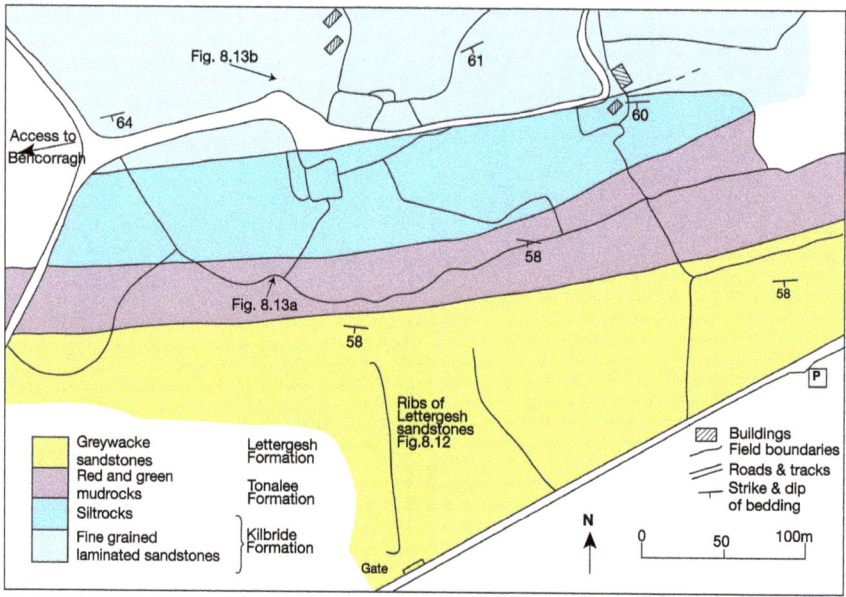

Fig. 11 Simplified geological map of Drin

2.5.1 Drin

Ribs of dipping beds of the Lettergesh Formation can be seen running down the hill from the west. Walk along the road for a further 300 m and access a field on the right via a gate at to examine the predominantly coarse-grained sandstones that are interpreted to have been deposited mainly by high-density turbidity currents. Massive, graded sandstones (Fig. 12a,b) commonly with vertical dewatering pipes, and stratified sandstones with parallel atenation and, less commonly, small scale cross stratification are present, with the massive graded portions consistently underlying the parallel laminated parts where they are part of the same bed (Fig. 12). Some of the coarser beds have erosive bases and/or are loaded into the beds below (Fig. 12c–e).

These greywacke sandstones are dark coloured on fresh surfaces and a variety of grain types are visible with a hand lens. Those with a distinctive reddish or pinkish tinge can be seen to be volcanic rock fragments in thin section (Fig. 12f) and grains of this type comprise a large proportion of the sandstones. They appear to be very fresh, first cycle detritus.

To examine a range of bed types and structures walk to the N-NW across the strike towards the line of a small stream and the edge of the small fields. On approaching the stream walk c.100 m upstream to larger exposures of red and green mudrocks of the Tonalee Formation in and adjacent to the stream (Fig. 13a) (53.565088°, −9.499031°). This formation forms a valuable marker horizon within the Killary-Joyce succession.

Fig. 12 Images from the lower part of the Lettergesh Formation: **a** Graded sandstone **b** Massive graded sandstone with thin layer of parallel lamination at the top and thin mud parting between this and the next sandstone **c** Amalgamated parallel laminated sandstones **d** Erosional scour filled with coarse sandstone showing size grading and parallel lamination in upper part beneath a sharp contact at the base of the next sandstone **e** Loading at the base of the coarse sandstone in the upper part of the image: the upper part of the graded sandstone beneath shows parallel lamination succeeded by small scale cross stratification produced by ripples **f** Thin section image (xpl0) from the Lettergesh sandstones at Drin. Note the dominance of feldspathic rock fragments; the clast marked by an arrow preserves some fiamme

Fig. 13 **a** Typical red Tonalee Formation mudrocks **b** Laminated sandstones from the Finny School Laminated Sandstone Member of the Kilbride Formation; Kilbride peninsula in the background

Where a small track crosses the stream walk eastwards down through the succession that beneath the Tonalee Formation consists firstly of siltrocks with scattered fossil debris and then fine-grained laminated quartzose sandstones that are particularly well exposed next to an old NW trending track (Fig. 13b). Mapping of this area shows that these laminated sandstones of the Finny School Laminated Sandstone Member of the Kilbride Formation directly overlie the coarser-grained *Skolithos* bearing sandstones seen at Loc. 8.2. From here there is a direct walking route along the ridge crest to Bencorragh.

Provenance exercise for the KH-JC succession (Finny Church and Drin)

(1) What information can you determine in the field concerning the provenance of the Lough Mask Formation?
(2) How do the sandstones with the *Skolithos* burrows in the Kilbride Formation differ compositionally from the sandstones of the Lough Mask Formation?
(3) To what do you attribute this difference?
(4) What can you determine in the field concerning the composition of the Lettergesh sandstones?
(5) How do you interpret the provenance changes seen to occur through the succession?
(6) What further information is needed to support your interpretation?

This exercise is best accompanied by examination of thin sections under the microscope. An indication of some of the major differences is given by comparing Figs. 7b and 12f.

2.5.2 Bencorragh

For the ascent of Bencorragh which has spectacular exposures and views allow at least an hour and a half. Continue straight up the indistinct old track as far as the curved boundary wall and cross the fence here (arrow on Fig. 11). Follow the crest of the ridge to the summit of Bencorragh. Exposures near the summit of Bencorragh provide the best examples of pillow lavas seen anywhere in Ireland (Fig. 14). The structure of the pillows seen here and in the crag faces just to the south show that the lavas consistently dip steeply and young to the north (Fig. 14a–c). The pillows are elongate, showing that they were lava tubes and locally have spectacular radial vesicles infilled with quartz and epidote. The pillows here record a greenschist facies assembly, whilst those at higher stratigraphic levels (Loc 6.15) are anchimetamorphic which may reflect burial metamorphism within the volcanic pile (Ryan et al. 1980).

There are fine views from here of the almost flat Maumtrasna Plateau to the north, the trace of the Derry Bay Fault on the north side of Lough Nafooey (Fig. 14d), and Lough Mask and the flat lying Carboniferous limestone lowlands to the east. Scattered exposures of basal Carboniferous conglomerates and sandstones with gentle dips are present on the surface of the plateau. There is considerable debate as to whether the Maumtrasna Plateau represents the result of Variscan faulting or whether it is due to later Tertiary uplift (Dewey 2000; Worthington and Walsh 2011). Return is by the same route.

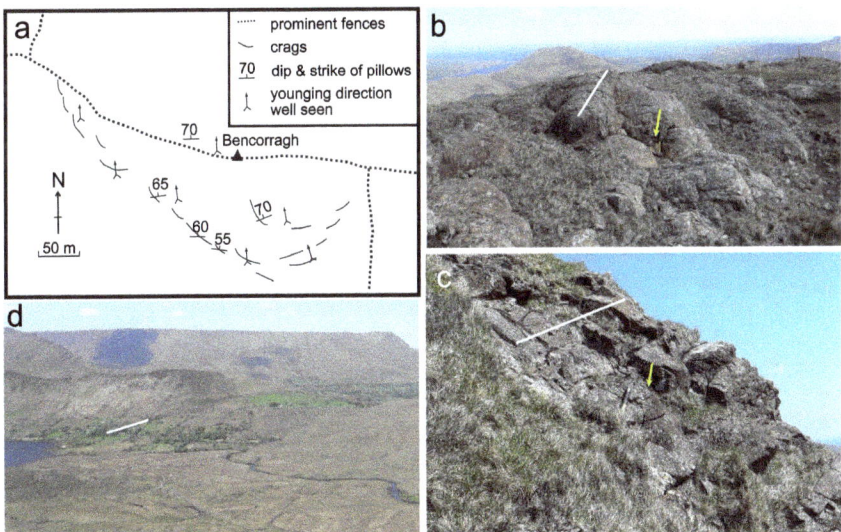

Fig. 14 **a** Sketch map of localities near Bencorragh summit **b** Pillow lavas on the summit of Bencorragh. Beds dip steeply to and young to the left (north) and the yellow line approximates dip and the white line marks strike on (b) and (c) **c** Pillow lavas seen in profile in the crags south of Bencorragh **d** Maumtrasna Plateau seen from Bencorragh. The scar in the middle ground (indicated by white line) marks the course of the Derry Bay Fault

2.6 Locality 8.5: Owenduff Bridge (53.573714°, −9.794439°)

Access and other Information Take the N59 west from Leenane and park at Owenduff bridge on the wide verge on the side of the road (53.573714°, −9.794439°). This is a main road on which traffic moves quickly so take great care and wear high visibility clothing.

This locality allows examination of the lower part of the Killary Harbour – Joyce Country succession and its relationship to the metamorphic rocks of Connemara. Walk west after parking across the bridge for about 40 m. Here, Dalradian psammites with common quartz veins are exposed on either side of the road. Now walk back to the exposures adjacent to the bridge. These are immature breccias, the lowermost being dominated by locally derived Dalradian clasts (Fig. 15a). Above this bedding is readily discernible and locally clasts show good imbrication indicating approximately westerly flow (Fig. 15b). The overabundance of vein quartz and the moderate sorting indicate that these are not simply local scree deposits but that the sediment had undergone a degree of sorting and segregation. These breccias are interpreted as infilling depressions on an irregular eroded surface of Dalradian rocks.

Continue eastwards along the road. Rocks exposed just to the north of the road are well-sorted, grey-green sandstone that contains some shell fragments. These are part of the Kilbride Formation, which, along with the Lough Mask Formation, is relatively thin in this area being together a maximum of a few tens of metres.

Continue towards Leenane to the next prominent crags adjacent to the road (53.574237°, −9.792715°), where there is a series of well-rounded cobble conglomerates and interbedded sandstones. The most common clast types are pale-weathering sandstones and pale porphyries. (Fig. 16a, b). Vein quartz is notably absent or rare and there is a lack of clasts that can be matched to the Dalradian basement. Despite their well-rounded nature, the clasts display imbrication (Fig. 16c) that suggests flow from northeast to southwest. These rocks have been assigned to the Gowlaun Conglomerate Member that forms the basal part of the Lettergesh Formation.

Fig. 15 a Basal breccias of the Lough Mask Formation dominated by Dalradian schist clasts **b** Breccias of the lower part of the Lough Mask Formation: note the bedding, the common presence of vein quartz, and the imbrication of elongate clasts that dip to the right at steeper angles than bedding

Fig. 16 a, b Close-packed texture of the Gowlaun Conglomerate **c** Imbrication of clasts dipping more steeply to the right (northeast) than bedding **d** Large, poorly rounded clast of Dalradian schist (just left of hammer)

Walk northwest across the bog to a large prominent crag at (53.574425°, −9.793559°). At the western end of this exposure, a large (>50 cm) clast of quartz-veined psammite is present that is not rounded and may have been acquired from the local basement (Fig. 16d). The source of the majority of the clasts is not known and they may represent a part of the succession that is not otherwise preserved.

2.7 Locality 8.6: Lough Fee: (53.592468°, −9.831395°)

Access and other information This is a viewing stop of the hillside south of Lough Fee and Lough Muck. There are numerous possible places to pull off at the side of the road, the largest being just after passing a small, wooded peninsula at 53.592468°, −9.831395°.

A series microgranitic intrusive rocks are present in the area from Lough Fee to the Atlantic coast. These have been described by Mohr (1998) who suggested they were intruded into sediments that were only moderately lithified. The thickest of these sills is the Benchoona sill that forms an impressive hill to the south of Lough Fee (Fig. 17) where it reaches a maximum thickness of 330 m. The sill thins markedly against the Letterfrack fault and Mohr (1998) suggests that intrusion overlapped with tilting of the Silurian succession and movement on the Letterfrack fault. An example of one of the thinner sills can be examined at locality 8.8.

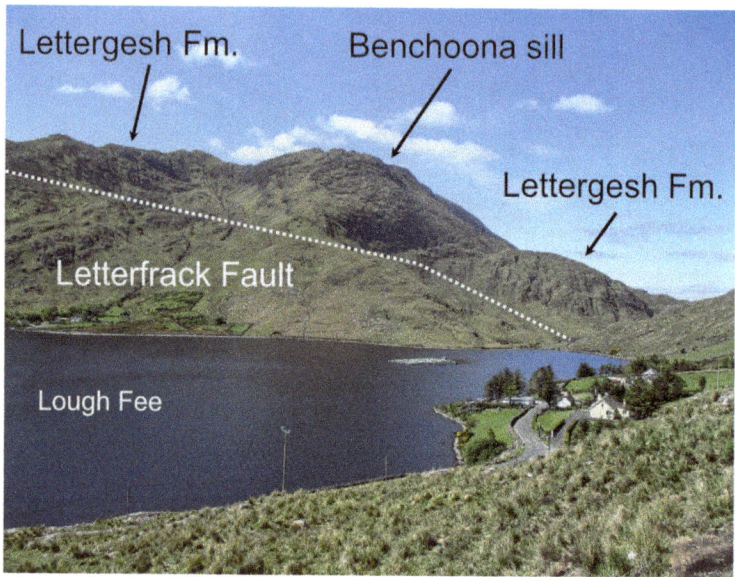

Fig. 17 View of the Benchoona sill and the surrounding Lettergesh Formation sediments at the south-west end of Lough Fee

2.8 Locality 8.7: Gortnaling (53.598628°, −9.912364°)

Access and other Information From Leenane, head west on the N59 and then turn right onto an unnamed road signposted for Tully Cross. Some parking is usually available outside the national school in Gowlaun (53.598628°, −9.912364°). Walk southwest along the road to a stone-faced bungalow with a track just beyond it. Enter the field opposite (53.597450°, −9.917221°) containing two large trees and walk down to the east side of Gortnaling Cove (Fig. 18). The section can be accessed by carefully descending a low grassy cliff at the seaward end of the field; take great care if it is wet.

This locality allows examination of the relation of the basal parts of the Killary Harbour – Joyce Country succession with the metamorphic rocks of Connemara.

Although there is considerable faulting in this area, the contact between the Dalradian and the base of the Silurian can be seen (Fig. 19a). Large blocks of heavily serpentinised ultramafic rocks containing bastite pseudomorphs after orthorhombic pyroxene suggesting a harzburgite origin occur at the contact (Fig. 19a, b) which also contains locally derived Dalradian material. The Dalradian beneath comprises upper greenschist-facies semi-pelitic psammites with thin foliation-parallel quartz sheets. The basal few decimetres of the Silurian sequence are dominated by locally derived Dalradian clasts (Fig. 19b). Above this are bedded breccias, richer in vein-quartz clasts (Fig. 19b, c), similar to those seen at Owenduff Bridge (Loc. 8.5). These breccias are the basal part of the Lough Mask Formation. There are two interpretations

Fig. 18 Geological map of Gortnaling Cove

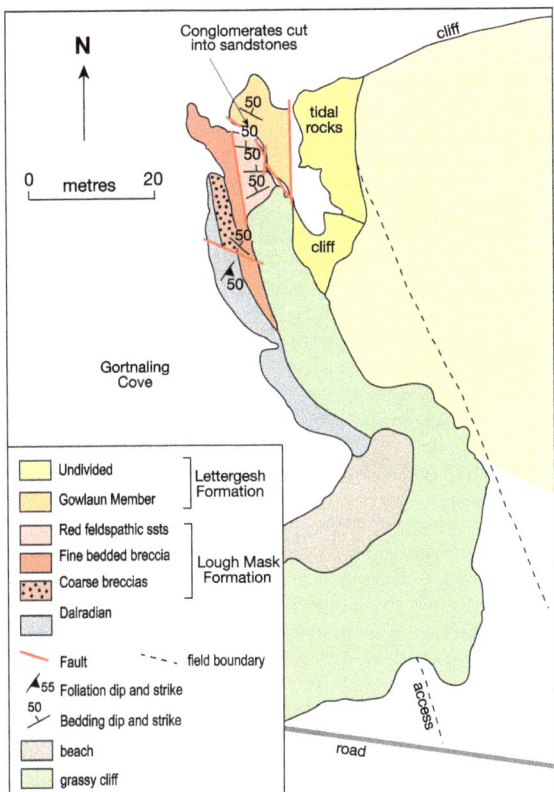

for this contact (see Basal Silurian Exercise below). Firstly, that it is an unconformity modified by subsequent deformation or secondly that this is a weathered low-angle extensional shear zone.

On most of the foreshore, the basal breccias are separated by a fault from a series of well-bedded medium- to coarse-grained feldspathic sandstones that are assigned to the Lough Mask Formation (Fig. 19d) but a conformable stratigraphic contact is visible just beneath the steep grassy cliff. These feldspathic sandstones are separated by further faults from a series of conglomerates that have a very different clast assemblage to the breccias of the basal Lough Mask Formation. The conglomerates have been assigned to the Gowlaun Member of the Lettergesh Formation (Fig. 19e, f). They continue to the edge of a steep inaccessible cliff, beyond which well-bedded sandstones and mudrocks of the Lettergesh Formation are visible in the cliff sections.

It is not clear if the Kilbride Formation has been excised by the numerous faults, was erosively removed in Gowlaun times, or was never deposited here. The faulting also means that the thickness of the Gowlaun Member here is not known; it reaches its maximum thickness on Altnageera, some 3 km inland, estimated at 370 m by Laird and McKerrow (1970), where it appears to have cut through the Lough Mask Formation to the Dalradian basement. The clast assemblage in the Gowlaun Member

Fig. 19 a Unconformity between Dalradian schists (bottom left) and breccias of the Lough Mask Formation. Note the large, brown-weathering blocks of serpentinite at the unconformity surface **b** Close up view of the unconformity looking west; Dalradian is on the left with prominent serpentinite blocks at the unconformity; coarse grained breccias of the lower part of the Lough Mask Formation are to the right **c** Well bedded fine grained breccias in the basal part of the Lough Mask Formation; note the large amount of vein quartz **d** Coarse, feldspathic sandstones of the Lough Mask Formation **e** Coarse conglomerates of the Gowlaun Member cutting into sandstones at side of fault gully shown on Fig. 18 **f** Conglomerates of the Gowlaun Member exposed on the side of the fault gully shown in Fig. 18 Note the angular clasts which are of Dalradian schists

here is very different to that seen near Owenduff Bridge, as clasts are more angular and include numerous Dalradian schist clasts. This could be interpreted as resulting from collapse of submarine canyon walls of schist as the Gowlaun currents flowed through the canyon that was cut into the basement. Unfortunately, there is no control, from what is essentially a two-dimensional exposure, of the overall orientation of this channel and the sediment transport direction from the northeast is only derived from exposures south of Lough Fee (Williams and O'Connor 1987).

The basal breccias at Gortnaling and Owenduff Bridge (loc. 8.5) are part of a carapace of clast-supported breccias that underlie all the Silurian deposits of South North Galway where they rest upon the Dalradian of Connemara (Williams and Rice 1989; Leake 2014). The high concentration of quartz suggests that these breccias may represent a lag deposit on an erosion surface. Williams and Rice (1989) identify sections at the local base of the Silurian, including Gortnaling, where fault rocks become progressively infiltrated with sediment up section. They attribute this to the breccia having developed along a regional low-angle extensional detachment associated with the exhumation of the Connemara Massif. Clift et al. (2004) suggest that this crustal scale extensional detachment entrained mafic and ultramafic blocks whilst it was active during the Mid-Late Ordovician and was then partially eroded prior to Silurian deposition. Leake (2014) offers an alternative explanation and argues that the quartz rich breccias are due to the preferential preservation of the more resistant material and that the deformation fabrics in parts of the breccia are due to subsequent intra-Silurian folding.

Basal Silurian exercise

The nature of the sub-Silurian unconformity is regionally important in assessing the post-Grampian history of Connemara and the origin of the Killary Harbour–Joyce Country depositional basin. The lowest Silurian sediments have been interpreted both as simple sedimentary breccias overlying an irregular but largely peneplaned surface and also as erosional remnants of a major extensional detachment within the Dalradian succession.

This exercise requires examination of both locality 8.5 (Owenduff Bridge) and locality 8.7 (Gortnaling Cove). The assessment of the field evidence can be enhanced by using information from some key published sources (Chew and Van Staal 2014; Clift et al. 2004; Friedrich and Hodges 2016; Geological Survey of Ireland—www.gsi.ie; Leake 2014; Williams and Rice 1989). What additional field data/samples would you collect to further answer questions you have raised in the field?

Owenduff Bridge

(1) What sedimentary structures are observed in the breccias?
(2) What information can be gleaned from these structures?
(3) What is the composition of the breccias?
(4) How do you produce a composition of this nature?

Gortnaling

(1) Describe the nature of the contact between the Dalradian and the Silurian. What type of contact is this?
(2) Why are serpentinites concentrated at or near the contact?
(3) What is the relationship of the serpentinites to the enclosing Dalradian?
(4) How far beneath the contact are the serpentinites found?
(5) What is the nature of the structure in the Dalradian beneath the contact; is there any evidence for high strain zones?

(6) Construct the lower part of the Silurian stratigraphy based on these exposures. Be careful to base the summation on demonstrable unfaulted contacts.
(7) How does this succession compare to that seen in the eastern part of the outcrop (Lough Mask and Lough Corrib areas) and at Owenduff Bridge?

Provenance of the Gowlaun Conglomerate

This exercise can be carried out safely at Owenduff Bridge where there are numerous large exposures to accommodate a student group. If safety conditions can be satisfied the exercise can be repeated at Gortnaling although available exposures are more limited.

(1) Should size and shape of clasts be measured? Why might this be important?
(2) What transport distance is needed to accomplish the level of rounding displayed by the clasts?
(3) How well can a maximum transport distance be constrained?
(4) Devise a sampling strategy for accurately estimating the clast composition of the Gowlaun Conglomerate?
(5) What criteria would you use in the field to distinguish between quartzose sandstones and metamorphic quartzites?
(6) In which direction did the source area lay?
(7) What is the best reconstruction you can make of the composition of the source area?
(8) Does the information from Owenduff Bridge give the same answer as that from Gortnaling?

2.9 Locality 8.8: Lettergesh Beach: (53.602154°, −9.908356°)

Access and other information From locality 8.7 take a short drive north to the signposted beach car park at Lettergesh (53.602154°, −9.908356°). From here walk 110 m right (north) to the first headland.

Sandstones of the Lettergesh Formation are intruded by one of the many microgranitic sills that are common in this area (Fig. 20a). There is some evidence of baking of the Lettergesh sediments, in the form of irregular green-weathering patches in the sandstones beneath the sill (Fig. 20b). There are no published details of the mineralogical changes involved.

The Silurian of North Galway and South Mayo

Fig. 20 a Sandstones of the Lettergesh Formation overlain by microgranitic sill **b** Contact metamorphic effects in Lettergesh Formation sandstones

2.10 Locality 8.9: Salrock and Salrock School (53.602185°, −9.857188°)

Access and other Information Parking is available on the Lough Muck side of the road near the old Salrock School (now a private house) at 53.602185°, −9.857188°. Walk back along the road to the east.

2.10.1 Salrock School

These localities allow examination of the shallowing upward part of the Killary Harbour – Joyce Country succession. Several good roadside exposure here show the essential characters of the Glencraff Formation (Fig. 21). This formation consists of

Fig. 21 Thin bedded sandstones and mudrocks of the Glencraff Formation, Salrock School

c. 60 m of parallel-sided thin sandstones interbedded with grey mudrocks. They are in marked contrast to the thick-bedded turbidites of the Lettergesh Formation below and the more lenticular interbedded sandstones and mudrocks of the Lough Muck Formation above. These sandstones have been interpreted as turbidites by Laird and McKerrow (1970) and as outer shelf storm deposits by Nealon (1989).

2.10.2 Salrock

Return a little to the east and turn left on an unnamed road to Rosroe Pier. Just after this road crosses the crest of a ridge and some large pine trees appear on the right hand side, there is parking for several vehicles on the left side of the road (53.602388°, −9.851000°). Access the open hillside to the left, keeping to the left of a telegraph pole and then keep near to the crest of the ridge.

This exposes both the mudrock-dominant (Fig. 22a) and sandstone-dominant (Fig. 22b) facies of the Lough Muck Formation and also some interesting large-scale structures. These consist of packages of beds that are discordant, often subtly, to those beneath. Examples where the younger beds are asymptotic to a basal surface and also those where the upper beds are steeper and markedly discordant are seen. In many cases the beds just above the discordances are deformed by soft-sediment deformation, but in others they are not (Fig. 23). These structures have variably been interpreted as rotational slump scars (Laird 1968) or drapes of tidal channels (Williams and Nealon 1987). Whichever interpretation is accepted basically defines where the slope to shelf transition occurs in what is generally a shallowing-upward sequence (Fig. 2).

After returning to the parking it is worth a short drive down the road for a panoramic view of the Salrock Thrust where Ordovician rocks of the Rosroe Formation are thrust over the Salrock Formation of the Killary Harbour–Joyce Country succession (Fig. 24).

Fig. 22 **a** Thin bedded facies of the Lough Muck Formation **b** 8.7.3 Sandstone rich facies of the Lough Muck Formation

Fig. 23 a Rotated package of sandstone dominant facies with clear truncation at its upper surface **b** Large gently discordant bed packages **c** Bed package showing asymptotic discordance at its lower surface **d** Large rotated package with curved profile; the beds on the left provide the regional horizontal

Fig. 24 Panoramic view of the Salrock Thrust. Killary Harbour is hidden from view but the gentle dips of the South Mayo Syncline are evident in Mweelrea to the north

2.11 Locality 8.10: Glassilaun (53.614860°, −9.874965°)

Access and other Information From Leenane take the N59 and then turn right onto the unnamed road towards Tully Cross. About 2 km beyond the western end of Lough Muck take the small, signposted road to Glassilaun Beach, where there is abundant parking. There are good sections at either end of the beach and in the small group of rocks in the central part approximately half tide or lower is needed to access and examine them.

This beach has excellent exposures of the Salrock Formation, the uppermost part of the Killary Harbour–Joyce Country succession. The Salrock Formation is dominated by red and purple mudrocks (Fig. 25a, b) and interbedded sandstones (Fig. 25b, d) but it also contains numerous green mudrocks, especially where the sandstones are thicker. This colour variation is thought to be secondary as it does not always follow bedding (Fig. 25c). It is probable that the whole section was initially red and that reduction of the iron-bearing minerals occurred during later fluid movement, which would have been greatest through the more permeable thicker sandstone horizons. The occasional presence of *Lingula* and gastropods without any truly open marine fossils suggests that these rocks were deposited in marginal marine mudflats.

The formation is intruded by several microgranite sills, a good example being visible in the rocks near the centre of the beach (Fig. 26).

An excellent example of antitaxial, extensional veins occurs in the first, low, smooth outcrops at the eastern end (53.615701°, −9.876259°) of the fine sandy beach,

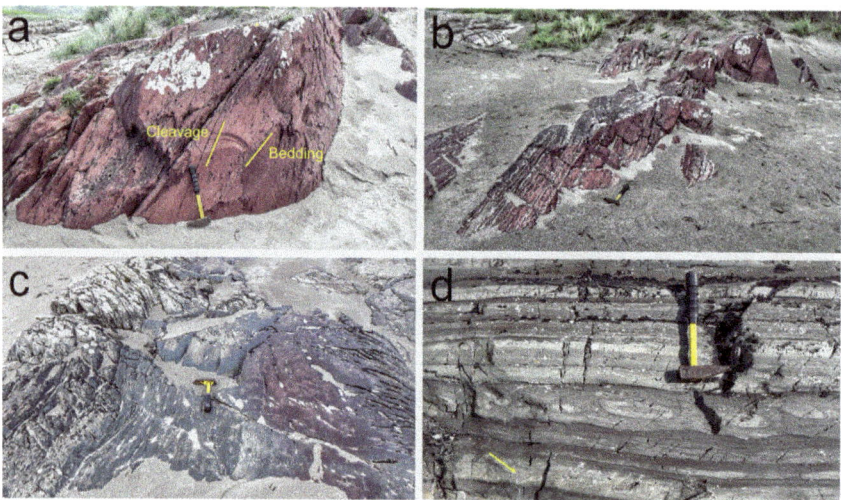

Fig. 25 **a** Homogenous red mudrocks of the Salrock Formation **b** Homogenous red mudrocks of the Salrock Formation interbedded with paler coloured thin bedded sandstones **c** Thick green sandstone (left) with homogenous red mudrocks to the right: the mudrocks on which the hammer rests have undergone secondary reduction **d** Thin bedded red mudrocks interbedded with thin non-red sandstones a thin horizon of soft sediment deformation is just below the hammer head

Fig. 26 Microgranite sill (hammer rests on this) cutting thin bedded mudrocks and sandstones of the Salrock Formation

125 m northwest of the carpark, where hard, grey-green, fine-grained sandstones grade into interbedded red–purple mudrocks of the Salrock Formation (Fig. 27a, d). A striking feature of the vein-fill mineralogy is that it is related systematically to the mineralogy of the alternating bands of quartz-rich sandstones and red–purple mudrocks, quartz in the sandstone veins and chlorite in the mudrock veins (Fig. 27a); veins are antitaxial. This indicates that the vein-fill "fed" on the chemistry of the immediate vein wall, an example of the "substrate effect" (Lander and Laubach 2015) and that the veins grew incrementally with no through-going gap, no consequent

Fig. 27 **a** Vein whose fill mineralogy is related to that of the bedrock, quartz in the sandstone and chlorite in the mudstone **b** Vein with antitaxial epidote rims and vuggy quartz later stage fill

fluid transport, and in the solid state. It is probable that these veins developed during regional transtension (see Loc 8.14).

Veins in the calcareous greywackes of the underlying Lettergesh Formation, are particularly well seen in a small quarry with a mobile telephone mast on the west side of the Leenane-Clifden road 200 yards south of the Lettergesh turnoff (53.579563°, −9.786943°). These show epidote rims and centre fills of quartz, antitaxial in the epidote stage and vug-fill in the later quartz stage (Fig. 27b).

3 Croagh Patrick Succession

3.1 Aim of Excursion

The aim of this excursion is to examine the stratigraphy of the strongly deformed Croagh Patrick succession and to show the relationship to the underlying Ordovician and to some of the major faults in the area.

3.2 Locality 8.11: Derryheagh and Cregganbaun (53.695560°, −9.799678°)

Access and other Information Take the R335 south from Louisburgh as far as (53.695560°, −9.799678°) where parking is available (Fig. 28). From the parking spot walk north along the road and enter via the gate on the right just before the rendered stone wall.

Exposures to the west of the large barn are of the Cregganbaun Conglomerate (Fig. 29a, b) which forms the basal unit of the Croagh Patrick succession. This contains well-rounded clasts up to 35 cm across of pink and pale quartzite and psammite. Rare interbedded sandstones (Fig. 29c) define bedding, which strikes about 080 and dips 42° north. This is discordant to bedding in the Sheeffry Formation to the east. There is an obvious cleavage in the sandier layers and the near-vertical alignment of clasts in parts of the conglomerate is tectonic rather than a primary sedimentary fabric. Exposures to the north are of compact quartzose sandstone, the texture of which is in part due to baking by the Corvock granite. These rocks are part of the Cregganbaun Quartzite Formation although they lack the very pale, almost white appearance seen elsewhere, e.g. Kilgeever (Loc. 8.12).

Return to the parking and drive northwards towards Cregganbaun School. Just beyond the road junction to Killeen there is a large parking area on the west side of the road (53.703535°, −9.808185°). In this area, fine grained, pale-weathering sandstones and siltrocks that are part of the Bouris Formation locally contain shell debris that contains some identifiable coral and brachiopods fragments, mainly preserved

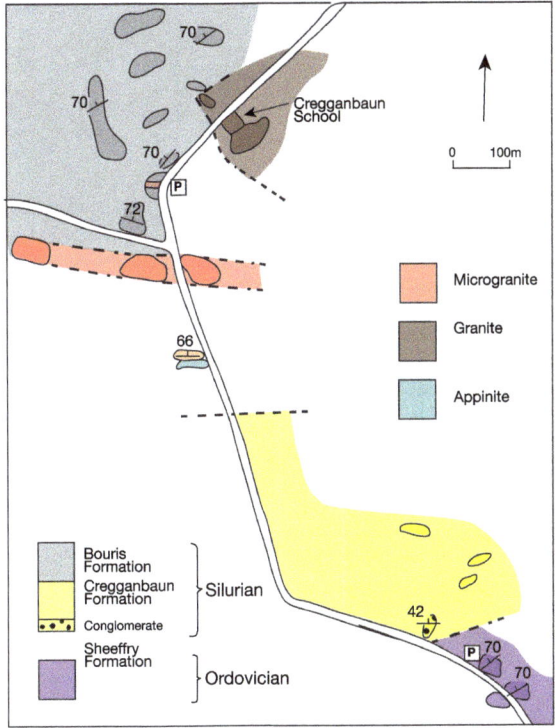

Fig. 28 Simplified map of the Derryheagh/Cregganbaun area

as moulds. A small felsitic intrusion is also present. From here, walk towards Cregganbaun School and, just before reaching it, enter the open moorland to the south of the school. There are abundant exposures of the Corvock Granite here (Fig. 28), which is even-grained and quartz-rich and has caused widespread thermal alteration in the area. The granite has been dated at 410 ± 4 Ma (Graham and Riggs 2021) (Lower Devonian) and at 387 ± 12 Ma, approximately Middle Devonian but with large error (O'Connor 1989). It clearly post-dates the Silurian rocks and has caused widespread thermal alteration in the area.

Just opposite the school (53.704526°, −9.806506°) a small trackway provides access to a large area of open moorland. Some granite is exposed near the entrance to the moorland. Walk westwards across numerous exposures of Bouris Formation until reaching a marked break in slope and then south across strike to see the typical features of the Bouris Formation (Fig. 29d–f). The formation is somewhat sandier here than in the exposures to be visited at Emlagh (Locality 8.14).

Fig. 29 a Cregganbaun Conglomerate at Derryheagh **b** Cregganbaun Conglomerate at Derryheagh with closely packed clasts of quartzite **c** Isolated cobbles in strongly deformed sandy matrix, Cregganbaun Conglomerate, Derryheagh **d** Fine grained sandstones in Bouris Formation **e** Typical muddy Bouris Formation **f** Thin bedded sandstones in muddy Bouris Formation bedding is subvertical and cleavage dips more gently to the left

3.3 Locality 8.12: Kilgeever (53.765071°, −9.765410°)

Access and other Information Entering Louisburgh on the R335 from Westport, turn left onto Chapel Street at the crossroads in the centre of the village. Go straight for about 2 km and then fork right at a brown sign for Clew Bay Archaeological Site 10 and Kilgeever cemetery. There is abundant parking opposite the cemetery

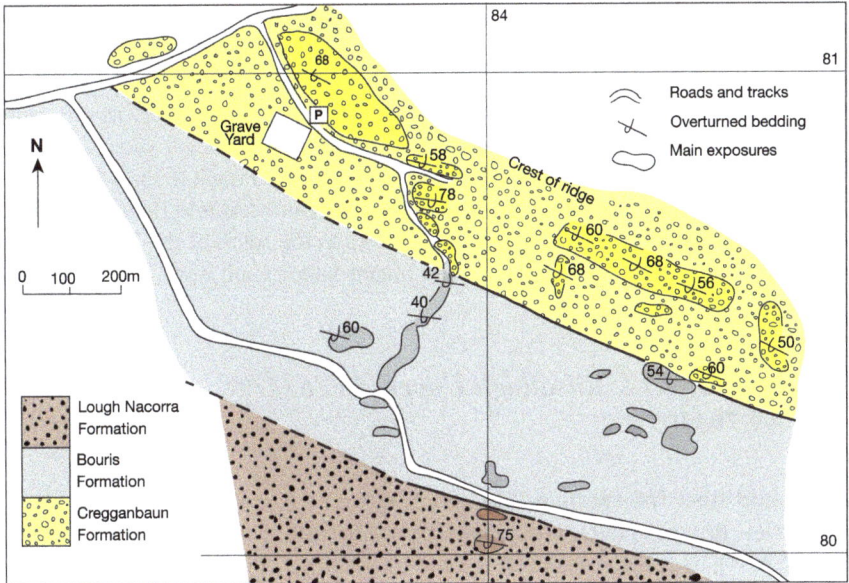

Fig. 30 Simplified geological map of Kilgeever area

(53.765071°, −9.765410°), where there is a large quarry exposure (Fig. 30. The ridge can be accessed via a small steep path opposite the eastern end of the cemetery.

The rock is quartzite that is strongly deformed such that it is difficult to determine original sedimentary structures. A few preserved cross-beds indicate that younging is to the south and that the bedding is overturned. The purest, white-weathering quartzites are seen near the top of the ridge (Fig. 31a). In general, thicker, more quartzose beds dominate the northern part of the exposure and to the south, younger beds are thinner.

Fig. 31 a Pure white quartzites on ridge of Kilgeever Hill b Flaggy quartzites, psammites and pelites in the Cregganbaun Quartzite Formation

From the main car park, continue east along the track to a second quarry on its north side. The upper part of the quarry face is mainly quartzite, with increasing admixture of finer-grained pelitic layers towards the base (younger beds) (Figs. 31b). The pelitic layers have a soapy feel but do not have any megascopic new mineral growth. This unit lies within the Cregganbaun Quartzite Formation as more quartzites are present to the south. A cross-section exposed in an old track to the south shows that the Cregganbaun Quartzite passes transitionally upward into the finer grained Bouris Formation (Fig. 30). There are fine views to the south of the Sheeffry hills, the Doo Lough gap through which run the Maam Valley Fault Zone, and Mweelrea.

3.4 Locality 8.13: Kilsallagh Upper (53.763179°, −9.705105°)

Access and other Information Take the L5880 from Lecanvey and after c. 1 km take the next left, driving uphill to Kilsallagh Upper. Park in the large car park (53.763179°, −9.705105°), or at the roadside at the end of the ridge to the east, which marks an alternative access route to Croagh Patrick (Fig. 32). The ridge is formed by the

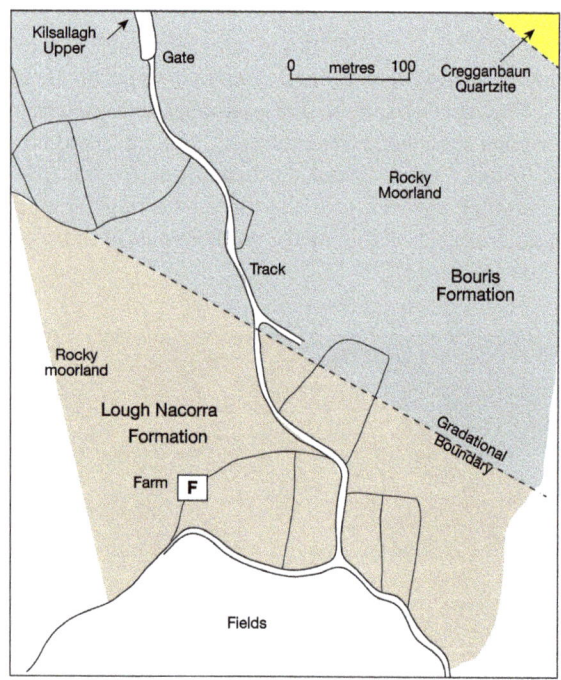

Fig. 32 Simplified geological map of Kilsallagh Upper

Cregganbaun Formation, which provides the pale debris for the extensive screes.

The purpose of this stop is to examine some easily accessible sections in the Lough Nacorra Formation as well as enjoying some spectacular views of South Mayo geology and scenery. Walk southwards from the car park past a few houses and then through gates and a farmyard, keeping to the metalled track. Be sure to shut the gates. Just through the farmyard, outcrops near a small stone shed on the east side of the track are of the mainly pelitic Bouris Formation. The dip is steeply northwards but the rocks are overturned and young to the south (Fig. 33a).

Continue up the track through further gates, as far as the ridge crest. Extensive exposures demonstrate the gradational passage from the Bouris Formation to the Lough Nacorra Formation. There is a consistent overall dip to the north in rocks that are overturned on the northern limb of the Croagh Patrick Syncline. The rocks are strongly deformed and contain numerous minor parasitic and kink folds. However, primary way-up evidence such as truncated small-scale cross-bedding can be found in the coarser horizons (Fig. 33b). On the southern side of the ridge, towards an isolated farmhouse, the succession is dominated by well sorted, fine- to medium-grained psammites (Fig. 33c) many of which are dominated by parallel lamination.

Fig. 33 a Bouris Formation exposure looking east, the bedding can be seen dipping moderately to the left (north) but is overturned and younging is to the south **b** truncated cross-beds giving primary way-up evidence, Lough Nacorra Formation **c** typical psammitic Lough Nacorra Formation dominated by flat lamination **d** Silurian stratigraphy exposed on the southern side of the Croagh Patrick ridge

3.5 Locality 8.14: Emlagh Point (53.744148°, −9.897374°)

Access and other Information From Louisburgh, take the R355 south heading for Killeen and the Lost Valley. After about 5 km turn right (signposted Doogmakeon) at 53.739195°, −9.851912° and continue to the end of the road where there is plentiful parking at 53.744148°, −9.897374°. Continue down track to beach. A low tide is necessary to study these outcrops. This locality affords a traverse from the polyphase, strongly sheared, Bouris Formation with subvertical foliation, through the north-dipping Emlagh Fault, to the non-metamorphosed sediments of the Louisburgh-Clare Island succession (Fig. 4). The Emlagh Fault is sinistral normal, so the Bouris Formation is in the footwall and the Louisburgh-Clare Island succession forms the hanging wall.

The footwall rocks are superbly exposed from Gortnagarryan Strand to the tip of Emlagh Point first in distributed outcrops then on a continuous wave-cut platform. Bedding is transposed into parallelism with the sub-vertical foliation to form a shear banding with stripes of pelite and fine psammite. On the headland just opposite the parking two small, irregular, tonalite intrusions and several aplite veins are affected by the foliation (53.744204°, −9.899294°); the Corvock granite 10 km to the east is the likely source of these intrusions. The banding is the first-phase foliation in the footwall. It is almost penetrative, but lenses of less-deformed fine psammite are preserved and, in a few places, show current cross-bedding that consistently indicates younging to the north.

On the north side of Emlagh Point, the north-dipping Emlagh Fault is exposed on the foreshore (Figs. 34 and 35). The rocks exposed north of the fault belong to the Strake Banded Formation characterised by red mudrocks and sandstones. In this area there are also three levels at which calcareous laminites occur and these form useful local markers (Maguire 1989). Detrital zircon from sandstones here are mainly rounded purple zircons of probable Grenville age and euhedral gains that indicate that volcanism continued into the Silurian (Dewey and Mange 1999), which has been confirmed by Riggs et al. (in press). We suggest that the best way to follow the geology here is to work westwards along the fore-shore outcrops and wave-cut platform in the Bouris formation, turn north to the north side of Emlagh Point, then eastwards to the outcrop of the Emlagh Fault and its hanging wall (Figs. 34 and 35a).

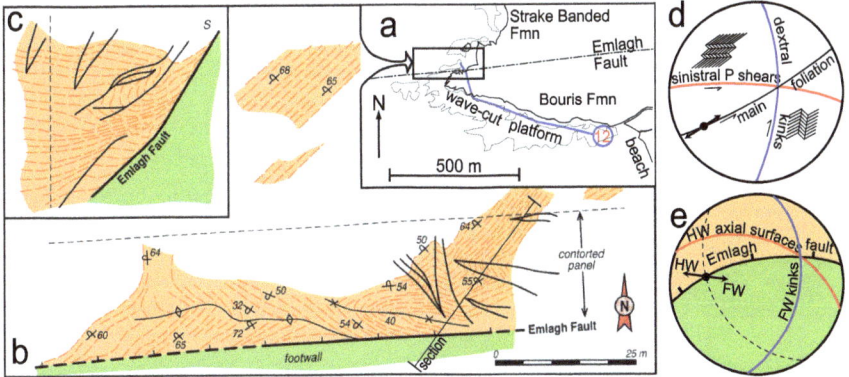

Fig. 34 a Location map for locality 8.14 showing traverse (blue line) **b** Structural map of the Strake Banded Formation in the hanging wall of the Emlagh Fault **c** NE–SW section through the Emlagh Fault. The dashed line shows the edge of the severely contorted panel. **d** main fabric elements of the Bouris Formation (footwall) **e** Main fabric elements of the Strake Banded Formation (hanging wall)

The Emlagh Point outcrops constitute a fine example of the effects of late Silurian–early Devonian sinistral transtension superimposed on a transpressional then a strike-slip milieu. Structures throughout the South Mayo Trough show this sinistral structural sequence of transpression with cleavage and early, gently-plunging open to tight folds at the meso-scale, evolving through strike-slip east–west shear zones to transtension with structures ranging from large steeply-plunging open folds to a variety of outcrop scale minor structures. We suggest that this sequence is a result of a changing slip direction across the sinistral collision zone between the Laurentian and Avalonian continents during the late Silurian–early Devonian. In the Bouris Formation which outcrops in the footwall of the Emlagh Fault, the vertical shear foliation is coincident with transposed bedding and an early cleavage, which gives the rock its stripy character. Sheets of laminated quartz with a sub-horizontal fibre lineation are parallel with foliation, and the foliation is folded by steeply-plunging sinistral tight to isoclinal folds (Fig. 35b). Dextral kink bands (Fig. 35c) and sinistral P folds (Fig. 35d) indicate sinistral transtension. The northeast-striking foliation suggests that it lies in the short limb of a transtensional fold typical of the South Mayo Trough. The absence of SC structures, so common in the Deer Park to the east, is typical of transtensional shear zones. SC structures can only form in a thinning transpressional shear zone, whereas kink bands and P structures can form only in a widening transpressional shear zone (Dewey and Ryan, in preparation). This is because shear zones thin in transpression and thicken in transtension. Late dextral kink bands are common through the South Mayo Trough in slates; they formed in the sinistral transtension final phase of regional deformation. They are especially well-developed in the slates of the Glenummera Formation and in the red slates of the Salrock Formation at Glassilaun beach.

Fig. 35 a General view looking east-northeast of the Emlagh Fault whose trace is marked with a dotted line **b** isoclinal folds in laminated quartz sheet, Bouris Formation **c** dextral kink bands and **d** sinistral 'P' shears in the main shear fabric of the Bouris Formation **e** increase in fold intensity in the Strake Banded Formation (left hand side) towards the Emlagh Fault (dashed line) **f** contorted panel (see Fig. 34b) in the Strake Banded Formation within 2 m of the Emlagh Fault

The Emlagh Fault is planar, dips north at about 65°, and is heavily silicified to a very tough rock in a band about ten centimetres wide. We have not seen slickensides along the Fault. The hanging wall Strake Banded Formation is mostly inverted with a super-posed east-plunging antiform and synform whose axial surfaces strike at about 20° clockwise to the Emlagh Fault. Intensity of folding in the hanging wall increases towards the fault (Fig. 35e) until, in an approximately 2 m zone along the Fault, there are very tight refolded folds arranged in phacoids and lozenges (Fig. 35f).

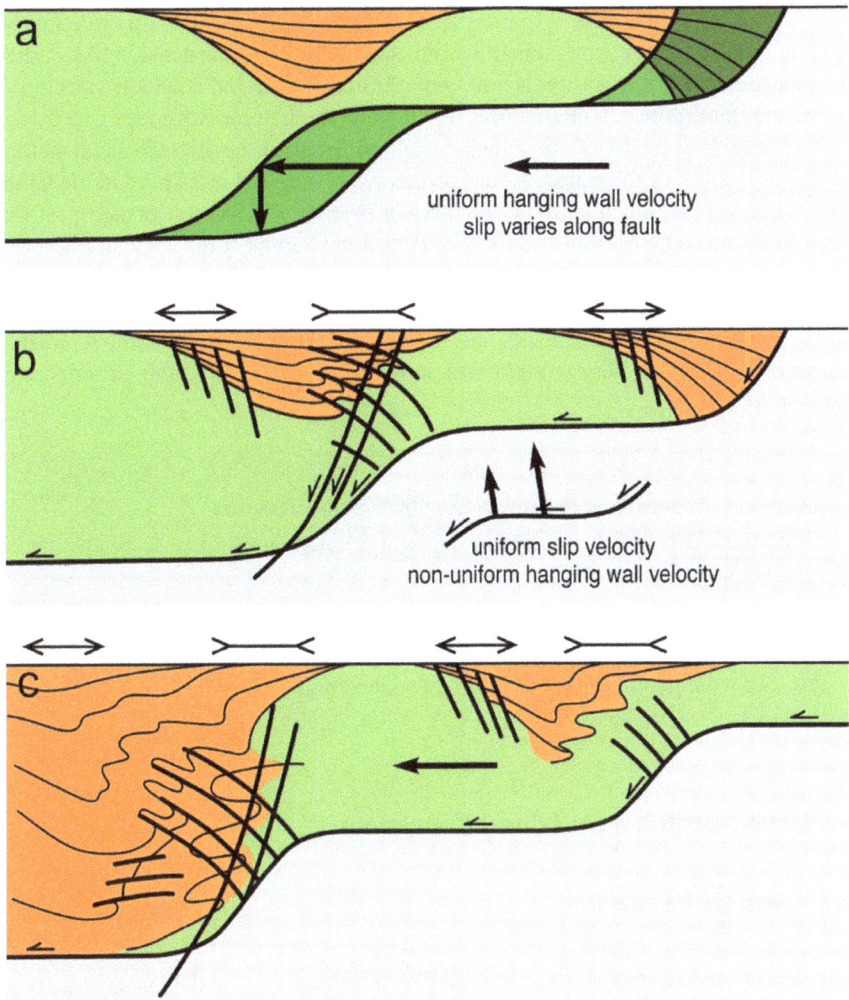

Fig. 36 Schematic diagram showing the structural development of basins in the upper part of the brittle lid above a normal fault with ramps and flats (see text for details)

The kinematics of the Emlagh structure is illustrated in Wulff Net stereo-projections in Fig. 34d, e. In the footwall (Fig. 34d), the sinistral P shears and dextral kink bands form a conjugate pair across the sinistral shear foliation with which they share a steeply-plunging intersection lineation. The slip direction on the shear foliation (black dot), from sinistral isoclines and quartz fibre, plunges gently to the southwest. In Fig. 35e, kink bands in the footwall in a zone some 10 m wide adjacent to the fault dip to the east at a slightly shallower inclination than those farther away from the Fault, and fold axial surfaces in the hanging wall dip at about 45°, are clockwise-transecting with respect to the Fault; kinks, folds, and Fault have a common intersection. This suggests an origin in sinistral transtension with a sinistral-normal slip direction on the Fault and a change from sinistral strike slip to sinistral transtension. An earlier regional sinistral transpression is not seen at Emlagh Point but is widely-developed across the South Mayo Trough where it generated the first cleavage (see Loc 6.11).

The stratal inversion of the hanging wall adjacent to the Emlagh Fault before the superposition of the transtensional structures is not seen elsewhere in the Clare Island-Louisburgh Basin and may be a result of a large-scale non-planarity of the Emlagh Fault. Where a normal fault has flats and ramps (Fig. 36a), asymmetric basins form above blind and surface-intersecting ramps. Because of the downward motion and slower forward velocity of basin fill above a blind ramp, counter-intuitive thrusting and overfolding develop, and normal faults develop at the ramp to normal transition (Fig. 36b). Very complicated superimposed fold and fault patterns may develop (Fig. 36c).

> **Exercise: Structure of the northern Silurian successions**
> Using the information acquired in the field at localities 8.12 to 8.18, locality 6.1 and Fig. 37.
> (1) Draw a series of sketch N-S sections to illustrate the structure of the Croagh Patrick and Louisburgh-Clare Island successions.
> (2) In what geological settings might such structural profiles develop?

4 Louisburgh–Clare Island Succession

4.1 Aim of Excursion

The aim of this excursion is to examine the stratigraphy and paleoenvironments of the Louisburgh–Clare Island succession, its relationship with the underlying rocks of the Clew Bay Complex and its subsequent deformation history. A summary map is shown in Fig. 37.

4.2 Locality 8.15: Shivlagh Rocks (53.777794°, −9.796047°)

Access and other Information Proceeding north-east from the centre of Louisburgh fork left at the Garda Station and then straight (northeast) until the road bends 90° left, then 90° right and descends to a small stream where there is limited parking at (53.777794°, −9.796047°). Just over the stream, a stile and short path access the coast near Shivlagh Rocks (Fig. 38). At least half tide is needed to study this section.

The Silurian of North Galway and South Mayo

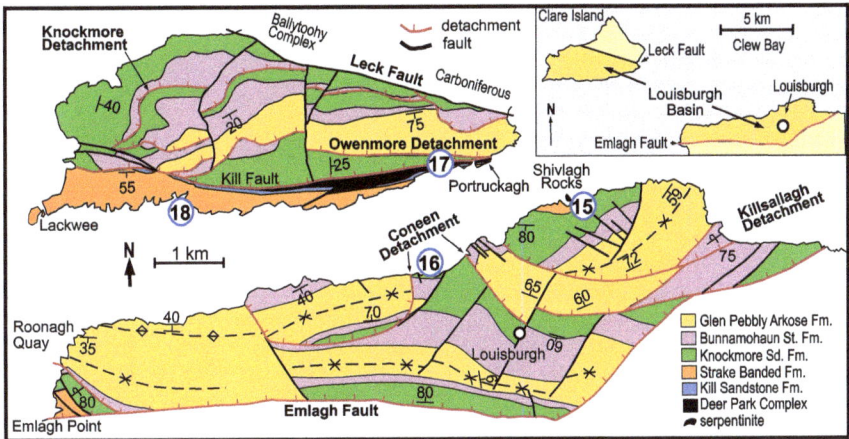

Fig. 37 Summary geological map of the Louisburgh-Clare Island succession with localities 8.15 to 8.18 shown by circled numbers '15' – '18' respectively. Abbreviations used: St. = Siltstone; Sd. = Sandstone; Fm. = Formation

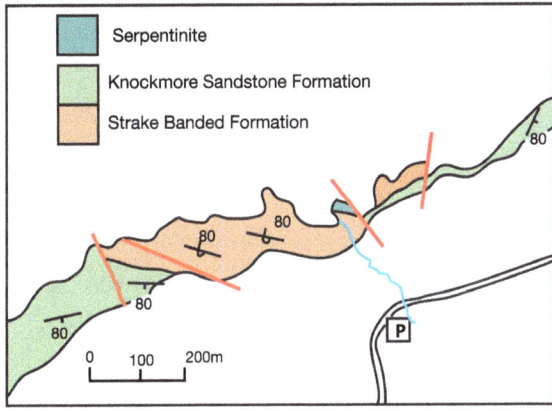

Fig. 38 Geological map of the Shivlagh rocks area

Examine the cliff section on the east side of the bay, which is formed of red mudrocks and thin sandstones of the Strake Banded Formation, the lowest part of the Louisburgh–Clare Island succession seen on the mainland. A prominent yellow-weathering bed in the cliff that is offset by a shallowly north-dipping normal fault is a tuffaceous sandstone (Fig. 39a). This is the only thick bed of this type seen on the mainland, in contrast to the outcrop further west on Clare Island where they are more common. The upstanding ridge of tidal rocks seaward of here is formed by a much-altered serpentinite intrusion that has caused discolouration of the adjacent Strake Banded Formation. The intrusion is cut off by a fault to the east, but this part of the coastline is not easily accessible.

Fig. 39 **a** Pale weathering tuffaceous sandstone within the mainly red mudrocks of the Strake Banded Formation. The white line indicates the trace of the low angle normal fault. The line is broken where the fault is obscured by material in the foreground. Beds are overturned and young to the right **b** Interbedded red mudrocks and grey sandstones, Strake Banded Formation **c** Contact between the red mudrocks of the Strake Banded Formation (left) and the granule rich grey-green sandstones of the Knockmore Sandstone Formation (right) **d** Typical field appearance of the Knockmore Sandstone Formation on the foreshore exposure. The hammer rests on a thin bedded portion with desiccated mudrocks. Only the major bedding planes stand out **e** Flat laminated sandstone on wave polished surface **f** Intraformational mud clasts in flat laminated sandstones on wave polished surface.

Return to the boulder beach and walk westwards through typical Strake Banded Formation. Beds are mainly overturned and dip to the north but young to the south (Fig. 39b). In the first rocky bay, small grey dykes can be seen cross-cutting the layering. These dykes are themselves faulted, very fine-grained and appear to lack chilled margins. These have also been described as being of serpentinite composition (Phillips et al. 1970).

At the next headland there are several small faults, but just beyond it a clear stratigraphic contact can be seen between the Knockmore Sandstone Formation and the Strake Banded Formation (53.778877°, −9.803201°). The base of the Knockmore Sandstone Formation is sharp and erosive, locally rich in quartz pebbles and cross-bedded (Fig. 39c). Although there is no obvious structural discordance, there may be a significant time gap here, as there is a major change in the style of sedimentation. Above the coarse basal bed, the sandstones are finer grained and dominated by flat and low-angle lamination with only occasional cross-beds. On most of the coastal exposure, it is only the major bedding planes that weather out (Fig. 39d) and the sandstones show little apparent internal structure, but fine lamination is visible on the wave polished rocks (Fig. 39e, f). Mudrock horizons are relatively thin and commonly show desiccation cracks, emphasising that the sediment surfaces were commonly exposed. The sandstones are interpreted as fluvial deposits.

4.3 Locality 8.16: Cooneen Harbour (53.770229°, −9.830518°)

Access and other Information Parking is available for a few cars at Cooneen Harbour (53.770229°, −9.830518°). For more or larger vehicles use the beach car park at Carrowmore Strand (53.769422°, −9.824973°) and walk across the beach. Walk down the slipway onto the beach.

The rocks exposed on the foreshore at the western end of the beach (Figs. 40 and 41a) are probably the best exposures of the Bunnamohaun Siltstone Formation on the mainland for demonstrating internal structures.

The formation here is dominated by red mudrocks consisting of deep-red silts thinly interbedded with laminae of coarser red silt to fine sand. These mudrocks are interbedded with packages dominated by fine grained red sandstones. The red mudrocks contain some ripple structures in the coarsest bands, some of which are quite symmetrical and may have been generated by wave action (Fig. 41b). There

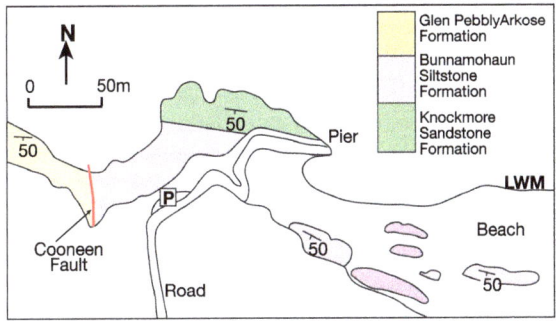

Fig. 40 Sketch map of the Cooneen Harbour coastal section

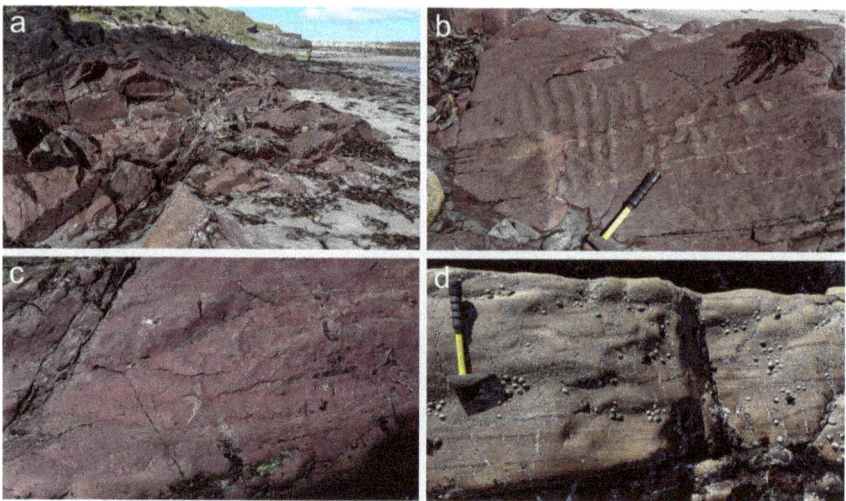

Fig. 41 a Bunnomohaun Siltstone Formation at Cooneen Harbour. The foreground is dominated by red mudrocks with thin sandstones more evident towards the back of the beach **b** Wave ripples exposed on a bedding plane surface **c** Typical bedding scale in red mudrocks cross-cutting sandier segregations are either desiccation crack fills or dewatering structures **d** Knockmore Sandstone Formation at Cooneen Pier with heavy mineral bands

are numerous desiccation-crack fills seen in cross-section as well as a variety of generally more irregular dewatering structures (Fig. 41c). The rocks are interpreted as flood plain deposits that repeatedly dried out; the few wave-generated ripples are interpreted to have formed when there was sufficient water depth after floods. The sandier horizons probably represent temporary influx of the channel systems present on the flood plain.

Either go around the end of the pier if the tide is low or cross the pier onto the rocks on the north side. The rocks are the upper part of the Knockmore Sandstone Formation and the contact with the overlying Bunnamohaun Siltstone Formation here runs mainly under the pier. The sandstone here differs slightly from the formation elsewhere in being brownish rather than green/grey but is similar in other aspects. Some heavy mineral bands are evident in the sandstones (Fig. 41d), which are mainly flat-laminated but locally cross-bedded. Thin grey laminite bands, which are common near the top of this formation elsewhere, are absent here although they are known from just behind Cooneen House some 300 m away. Follow these sandstones around into the cove to the west of the pier, where the contact with the Bunnamohaun Siltstone Formation can be seen.

A marked cleft on the west side of this cove marks the trace of the Cooneen Fault (53.770109°, −9.832105°). The fault dips west at about 60° and younger rocks of the Glen Pebbly Arkose Formation are present in the hanging wall (west side) (Fig. 42). The simplest explanation is that this is a normal fault, although it has previously been referred to as the Cooneen Thrust (Phillips et al. 1970).

The Silurian of North Galway and South Mayo   293

Fig. 42 Gently right dipping trace of the Cooneen Fault; the footwall rocks to the left belong to the Bunnomohaun Siltstone Formation and the hanging wall rocks to the right belong to the younger Glen Pebbly Arkose Formation

4.4 Locality 8.17: Clare Island Southeast (53.795999°, −9.957834°)

Access and other Information From the harbour, walk west towards Kill. Pass three closely spaced houses on the right and then just before the next house on the left enter through a gate (53.799131°, −9.959510°) to follow a track diagonally left upslope, then walk due south, parallel to the field boundary to the headland of Portruckagh

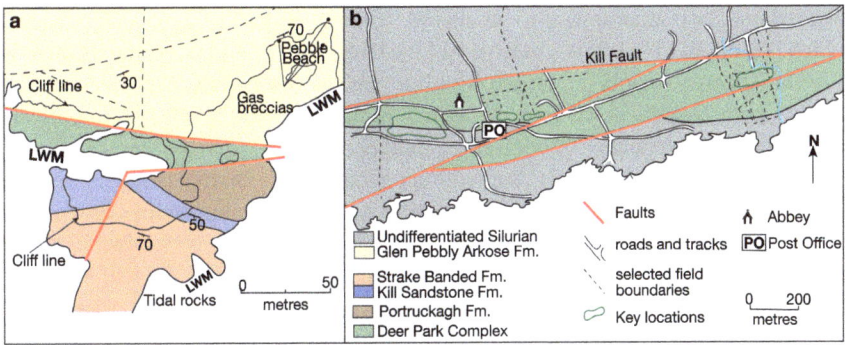

Fig. 43 **a** Detailed map of the Portruckagh locality **b** Simplified map of the area between Portruckagh and Kill showing the key exposures described in the text

Fig. 44 **a** Kill Fault zone, Portruckagh. The hammer rests on reddened Glen Pebbly Arkose Formation; the lower part of the image is of sheared Portruckagh Formation **b** Sheared Portruckagh Formation in Kill Fault zone, Portruckagh **c** Gas breccia in Glen Pebbly Arkose Formation, Portruckagh **d** Strongly discordant gas breccias in Glen Pebbly Arkose Formation, Portruckagh; the white line represents bedding

(Fig. 43a). Access the cove on the east side with fine views over Clew Bay, Croagh Patrick, Sheeffry Hills and Mweelrea.

These localities expose fragments of the Clew Bay and Deer Park complexes (see chapter "The Ordovician Fore-arc and Arc Complex") and the lower part of the Louisburgh-Clare Island succession, within a major fault zone. The north side of the cove has magnificent exposure of the Kill Fault (53.795999°, −9.957834°) (Fig. 44a, b). Broken grey sandstone clasts, some containing quartz veining, in a sheared muddy matrix appear to be altered versions of the Portruckagh Formation, which is thought to be a low-grade part of the Clew Bay Complex. At the base of the cliff this sheared rock is in contact with brecciated Glen Pebbly Arkose Formation that is locally reddened. There is no veining in the Kill Fault Zone, presumably due to the impermeable and easily deformed muddy matrix.

Walk east to the next small cove with a pebbly beach. On this traverse there are numerous examples of brecciated Glen Pebbly Arkose Formation, with many breccia zones strongly discordant to bedding (53.796282°, −9.957078°) (Fig. 43c, d). These breccia zones are not confined to areas immediately adjacent to the Kill Fault, but can be found some distance from it. These unusual rocks have been interpreted as gas breccias (Phillips 1974), with the Kill Fault zone acting as the conduit from depth for the volatiles.

Return to the beach at Portruckagh, where the northernmost exposures are of low grade sandstones and mudrocks of the Portruckagh Formation. To the south

Fig. 45 **a** Veined serpentinite, Portruckagh **b** Dismembered sandstones and mudrocks of Portruckagh Formation, Portruckagh **c** Near vertically dipping Kill Sandstone Formation (left) resting unconformably on Portruckagh Formation (beneath rucksack), Portruckagh **d** Vertically dipping Kill Sandstone Formation (right) passing upwards into red mudrocks of the Stake Banded Formation (left), Portruckagh; beds young to the left (south)

of these is a sliver of strongly veined and altered serpentinite (Fig. 45a), which has a sharp contact to the south with a more extensive outcrop of the Portruckagh Formation. The latter consists of dark-coloured sandstones and grey mudrocks of low metamorphic grade (Fig. 45b) The rocks are cleaved but not schistose and bedding is readily distinguished. Access the headland on the south side of the beach, where more of the Portruckagh Formation is exposed. Finer grained portions show small-scale east-plunging folds and numerous small faults.

On the east-facing headland, a zone of sandstones with extensive quartz veining marks the base of the Kill Sandstone Formation, with the location of the quartz veins presumably determined by the more permeable nature of the strata (Fig. 45c). A thin conglomerate, less than a metre thick, is locally present at this steep, slightly overturned unconformity. In the cliff section to the west, the Kill Sandstone Formation passes gradationally into thin-bedded red mudrocks and sandstones of the Strake Banded Formation (Fig. 45d).

Return across the headland to the narrow isthmus connecting it to the main part of the island. In situ serpentinite can be examined on a thin ridge that juts into the bay to the west. Glen Pebbly Arkose Formation is visible in the cliffs on the north side of this cove and the Kill Fault can be seen in the low tidal rocks although it is not safe to access here. Return to the road by the same route used to get to Portruckagh.

Walk west along the road for 800 m, passing a black barn and then one more house on the right and then look for a large (c. 80 m long) exposure below the road towards the sea at Gurtin (Fig. 43b) (53.794322°, −9.974841°). Access it through a gate at 53.795776°, −9.975029°. The exposure is composed of amphibolite that in places has a strong tectonic fabric produced by mineral alignment. There are also parts that were originally sediments that are now metamorphosed to pelites with some quartzose sandy layers Phillips (1973).

Return to the road and continue west for c. 900 m towards the abbey. Enter the last field on the right before the fork right to the shop and abbey. The lowest part of the exposure is formed of metasedimentary rock with obvious large muscovites, quartz and locally garnets. The upper crags are of amphibolite. Return to the road and enter the next field, through the gate, to see further examples of amphibolite, most showing a clear tectonic fabric but with more massive parts.

Collectively these three localities show rocks that are interpreted to be part of an ophiolite and its associated sedimentary envelope referred to as the Deer Park Complex.

Return to the main road and walk west past the turn to the abbey and a cross-roads and then enter the large field on the right hand side through a gate near its western end (53.792413°, −9.993151°). The small ridge of exposures closest to the road is of veined, quartzose sandstone of the Kill Sandstone Formation. The next ridge of exposures to the north comprises spectacular examples of the Kill Conglomerate, which forms the basal part of the Silurian succession (Fig. 46a). The conglomerate is dominated by well-rounded clasts of quartzite up to 40 cm across, but there are also large clasts of vein quartz and, less commonly, jasper. A few small clasts of amphibolite are also present. The quartzite clasts are interpreted as fragments of the Dalradian metasedimentary rocks that are currently exposed along the northern side of Clew Bay. The next ridge to the north is an excellent example of amphibolite, parts of which have a strong tectonic fabric (Fig. 46b) that is steeply dipping and

Fig. 46 **a** Coarse-grained Kill Conglomerate, west of Abbey, Kill **b** Fine scale striping in amphibolite, Kill

striking roughly east–west, but other parts are less deformed and have more of the appearance of metabasites.

4.5 Locality 8.18: Clare Island Southwest (53.792903°, −10.018950°)

Access and other Information From the harbour, walk west for just over 4 km towards the main boundary fence, which is marked by a large prominent wall across the hillside to the north built as part of the famine relief project and is marked on the road by a cattle grid. Just before this, where a stream crosses the road, enter the fields to the south via a gate (53.792903°, −10.018950°) and walk parallel with the wall on the west side as far as the coast (Fig. 47).

Extensive coastal exposure here is of steeply dipping Strake Banded Formation. There is abundant evidence from truncated cross-strata and graded beds that the rocks consistently young to the south (Graham 2001). The formation consists of three main rock types: purple mudrocks; fine grained, mainly grey or purple sandstones; and pale-weathering, coarser grained sandstones. The pale-weathering sandstones contain abundant tuffaceous fragments and were derived from the reworking of tuffs; some of these have associated ash-fall tuffs that are commonly graded (Fig. 48a). U–Pb zircon dating of one of these tuffs (53.790180°, -10.018412°) has yielded an age of 423 ± 4 Ma (Graham et al. 2020), a late Silurian (Ludfordian to Pridoli) age. Palaeocurrent indicators from these tuffaceous sandstones (Fig. 48b) indicate transport from west to east, suggesting that the volcanic centre lay to the west. In contrast, the grey/purple sandstones show palaeocurrents from the east (Fig. 48c), as do currents in the finer grained sediments.

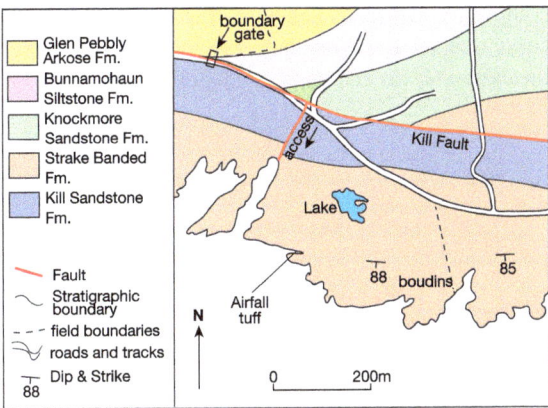

Fig. 47 Sketch geological map of Brockal

Fig. 48 a White weathering, graded, airfall tuff in Strake Banded Formation, Brockal. Beds are near vertical and young to the left **b** Erosive base to pale weathering tuffaceous sandstone, Strake Banded Formation, Brockal **c** Grey weathering, non-tuffaceous sandstone, Strake Banded Formation, Brockal: beds young to the right **d** Prominent, clockwise-transecting cleavage in mudrocks and thin sandstones of the Strake Banded Formation, Brockal, looking west: beds young to the left

The rocks here are probably sandwiched between two major faults, the Kill Fault to the north and one lying offshore to the south. This is the likely reason for the very steep dips. Although folds are not seen, the rocks contain a strong cleavage and the intersection lineation plunges moderately to steeply east. In plan view, cleavage is consistently clockwise of bedding (Fig. 48d). The structures are interpreted as due to compression that was oblique to the main faults with an overall sinistral sense of displacement. There is also strong boudinage of sandstone beds, best seen in the more competent tuffaceous sandstones (Fig. 49a, b). These boudins indicate both along strike and vertical extension, indicating that the strong compression that produced the cleavage was accompanied by extension in the other two dimensions (Fig. 49c).

Fig. 49 a Boudins in sandstones of Strake Banded Formation showing along strike extension b Boudins in sandstones of the Strake Banded Formation showing vertical extension c Boudinage in sandstones of Strake Banded Formation, Brockal. Yellow line indicates the direction of along-strike extension; orange line indicates vertical extension

References

Anderson JC (1960) The Wenlock Series of South Mayo. Geol Mag 97:265–275

Bickle M, Kidd RGW, Nisbet E (1972) The Silurian of the Croagh Patrick range, Co. Mayo. Sci Proc R Dub Soc A4:231–249

Bishopp DW (1950) Some upper and lower limits of the Devonian system in Ireland, with proposed corrections to maps. P Roy Irish Acad B 53:25–32

Bluck BJ, Leake BE (1986) Late Ordovician to Early Silurian amalgamation of the Dalradian and adjacent Ordovician rocks in the British Isles. Geology 14:917–919

Boydell DR (2012) Graptolite biozone correlation charts. Geol Mag 149:124–132

Chew DM, van Staal CR (2014) Examples of large-scale hyperextension during the opening of the Iapetus ocean. Geosci Can 41:165–185. http://dx.doi.org/10.12789/geocanj.2014.41.040

Clift PD, Dewey JF, Draut AE, Chew DM, Mange M, Ryan PD (2004) Rapid tectonic exhumation, detachment faulting and orogenic collapse in the Caledonides of western Ireland. Tectonophysics 384:91–113

Dewey JF (1963) The stratigraphy of the Lower Palaeozoic rocks of central Murrisk, Co., Mayo, Eire, and the evolution of the South Mayo Trough. Q J Geol Soc Lond 119:313–344

Dewey JF (1967) The structural and metamorphic history of the Lower Palaeozoic rocks of central Murrisk, County Mayo, Eire. Q J Geol Soc Lond 123:125–155

Dewey JF (2000) Cenozoic tectonics of western Ireland. P Geologist Assoc 111:291–306. https://doi.org/10.1016/S0016-7878(00)80086-3

Dewey JF, Mange M (1999) Petrography of Ordovician and Silurian sediments in the western Irish Caledonides: tracers of a short-lived Ordovician continent-arc collision orogeny and the evolution of the Laurentian Appalachian-Caledonian margin. Geol Soc Lond Spec Pub 164:55–107

Dewey JF, Strachan R (2003) Changing Silurian-Devonian relative plate motion in the Caledonides: sinistral transpression to sinistral transtension. J Geol Soc Lond 160:219–229

Donovan SK, Harper DAT (2003) Llandovery Crinoidea of the British Isles, including description of a new species from the Kilbride Formation (Telychian) of western Ireland. Geol J 38:85–97

Donovan SK, Doyle EN, Harper DAT (1992) A flexible crinoid from the Llandovery (Silurian) of western Ireland. J Paleontol 66:262–266

Doyle EN (1994) Storm-dominated sedimentation along a rocky transgressive shoreline in the Silurian (Llandovery) of Western Ireland. Geol J 29:193–207

Doyle EN, Harper DAT, Parkes MA (1990) The Tonalee fauna: a deep-water shelly assemblage from the Llandovery rocks of the west of Ireland. Irish J Earth Sci 10:129–143

Doyle EN, Hoey AN, Harper DAT (1991) The rhynconellid brachiopod Eocelia from the Upper Llandovery of Ireland and Scotland. Palaeontology 34:439–454

Friedrich AM, Hodges KV (2016) Geological significance of 40Ar/39Ar mica dates across a mid-crustal continental plate margin, Connemara (Grampian orogeny, Irish Caledonides), and implications for the evolution of lithospheric collisions. Can J Earth Sci 53:1258–1278. http://dx.doi.org/10.1139/cjes-2016-0001

Gardiner CI, Reynolds SH (1912) The Ordovician and Silurian rocks of the Kilbride peninsula (Mayo). Q J Geol Soc Lond 68:75–102

Gilmore B (1992) Scroll coprolites from the Silurian of Ireland and the feeding of early vertebrates. Palaeontology 35:319–333

Graham JR (2021) Geology of South Mayo: a field guide. Geological Survey of Ireland, Dublin, p 158

Graham JR, Leake BE, Ryan PD (1989) The geology of South Mayo. Scottish Academic Press

Graham JR, Wrafter JP, Daly JS, Menuge J (1991) A local source for the Ordovician Derryveeny Formation, western Ireland; implications for the Connemara Dalradian. Geol Soc Lond Spec Pub 57:199–213

Graham JR, Riggs NR, McConnell B (2020) A reliable age for the Louisburgh-Clare Island succession and its significance in the stabilisation of the Laurentian margin in Ireland. Irish J Earth Sci 38:5–14

Graham JR, Badley ME (2021) The geology of the Kilbride Peninsula and adjacent areas County Mayo. Irish J Earth Sci 39 https://doi.org/10.3318/IJES.2021.39.3

Graham JR, Evans DH (2001) Peltoclados Clarus Gen. and Sp. Nov., an enigmatic fossil from Clare Island with a probable Charophyte relationship. In: New Survey of Clare Island. Royal Irish Academy, pp 49–61

Graham JR, Riggs NR (2021) A Lower Devonian age for the Corvock granite and its significance for the structural history of South Mayo and the Laurentian margin of western Ireland. Can. J Earth Sci. doi. https://doi.org/10.1139/cjes-2021-0029.

Graham JR (2001) The Silurian rocks of southern Clare Island. In: New Survey of Clare Island. Royal Irish Academy, pp 35–48

Griffith R (1838) Geological map of Ireland to accompany the report of the Railway Commissioners 1837

Harper D, Scrutton C, Williams D (1995) Mass mortalities on an Irish Silurian seafloor. J Geol Soc Lond 152:917–922

Kelly T, Max M (1979) The geology of the northern part of the Murrisk Trough. P Roy Irish Acad B 79:191–206

Kilroe JR (1907b) The Silurian and metamorphic rocks of Mayo and North Galway. P Roy Irish Acad B 26:129–160

Kilroe JR (1907a) A Description of the Soil-geology of Ireland: Based Upon Geological Survey Maps and Records, with Notes on Climate. HM Stationery Office

Kinahan GH, Symes RG, Wilkinson SB, Nolan J, Leonard H (1876) Explanatory Memoir to accompany Sheets 73 and 74 (in part), 83 and 84. Geological Survey of Ireland

Laird MG (1968) Rotational slumps and slump scars in Silurian rocks, western Ireland. Sedimentology 10:111–120

Laird MG, McKerrow W (1970) The Wenlock sediments of north-west Galway, Ireland. Geol Mag 107:297–317

Lander R, Laubach S (2015) Insights into rates of fracture growth and sealing from a model for quartz cementation in fractured sandstones. Geol Soc Am Bull 127:516–538

Leake BE (2014) A new map and interpretation of the geology of part of Joyces Country, Counties Galway and Mayo. Irish J Earth Sci 32:1–21

Maguire CK, Graham JR (1996) Sedimentation and palaeogeographical significance of the Silurian rocks of the Louisburgh-Clare Island succession, western Ireland. T Roy Soc Edin-Earth 86:123–136

Maguire CK (1989) The sedimentology and stratigraphy of the Silurian succession of Louisburgh and Clare Island. PhD Thesis, University of Dublin

McKerrow W, Campbell C (1960) The stratigraphy and structure of the Lower Palaeozoic rocks of north-west Galway. Sci Proc R Dub Soc A 1:27–51

Melchin M, Sadler PM, Cramer BD (2020) The silurian period. In: Geologic Time Scale 2020. Elsevier, pp 695–732

Menuge J, Williams D, O'Connor P (1995) Silurian turbidites used to reconstruct a volcanic terrain and its Mesoproterozoic basement in the Irish Caledonides. J Geol Soc Lond 152:269–278

Mohr P (1990) Late Caledonian dolerite sills from SW Connacht, Ireland. J Geol Soc Lond 147:1061–1069

Mohr P (1998) Late-Caledonian Syn-Tectonic Magmatism in North-West Connemara: the Binn Chuanna Sill. Irish J Earth Sci 16:1–18

Nealon T, Williams DM (1988) Storm-influenced shelf deposits from the Silurian of Western Ireland: a reinterpretation of deep basin sediments. Geol J 23:311–320

Nealon T (1989) Deep basinal turbidites reinterpreted as distal tempestites: the Silurian Glencraff Formation of north Galway. Irish J Earth Sci 55–65

O'Connor PJ (1989) Chemistry and Rb-Sr age of the Corvock granite, western Ireland. Geol Surv Ireland Bull 4:99–105

O'Connor PD, Williams DM (1999) Tectonic control on sedimentation during transgression: a case study from Silurian successions in Ireland and Scotland. Geol Mag 136:75–82

Palmer D, Johnston JD, Dooley T, Maguire K (1989) The Silurian of Clew Bay, Ireland: part of the Midland Valley of Scotland? J Geol Soc Lond 146:385–388

Peach BN, Horne J, Teall JJH (1899) The Silurian Rocks of Britain: Scotland, by BN Peach and John Horne, with petrological chapters and notes by JJH Teall. HM Stationery Office

Phillips WEA (1973) The pre-Silurian rocks of Clare Island, Co., Mayo, Ireland, and the age of metamorphism of the Dalradian in Ireland. J Geol Soc Lond 129:585–606

Phillips WEA (1974) The stratigraphy, sedimentary environments and palaeogeography of the Silurian strata of Clare Island, Co Mayo, Ireland. J Geol Soc Lond 130:19–41

Phillips WEA, Rickards R, Dewey J (1970) The lower Palaeozoic rocks of the Louisburgh area, Co. Mayo. P Roy Irish Acad B 70B:195–210

Phillips WEA (1966) The geology of Clare Island, .5. Mayo. Unpublished PhD Thesis University of Dublin

Piper DJ (1967) A new interpretation of the Llandovery sequence of north Connemara, Eire. Geol Mag 104:253–267

Piper DJ (1970) A Silurian deep sea fan deposit in western Ireland and its bearing on the nature of turbidity currents. J Geol 78:509–522

Piper DJ (1972) Sedimentary environments and palaeogeography of the late Llandovery and earliest Wenlock of North Connemara, Ireland. J Geol Soc Lond 128:33–51

Rickards R (1973) On some highest Llandovery red beds and graptolite assemblages in Britain and Eire. Geol Mag 110:70–72

Rickards R, Smyth W (1968) The Silurian graptolites of Mayo and Galway. Sci Proc R Dublin Soc A 12:129–134

Riggs NR, McConnell B, Graham JR (2021) Sedimentary provenance of Silurian basins in western Ireland during Iapetus closure. In: Kuiper YD, Murphy JB, Nance RD, Strachan RA, Thompson MD (Eds) New Developments in the Appalachian-Caledonian-Variscan Orogen: Geol Soc Am Spec Paper 554, https://doi.org/10.1130/2021.2554(16) (in press).

Ryan PD, Floyd PA, Archer JB (1980) The stratigraphy and petrochemistry of the Lough Nafooey Group (Tremadocian), western Ireland. Q J Geol Soc Lond 137:443–458

Smethurst MA, MacNiocaill C, Ryan PD (1994) Oroclinal Bending in the Caledonides of western Ireland. J Geol Soc Lond 151:315–328

Upton B, Phillips WEA, Stillman CJ (2001) Late Palaeozoic ultramafic magmatism on Clare Island. In: New survey of Clare Island. pp 63–73

Waldron JW, Schofield DI, Dufrane SA, Floyd JD, Crowley QG, Simonetti A, Dokken RJ, Pothier HD (2014) Ganderia-Laurentia collision in the Caledonides of Great Britain and Ireland. J Geol Soc Lond 171:555–569

Williams DM, Harper D (1991) End-Silurian modifications of Ordovician terranes in western Ireland. J Geol Soc Lond 148:165–171

Williams DM, Nealon T (1987) The significance of large-scale sedimentary structures in the Silurian succession of western Ireland. Geol Mag 124:361–366

Williams DM, O'Connor P (1987) Environment of deposition of conglomerates from the Silurian of North Galway, Ireland. T Roy Soc Edin-Earth 78:129–132

Williams DM, Rice AHN (1989) Low-angle extensional faulting and the emplacement of the Connemara Dalradian, Ireland. Tectonics 8:417–428

Williams DM, O'Connor P, Menuge J (1992) Silurian turbidite provenance and the closure of Iapetus. J Geol Soc Lond 149:349–357

Worthington RP, Walsh JJ (2011) Structure of Lower Carboniferous basins of NW Ireland, and its implications for structural inheritance and Cenozoic faulting. J Struct Geol 33:1285–1299

Ziegler AM, McKerrow W (1975) Silurian marine red beds. Am J Sci 275:31–56

Ziegler AM, Cocks LRM, Bambach RK (1968) The composition and structure of Lower Silurian marine communities. Lethaia 1:1–27

The Late Silurian to Upper Devonian Galway Granite Complex (GGC)

Martin Feely, William McCarthy, Alessandra Costanzo, Bernard E. Leake, and Bruce W. D. Yardley

Abstract The Late Silurian to Middle Devonian calc-alkaline Galway Granite Complex (GGC) occupies a key location in the Irish and UK Caledonides. The GGC consists of a suite of Earlier Plutons (~424 Ma) i.e., the Roundstone, Inish, Omey and Letterfrack Plutons and the later Galway Batholith (~410–380 Ma). The latter comprises the Carna and Kilkieran Plutons. Zircon U-Pb and Molybdenite Re-Os chronometry of the component granites indicates that the formation of the GGC reflects four main stages of magma emplacement into the Grampian Connemara Metamorphic Complex and the Lower Ordovician South Connemara Group ranging from ~424 Ma to at least ~380 Ma. The five stages are: stage (1) The emplacement at ~424 Ma of the Omey, Roundstone and Inish Plutons and possibly the Letterfrack Pluton–collectively termed the Earlier Plutons; stage (2) emplacement of the Carna Pluton at ~410 Ma followed by stage (3) emplacement of the Kilkieran Pluton at ~400 Ma and stage (4) later granite intrusions within the Kilkieran Pluton e.g., the Shannapheasteen and Costelloe Murvey Granites, the latter was emplaced at ~380 Ma. A regional composite dolerite–rhyolite diking is also recorded and represents an Upper Devonian event in the GGC. These magmatic stages span the changing tectonic regimes i.e., from regional transpression during emplacement of the Earlier Plutons and the Carna Pluton to regional transtension during emplacement of the Kilkieran Pluton. The ~380 Ma Costelloe Murvey Granite is ~20 Ma younger than all other granite intrusions of the GGC and postdates Caledonian tectonism. This prolonged and episodic history of granite emplacement occurs in tandem with

---

M. Feely (✉) · A. Costanzo
School of Natural Sciences, Earth and Ocean Sciences, National University of Ireland Galway, University Road, Galway, Ireland
e-mail: martin.feely@nuigalway.ie

W. McCarthy
School of Earth and Environmental Sciences, Bute Building, Queen's Terrace, St. Andrews K16 9AL, UK

B. E. Leake
School of Earth and Ocean Sciences, Main Building, Cardiff University, Cardiff, Wales C10 3AT, UK

B. W. D. Yardley
School of Earth & Environment, University of Leeds, Leeds LS2 9JT, UK

magmatic molybdenite and chalcopyrite mineralisation in the GGC and forms an integral part of the granite related molybdenite mineralisation corridor located along the Caledonian orogen of the North Atlantic Massif. Regional scale vein fluorite with base metal sulphides in the GGC is correlated with Triassic-Jurassic hydrothermal vein mineralization throughout Europe and the North Atlantic margins.

1 Introduction

The Late Silurian to Middle Devonian Galway Granite Complex (GGC) comprises a suite of Earlier Plutons i.e., the Roundstone, Inish, Omey and Letterfrack Plutons and the later Galway Batholith consisting of two major plutons i.e., the Carna and Kilkieran Plutons (McCarthy et al. 2015a, b). The granites of the GGC were emplaced into the Connemara Metamorphic Complex to the north, west and east, and into the Lower Ordovician South Connemara Group in the south (Max et al. 1975; Leake and Tanner 1994; Ryan and Dewey 2004 and this volume). The younger Carboniferous limestones (~350 Ma) are in fault contact with the Galway Batholith to the east (Lees and Feely 2016, 2017). Gravity and aeromagnetic data indicate that the batholith extends offshore to the south and southeast, beneath the Carboniferous rocks of the Galway Bay area (Murphy 1952; Fairhead and Walker 1977; Max et al. 1983) beneath Galway Bay (Fig. 1).

The results of U-Pb zircon and Re-Os molybdenite chronometry have shown that the formation of the GGC reflects four major episodes of magma emplacement that extended from ~424 Ma to at least ~380 Ma (Buchwaldt et al. 2001; Feely et al. 2003, 2007, 2010, 2018, 2020; Selby et al. 2004; McCarthy 2013)—see Fig. 2. The four episodes are: (1) emplacement at ~424 Ma of the Omey, Roundstone and Inish Plutons and possibly the undated Letterfrack Pluton; (2) emplacement of the Carna Pluton at ~410 Ma followed by (3) emplacement of the Kilkieran Pluton at ~400 Ma and (4) later granite intrusions within the Kilkieran Pluton e.g., Shannaphaesteen and Costelloe Murvey Granites the latter was emplaced at ~380 Ma. A suite of late Devonian composite dolerite–rhyolite dikes are also recorded in the GGC (Mohr et al. 2018).

2 Tectonic Controls on the Generation and Emplacement of the Galway Granite Complex (GGC)

2.1 Magma Generation

Closure of the Iapetus Ocean triggered magma generation from the mid Silurian to mid Devonian. Magma generation was due to subduction related slab dehydration, slab break-off and crustal thickening during the convergence of Avalonia, Baltica

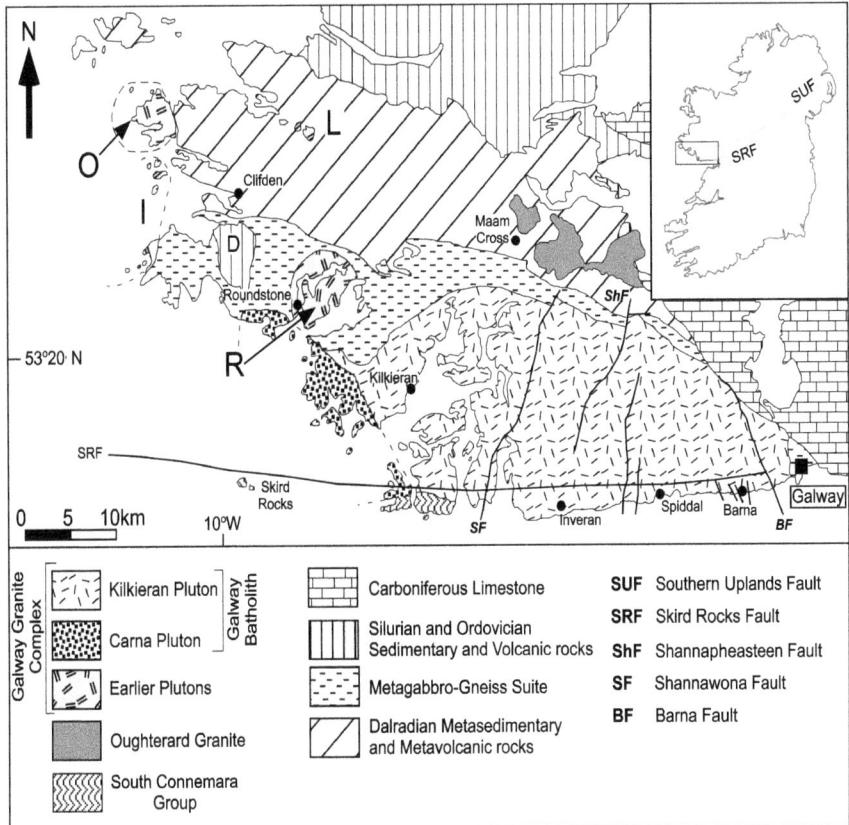

Fig. 1 The geological setting of the Galway Granite Complex. The spatial distribution of the Earlier Plutons i.e., Omey (O), Inish (I), Roundstone (R) and Letterfrack (L) Plutons, is indicated. The later Galway Batholith consists of the Carna and Kilkieran Plutons. The Metagabbro-Gneiss Suite and the Dalradian (metasedimentary and metavolcanic rocks) are collectively termed the Connemara Metamorphic Complex (CMC). The ~463 Ma Oughterard Granite intruded the rocks of the CMC. To the north lie the Silurian and Ordovician sedimentary and volcanic rocks of the South Mayo Trough. The Lower Ordovician South Connemara Group (metasedimentary and metavolcanic rocks) form a roof pendant along a section of the exposed southern margin of the Galway Batholith. The Lower Ordovician Delaney Dome (D) comprised of metarhyolites is located to the west of the Roundstone Pluton (R). To the east the GGC is in faulted contact with Lower Carboniferous limestones (~350 Ma). Map adapted from Townend (1966), Max et al. (1978), Leake and Tanner (1994), Pracht et al. (2004), Leake (2006, 2011) and Lees and Feely (2016, 2017)

and Laurentia (Atherton and Ghani 2002; Neilson et al. 2009; Archibald and Murphy 2021; Archibald et al. 2021) and later decompression due to Devonian transtension (Brown et al. 2008). An abundance of granite plutons, including those that make up the GGC, were emplaced into the Laurentian crust of Ireland and Scotland, during Late Silurian to Middle Devonian times i.e., during the latter part of the Caledonian Orogeny. During the late Silurian (~430–418 Ma), the collision between Baltica

Fig. 2 Schematic showing the temporal formation of the Galway Granite Complex's constituent plutons

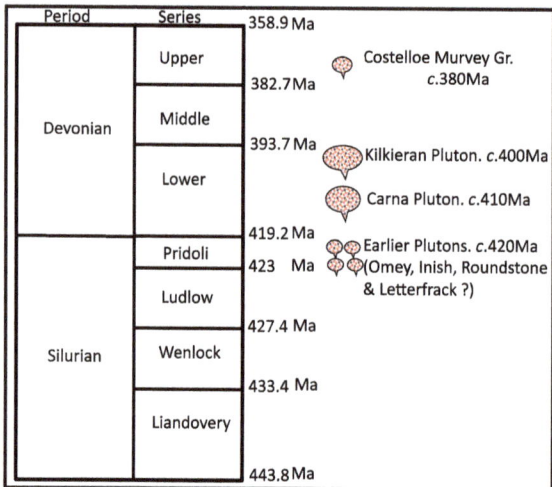

and Laurentia closed the northern part of the Iapetus Ocean initiating the Scandian Orogeny. The final closure of the Iapetus Ocean occurred when Laurentia collided with Avalonia (Chew and Strachan 2014). The generation of the granite magmas that formed the GGC can be accounted for by partial slab breakoff during the collision of eastern Avalonia with Laurentia (Fig. 3). As the slab foundered, the asthenosphere moved upwards advecting heat into the crust causing deep crustal melting that produced the magmas that formed the GGC (Archibald et al. 2021). Therefore, episodic emplacement must have occurred over a time period of ~24 Ma from ~424 Ma to ~400 Ma to account for the generation of the Earlier Plutons and the Galway Batholith (Fig. 4a, b).

The collision of Avalonia with Laurentia underpins the tectonic framework that not only triggered the generation of magma but also the subsequent ascent and emplacement of the Late Silurian to Middle Devonian granites in Ireland and Scotland.

2.2 *Magma Ascent During the Construction of the GGC*

Regional scale sinistral strike-slip faults, that parallel the Iapetan suture (e.g., the SUF and the HBF), were active during closure of the Iapetus Ocean (Fig. 5). There are many examples of Irish and UK late-Caledonian granites, including those of the GGC, whose emplacement was controlled by transtension or transpression (Dewey et al. 1998) acting across this sinistral SW-NE strike-slip regime.

Construction of the GGC was controlled by changes in the regional late-Caledonian transcurrent stress field, the most significant of which was the termination of transpression and the onset of regional transtension (McCarthy 2013). Two key

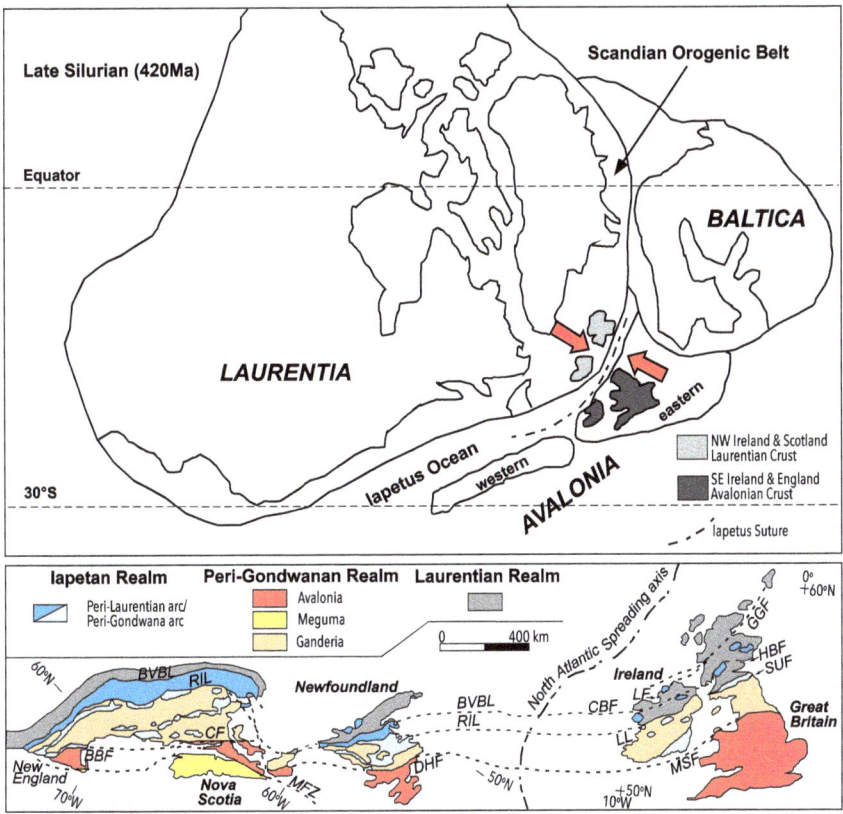

Fig. 3 A summary of the tectonic framework during the final closure of the Iapetus Ocean in the Late Silurian after the convergence of Baltica and Laurentia (Scandian Orogeny) and before Avalonia collides with Laurentian (Acadian Orogeny) to form the Iapetus suture (upper panel, adapted from Chew and Strachan 2014). A palaeogeographic reconstruction of the Appalachian-Caledonian Orogen (lower panel, from Archibald et al. 2021) showing the distribution of major terranes and structures. The last stage of these tectonic collisional events caused the Acadian orogeny (Early Devonian) as Avalonia collided with Laurentia. Abbreviations: BBF = Bloody Bluff Fault; BVBL = Baie Verte–Brompton Line; CF = Caledonian Fault; CBF = Clew Bay Fault; DHF = Dover–Hermitage Bay Fault; GGF = Great Glen Fault; HBF = Highland Boundary Fault; LF = Leannan Fault; LL = Lead Hills Fault; MFZ = Minas Fracture Zone; MSF = Menai Strait Fault; RIL = Red Indian Line; SUF = Southern Uplands Fault

sets of fabrics are identified within the GGC, in the Earlier Plutons (~424–410 Ma) a NNW-SSE subvertical magmatic fabric, whose orientation is consistent with regional transpression is recorded, while in the batholith a WNW-ESE sub-horizontal magmatic fabric whose orientation is consistent with regional transtension is recorded (McCarthy 2015a, b). The schematic diagram (Fig. 6) summarises a tectonic kinematic model for the construction of the GGC that occurred during the Late Caledonian sinistral transpression and later transtension. In this model, the Connemara terrane represents a deformation zone bound by the Highland Boundary Fault (HBF)

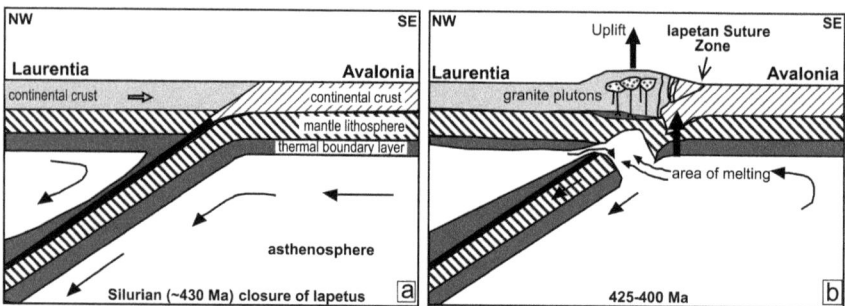

Fig. 4 Schematics showing slab breakoff on collision of Avalonia with Laurentia and the generation of an abundance of granite magmas from Late Silurian to Middle Devonian times including those of the GGC in Connemara. **a** Subduction of Avalonia under Laurentia during final closure of the Iapetus Ocean, arrows indicate plate movement directions **b** Avalonian slab breaks away and sinks. Melting, due to the rise of the asthenosphere under the lithosphere in the vicinity of the collision zone, generates granite magmas that ascend to their emplacement zones in the upper part of the Laurentian crust. Adapted from Atherton and Ghani (2002)

and the Skird Rocks Fault (SRF), orogen oblique structures are represented by the Clifden-Dog's Bay Fault and the Barna Fault (Fig. 6).

During late Caledonian transpression, clockwise block rotation promoted dilation along NNW-SSE faults. Ascending magma exploited these dilation zones during the construction of the Earlier Plutons of Omey, Roundstone and Inish that all yield U-Pb and Re-Os ages of between ~424 and 420 Ma and exhibit steeply inclined northwest-southeast striking magnetic foliations (McCarthy et al. 2015a, b). The undated Letterfrack pluton, is inferred to be a member of the Earlier Plutons. The Carna Pluton also occurs along a NNW-SSE fault, contains a prominent NNW-SSE subvertical fabric and intruded slightly later than the Earlier Plutons at 410 Ma. It is therefore interpreted to have intruded during the sinistral transpressive tectonic regime leading to the emplacement of the Carna Pluton along the intersection of the Clifden-Dog's Bay Fault and the Skird Rocks Fault. The first evidence for a change to regional transtension is recorded at ~405 Ma during the emplacement of the Kilkieran Pluton. Evidence for this transition is provided by field, petrographic and magnetic data which reveal magmatic state subhorizontal foliations and subhorizontal east-southeast-west-northwest lineations across the center of the Kilkieran Pluton (Baxter et al. 2005 and McCarthy 2013). These fabrics are interpreted by McCarthy (2013) to represent the switching of σ_1 and σ_2 that brought σ_1 from the horizontal, i.e., regional transpression, to the vertical plane, i.e. regional transtension. The larger size of the Kilkieran Pluton coupled with granite heterogeneity e.g. the magma mingling zone in the Kilkieran Pluton (El-Desouky et al. 1996; Crowley and Feely 1997; Baxter and Feely 2002; Feely et al. 2010) are interpreted to reflect increased volumes of magma emplacement as it migrated along the stitching Skird Rocks Fault (see Leake 1978), a deep seated, orogen parallel structure which was within the transtensional regime at ~400 Ma see introduction to the Kilkieran Pluton (Excursions 3 and 4).

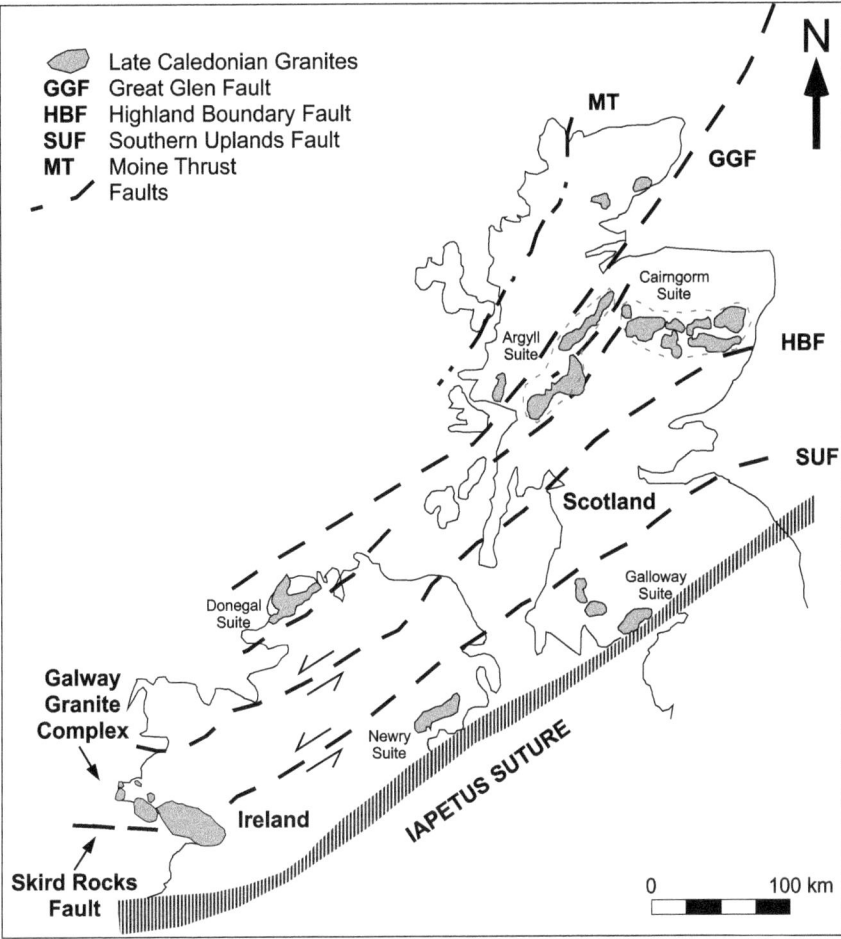

Fig. 5 Map of NW Ireland and Scotland showing the outcrops of the Late Caledonian granites (425–395 Ma) northwest of the Iapetus suture and their spatial relationships to the major orogen parallel faults. The GGC's Galway Batholith is spatially related to the sinistral strike-slip Southern Uplands Fault (SUF) that extends into Galway Bay locally termed the Skird Rocks Fault. Adapted from Atherton and Ghani (2002) and Dewey and Strachan (2003)

The Costelloe Murvey Granite marks the latest known phase of granitoid magmatism in the GGC at 380.1 ± 5.5 (Feely et al. 2003), this intrusion postdates all other granites in the Kilkieran Pluton by ~20–15 Ma. Buchwaldt et al. (2001) presented a binary isotopic model (Sr-Nd) which suggests the CMGr was generated by melting of older granite of the Kilkieran Pluton and notes that this geochemical fingerprint is quite distinct when compared to other members of the GGC. This intrusion is not related to the Caledonian Orogeny and is most likely related to some later magmatic event, possibly the closure of the Rheic Ocean (see Domeier 2016).

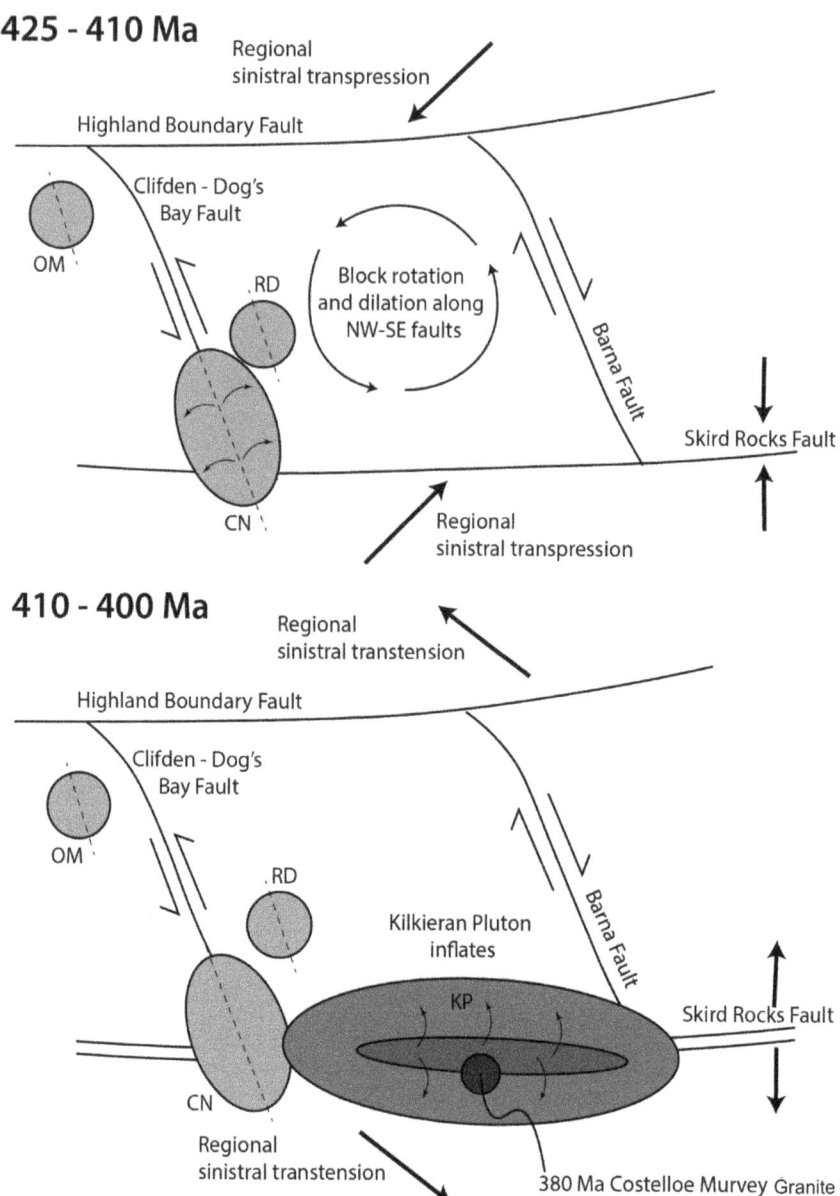

Fig. 6 Schematic representation of construction of the GGC in a tectonic framework (modified from McCarthy (2013)). The Skird Rocks Fault (SRF) is the western extension of the Southern Uplands Fault (Fig. 5). OP: Omey Pluton; RD: Roundstone Pluton; CN: Carna Pluton; KP: Kilkieran Pluton

2.3 Magma Emplacement During the Construction of the GGC

Petrographic fabrics identified in both the Omey and Roundstone Plutons are closely comparable and, although they are best reflected in rock magnetic data (McCarthy 2015a, b), these subtle mineral alignments are observable in the field. In the Omey Pluton two fabrics are described: (1) a weak magmatic fabric defined by locally aligned K-feldspars and euhedral platy biotite that define an overall concentric foliation and (2) a subtle foliation of elongate quartz, biotite and feldspar that define ~5 m wide subvertical NNW-SSE shear zones that are parallel to and extend from prominent NNW-SSE faults observed in the host rock. These shear zones display elongate quartz ribbons, aligned biotite and partially aligned, fractured feldspars. In thin section, quartz displays undulose extinction, biotite is kinked and envelopes plagioclase feldspar and fractures in plagioclase feldspar are infilled with residual melt pockets (see McCarthy et al. 2015b, Fig. 7). These data are interpreted to reflect magma ascent along a NNW-SSE conduit followed by lateral emplacement along the hinge line of the Connemara Antiform (see Omey Pluton, Excursion 5, Fig. 41a) forming a partially discordant phaccolith (McCarthy et al. 2015b).

The circular Roundstone Pluton has steeply inclined faulted contacts, within 100 m of this contact a subtle inflation fabric, defined by 3–6 mm elongate quartz ribbons (axial ratio c.1:3), smeared biotite and fractured feldspar occurs, this fabric is conveniently observed on the shoreline at Roundstone Village. A suite of microgranite sheets (G2) which are horizontal or subvertical and strike NNW-SSE dykes, crosscut the main granite facies (the G1 facies) of the Roundstone Pluton. Petrographic and magnetic fabric data (McCarthy et al. 2015a) show distinct magmatic fabric in the center of the pluton that is steeply inclined and distributed along a NNW-SSE girdle, parallel to G2 dykes. McCarthy et al. (2015a) interpret the Roundstone Pluton to be a punched laccolith and that ascent of magma occurred along a sub-plutonic NNW–SSE trending vertical fault and emplacement below the subhorizontal Mannin Thrust.

Only the northern and eastern edges of the Inish Pluton are exposed, the former on the islands of Inishturk and Turbot, and the latter along the western shoreline of Errismore peninsula. Two granite varieties are mapped (Leake 1986) i.e., Main Inish Granite and the older Marginal Granite. The latter contains many disoriented stoped blocks of country rocks passing outwards into a granite vein network that intrudes the country rocks. The Letterfrack Pluton (Townend 1966) comprises three NW–SE trending small intrusions with shallow outward-dipping contacts. They probably represent one large composite pluton. A coarse and fine-grained granite is described by Townend (1966). The fine-grained facies is confined to the middle body. Stoped blocks of country rocks are common especially in the contact zone of the largest, southeastern, coarse grained granite body.

The Galway Batholith's Carna and Kilkieran Plutons display evidence for forceful (ballooning) and passive (block stoping) emplacement. The Carna Pluton is elliptical in shape with a long axis orientated NNW–SSE, the intrusion stitches the Clifden-Dog's Bay Fault. Petrographic and Anisotropy of Magnetic Susceptibility (AMS) data from the Carna Pluton indicate that this fault structure was active during pluton construction and that the fault most likely directed magma ascent and emplacement (McCarthy 2013). Leake (2011) interpreted the pluton as a central ring complex emplaced in part by stoping. For example, the southern slopes of Errisbeg Hill, west of Roundstone expose kilometer-scale stoped blocks of the Metagabbro Gneiss (Leake 1974, 2011)—(see Excursion 2).

The contact zones of the later Kilkieran Pluton also display numerous examples of arrested stoped blocks of country rock. Kilometre-scale stoped blocks of rocks belonging to the Grampian Metagabbro Gneiss Suite are commonplace along sections of the Kilkieran Pluton's northern contact (Fig. 1 and Excursion 3). In addition to block stoping, the Kilkieran Pluton's emplacement was facilitated by ballooning (Baxter et al. 2005). A marginal deformation zone (~500 m wide) with planar fabrics e.g., quartz ribbons in the granite, is contact-parallel and is most intense at the pluton's northern contact zone e.g. at Glentrasna (see introduction to the Kilkieran Pluton in Excursions 3 and 4). Furthermore, planar fabrics are also developed parallel to internal, intrusive granite contacts. These magmatic fabrics are considered to have formed due to lateral expansion operating as successive magma batches were emplaced.

3 Contact Metamorphism of the Country Rocks

The Earlier Plutons (Omey, Roundstone, Inish and Letterfrack Plutons) were emplaced into the rocks of the Connemara Metamorphic Complex (CMC). Contact metamorphism overprints the Dalradian pelitic and carbonate rocks of the CMC. The Letterfrack Pluton intruded the Streamstown Schist and Lakes Marble Formations (Leake and Tanner 1994). Close to the granite contacts andalusite porphyroblasts are developed in the pelitic rocks. The aureole rocks of the Inish Pluton display cordierite and andalusite porphyroblasts in pelitic hornfels (Townend 1966; Cobbing 1969). The Roundstone Pluton intruded the MGGS and contacts do not display significant contact metamorphism (Leake and Tanner 1994). The contact metamorphism of the Omey Pluton's aureole rocks is well documented in Ferguson and Harvey (1979) and Ferguson and Al-Ameen (1985). These authors document, thermally overprinted pelites and carbonates of the Streamstown Schist and Lakes Marble Formations displaying andalusite hornfels and spectacular wollastonite + grossular + vesuvianite + diopside bearing skarns (Excursion 5).

According to Leake and Tanner (1994), there is insignificant contact metamorphism of the country rocks along the northern contact of the Galway Batholith is in contact with the rocks of the Metagabbro Gneiss Suite (MGGS; Fig. 1). Evidently, granite temperatures were insufficient to trigger metamorphic reactions in the MGGS. Given the bulk composition of the MGGS, temperatures >1000 °C would be required for such reactions to occur, the intruding granite magma was at temperatures <800 °C.

However, along the northern contact zone, andalusite and cordierite hornfels are locally developed in pelitic horizons within the Dalradian country rocks. In the southern contact zone, andalusite and cordierite hornfels occur in the pelitic rocks of the Lower Ordovician South Connemara Group. Interbedded volcanic horizons are amphibolites and contain hornblende, plagioclase, diopside and calcic garnet in amphibolites.

4 Petrologic and Geochemical Aspects of the GGC

The Inish and Omey Plutons are predominantly biotite-hornblende granite and the Letterfrack and Roundstone Plutons are predominantly granodiorite (Townend 1966; McCarthy et al. 2015a, b). A selection of discriminant diagrams highlights the mineralogical, petrological and geochemical characteristics of the GGC's granites (Fig. 7). On a modal Quartz–Alkali Feldspar–Plagioclase Feldspar (QAP) Streckeisen (1976) ternary diagram the range of rock types encountered in the GGC define a generally continuous field from the diorite to alkali-feldspar granite fields. The alkali-feldspar granites are collectively termed the Murvey Granite which includes the Costelloe Murvey Granite (Excursion 4). The latter is a radioelement rich (U ~ 15 ppm, Th ~ 40 ppm, K ~ 4.0%) High Heat Production (HHP) leucogranite (Feely and Madden 1986, 1987, 1988; Feely et al. 1989, 1991). GGC major element trends using AFM ($Na_2O + K_2O$–FeOt–MgO wt%) and CNK ($CaO–Na_2O–K_2O$ wt%) ternary diagrams display typical calc-alkaline variation trends. The arrows in Fig. 7b, c reflect the differentiation trends from mafic diorites to the felsic granites. A tectonic discrimination diagram (Fig. 7d) using relative abundances (ppm) of Hafnium (Hf), Tantalum (Ta) and Rubidium (Rb) places the batholith in the late and post-collisional tectonic field. Excursions 1–5 now follow, their spatial distribution are shown in Fig. 8.

5 The Carna Pluton (Excursions 1 and 2)

5.1 Introduction to the Carna Pluton: The Spatial Distribution of Its Major Granite Types

The Carna Pluton consists of the central Mace-Ards Granite surrounded by the Carna and Cuilleen granites. The Cuilleen granite occurs as a sheet within the more dominant Carna Granite (Fig. 9).

The Carna Granite displays a gradational contact (<300 m) with the surrounding marginal K-feldspar porphyritic Errisbeg Townland Granite (ETG). This contact displays a gradual southward decrease in size and abundance of pink K-feldspar

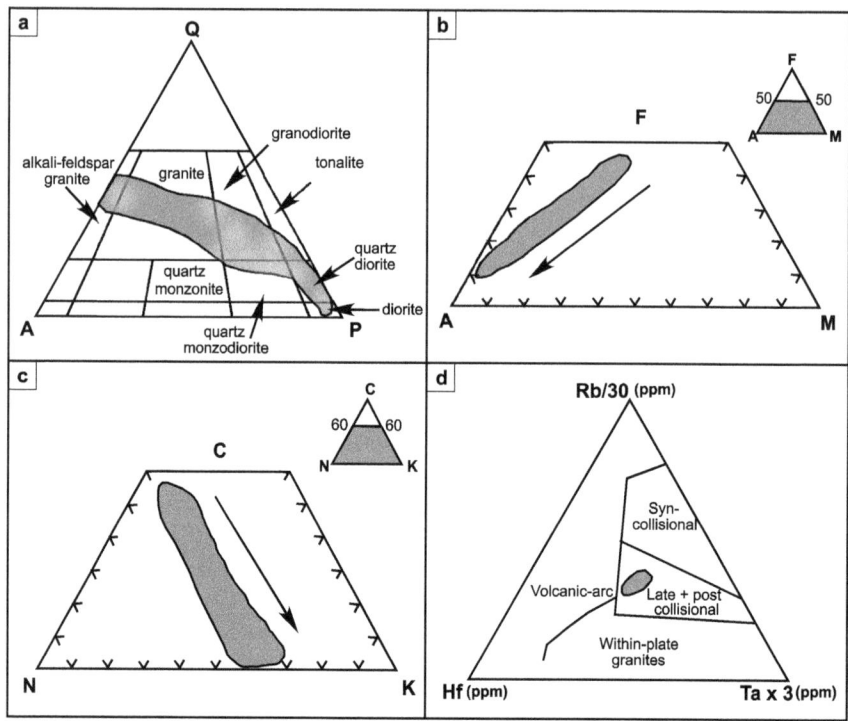

Fig. 7 a QAP ternary diagram for plutonic igneous rocks (Streckeisen 1976). Granite classification is based upon the relative abundances of quartz (Q), alkali feldspar (A) and plagioclase feldspar (P) **b** and **c** AFM ($Na_2O + K_2O$–FeOt–MgO wt%) and CNK ($CaO–Na_2O–K_2O$ wt%) ternary diagrams showing the calc-alkaline trend (black arrow) defined by the granites of the GGC and **d** Tectonic discriminant diagram showing that the granites define a field within the Late- and Post-collisional field (shaded area). Adapted from Feely et al. (2006)

phenocrysts towards the centrally located Carna Granite, SW of Gorteen Bay (Excursion 2: Figs. 18 and 19). A similar gradational contact is exposed to the southeast on Furnace Island, in South Connemara (Fig. 9). These gradational contacts indicate that both granite magmas were essentially coeval and solidified side by side without the formation of a chilled margin. Igneous layering is a common feature of the ETG. Alternating layers (10 cm to 2 m thick) are biotite-rich or K-feldspar -rich and were formed by gravity settling before the ETG magma fully crystallized. The layers dip at angles of ~20° outwards from the central Carna Granite. This radial, outward dipping mineral layering indicates that the layers were tilted after formation by the later, central intrusion of the Carna Granite.

Light pink coloured alkali feldspar (albite and K-feldspar) granite (i.e., the Murvey Granite) forms the outer margin to the ETG. There is a sharp transition (<40 m) between the two granites, the Murvey Granite lies above the ETG. A fine-grained garnet-bearing facies of the Murvey Granite is locally present near the contact west of Roundstone (Fig. 9). The contact between these two granite varieties is gradational

The Late Silurian to Upper Devonian Galway Granite Complex (GGC) 315

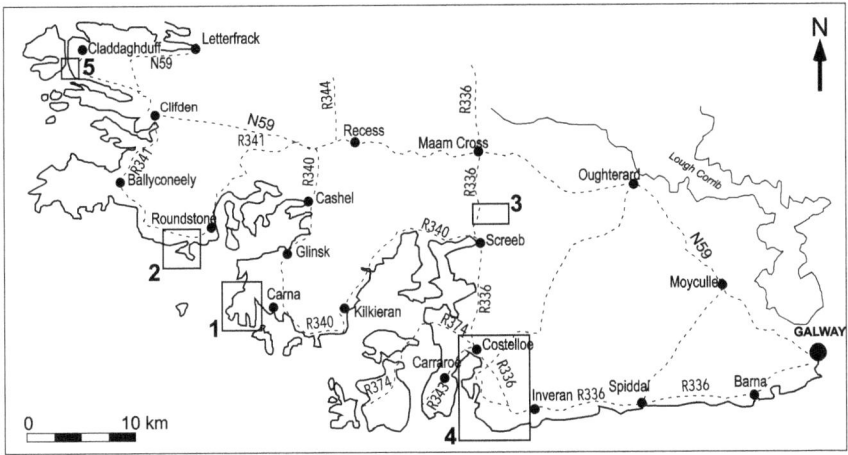

Fig. 8 Roadmap of Connemara showing the Excursions 1–5: 1 and 2 are in the Carna Pluton, 3, and 4 are focused on the Kilkieran Pluton and Excursion 5 is to the Omey Pluton and its aureole rocks. Detailed geological maps for each excursion that include the access roads to each locality are presented below

over ~10 m. Re-Os molybdenite age determinations indicate that the Murvey granite (and by inference the ETG) is ~410 Ma and were intruded by the slightly younger Carna, Cuilleen and Mace-Ards Granites at ~407 Ma (Feely et al. 2010) see Table 1.

The modal Quartz-Alkali Feldspar-Plagioclase Feldspar (QAP) Streckeisen (1976) ternary diagram (Fig. 10) shows that the rock types encountered in the Carna Pluton range in composition from granodiorites through granites to alkali feldspar granite (the marginal Murvey Granite). The Carna, Cuilleen and Ards Granites straddle the granodiorite-granite fields.

At Mace Head, west of Carna, the Mace-Ards Granite with 5% modal biotite + muscovite intrudes the Carna Granite (Leake 2011). Southeast of Mace Head the granite contains micaceous orbicules (50% biotite + muscovite + chlorite), 40–70 mm in diameter, surrounded by pale granite rims (10–40 mm wide) containing only trace amounts of biotite. The orbicule cores contain quartz and pseudomorphs of muscovite after cordierite. The Mace Head Mo-Cu mineralization zone has disseminated and quartz vein-hosted molybdenite and chalcopyrite orbicules –see Excursions 1 and 2 below.

Leake (1974 and 2011) reports the following K-feldspar and plagioclase compositions for the Carna Pluton: the Murvey Granite has K-feldspar ($\sim Or_{93-97}$) and plagioclase ($\sim An_{13-8}$). The Garnetiferous Murvey Granite has plagioclase ($\sim An_2$) and perthite ($\sim Or_{94}$). In both the ETG and the Carna Granite the K-feldspar is perthitic ($\sim Or_{85-97}$) and plagioclase is normally zoned ($\sim An_{35-15}$). The Cuilleen Granite has normally zoned plagioclase ($\sim An_{31-19}$) and K-feldspar ($\sim Or_{90}$). The Mace Ards Granite has normally zoned plagioclase ($\sim An_{40-17}$) and K-feldspar ($\sim Or_{86-97}$). Wright (1964) noted that the K-feldspar in the Carna Pluton is intermediate in structure between microcline perthite and orthoclase perthite.

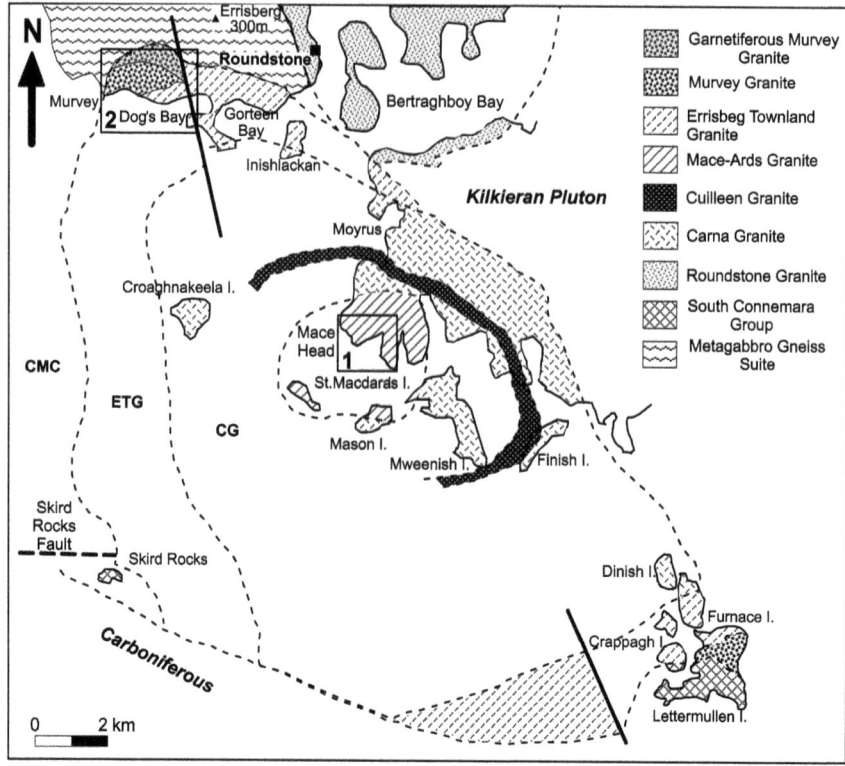

Fig. 9 The geology of the Carna Pluton adapted from Wright (1964), Max et al. (1978) and Leake (2011). Excursions 1 and 2 are shown. Heavy black lines are faults. Offshore extensions of individual plutons are shown as dotted lines

Table 1 Geochronology of the Carna Pluton. Chronometric data from [1]Pidgeon (1969) and [2] Selby et al. (2004)

Carna pluton		
Zircon (bulk sample) (U-Pb)	Molybdenite in granite hosted quartz vein (Re-Os)	
[1]Carna Gr	[2]Carna Gr (Mace Head) 407.3 ± 1.5 Ma	
412 ± 15 Ma	[2]Marginal Murvey Gr (W of Roundstone) 410.5 ± 1.5 Ma And 410.8 ± 1.4 Ma	

Fig. 10 QAP ternary diagram for plutonic igneous rocks (Streckeisen 1976) showing the granites of the Carna Pluton. Data from Wright (1964), Leake (1974) and (2011)

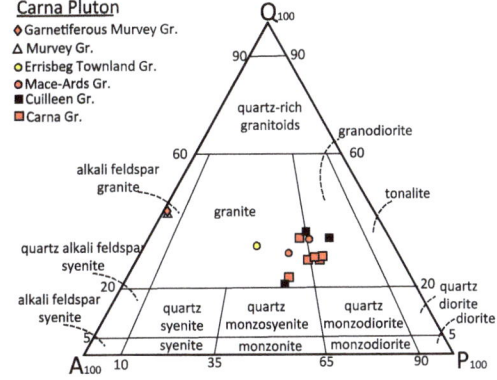

Excursions 1 and 2 where prepared by Martin Feely, Bernard E Leake and Alessandra Costanzo.

5.2 Excursion 1: The Central Sector of the Carna Pluton, West of Carna

Access and other Information From Carna post office travel NW for 2.7 km then turn left onto the narrow road leading south to Mace Pier (Fig. 11). At the first fork in the road turn right then proceed for 1.3 km to reach a T-junction, here take the road to the right and continue for 1.3 km to reach Mace Pier. Park here and proceed to location 1 (Figs. 11 and 12).

This coastal traverse SW of Mace Pier provides good examples of granite-related disseminated and quartz vein hosted molybdenite, chalcopyrite and pyrite mineralisation. In 2015 a diamond drilling program by MOAG Copper Gold Resources Inc. defined a Mo-Cu mineralised zone of ~1400 × 300 m with mineralisation to depths of 200 m. The zone is controlled by SSW-NNE trending faults in which Mo-Cu bearing quartz vein arrays occur. Later NNW-SSE faults cut the deposit. Diamond drill holes intersected mineralised intervals (~100+ m) averaging 0.04–0.07% Mo and similar for Cu. Locally values of up to 0.36% Mo and 0.32% Cu have been recorded in drill core. Quartz–magnetite segregations, magnetite-pyrite orbs, an orthoclase-feldspar breccia and an orbicular granite also occur at Mace Head (Fig. 12).

5.2.1 Locality 9.1: The Pier Granite (53.319419°, −9.889636°)

The Pier Granite is a medium grained, pink leucogranite containing 73–76 wt% SiO_2 and very similar in appearance and chemistry to the leucocratic Murvey Granite that occurs elsewhere in the Carna and Kilkieran Plutons. The Pier Granite was subjected to intense hydrothermal alteration. SW of the pier, distinctive iron oxide stains <2 cm

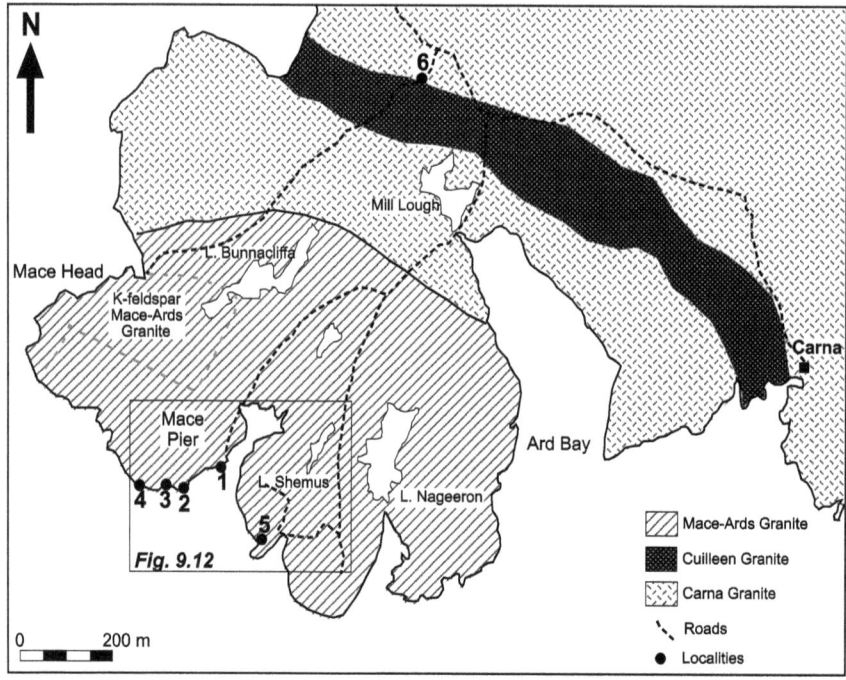

Fig. 11 Geologic map showing the distribution of the Mace-Ards, Carna and Cuilleen Granites in the Mace Head area. The boundary of the K-feldspar phenocrystic Mace-Ards Granite, NW of Mace Pier with the Mace-Ards Granite is indicated by the broken blue line. The six field locations, 9.1–9.6, are also shown (1–6) along with the main access roads in the area. Map adapted from Leake (2011). Mace Head area is outlined (Fig. 12)

across are clearly visible on all granite exposures (Fig. 13). These alteration zones weather proud and occur as an array of vertical tubular-like structures on granite outcrops. They are locally vuggy and contain pyrite, chalcopyrite ± molybdenite and probably represent fluid flow pathways during mineralisation. Disseminations of molybdenite, chalcopyrite and pyrite also occur in the granite. Dacite dikes cut the granite and do not display the same degree of hydrothermal alteration.

5.2.2 Locality 9.2: Mo-Cu Bearing Quartz Veins (53.318923°, −9.892080°)

Mo-Cu bearing quartz veins are common along this part of the coastline. They form part of the NNE-SSW fault-controlled network of mineralised quartz veins that define the Mace Head deposit. In general, the veins strike NNE-SSW and in general dip steeply to the NW, however there are several phases of vein emplacement and a range of orientations have been mapped. Muscovite commonly occurs along the

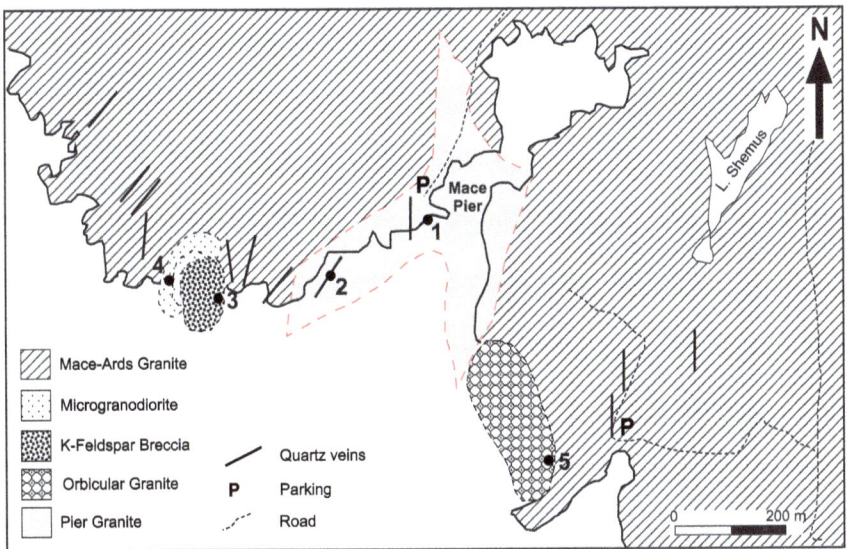

Fig. 12 Geologic map of the Mace Pier area showing the spatial distribution of the main rock units described in the text. Excursion locations 9.1–9.5 (1–5) are also indicated. Map adapted from Feely et al. (2006) and Leake (2011)

Fig. 13 The Pier Granite at Mace Pier displaying the centimetric scale iron oxide staining (from Feely et al. 2006, reproduced with permission from Geological Survey Ireland)

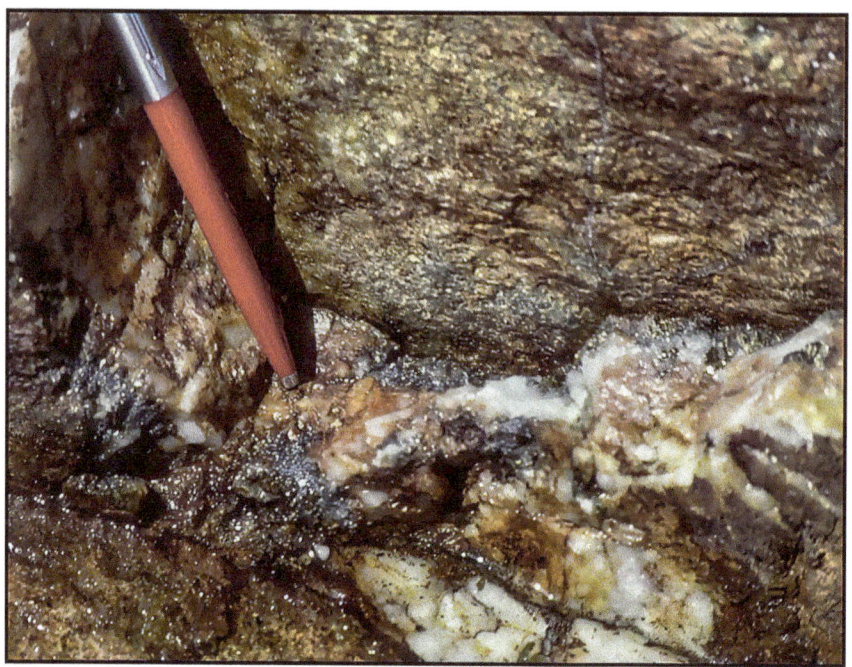

Fig. 14 A 10 cm wide NE trending vertical Mo-Cu bearing quartz vein. (From Feely et al. 2006, reproduced with permission from Geological Survey Ireland)

interface between the vein wall and the granite. At this locality, a 10 cm wide NE-SW trending vertical quartz vein is exposed along a prominent NE trending inlet. Mo-Cu mineralisation is visible along its marginal and central zones (Fig. 14). Molybdenite from this vein yields a Re-Os age of 407 ± 1.5 Ma.

5.2.3 Locality 9.3: The K-Feldspar Breccia (53.318198°, −9.896225°)

The host granite is the Mace-Ards Granite which contains ~5% biotite and is generally aphyric however, it does contain a zone characterised by K-feldspar phenocrysts (<2 cm long) to the NW of here (Fig. 11). The clast-supported perthitic-orthoclase feldspar breccia is exposed over an area of ~6000 m^2 and comprises a network of veins and pods (Fig. 15). The clasts comprise sub-angular fragments of perthitic orthoclase megacrysts (up to 10 cm long) and the Mace-Ards Granite (up to 25 cm long) set in a matrix of granular quartz (<1 cm), biotite (<5 cm) and apatite (<3 mm). Furthermore, the breccia intrudes a microgranodiorite plug (see Fig. 12, Locality 9.4), angular clasts of which are also entrained in the breccia. Neither breccia pods nor veins display any preferred orientation. Vertical exposures show that breccia veins are feeders to pods exposed on horizontal surfaces. Central parts of the breccia pods commonly contain milky quartz segregations <0.5 m across, some contain magnetite

Fig. 15 a The K-feldspar (orthoclase perthite) breccia. Note the shallow dipping quartz vein that intrudes the breccia **b** Magnetite clot within a milky quartz segregation spatially related to the breccia. (From Feely et al. 2006, reproduced with permission from Geological Survey Ireland)

clots <20 cm (Fig. 16). Disseminated molybdenite and chalcopyrite occur within the breccia. Contacts between the Mace-Ards Granite and the breccia are sharp and rectilinear. Locally however, the contact is marked by a biotite-rich foliated zone (10–20 cm wide) in which centimetric scale clasts of granite (<2 cm) and orthoclase feldspar (<5 cm) enveloped by the biotite foliation.

Derham and Feely (1988) argue that the orthoclase megacrysts and granite clasts were brecciated and silicified by P, K and Si bearing hydrous fluids as the feldspar and granite clasts were transported upwards, due to fluid overpressure, to their present level. The fluids crystallised forming the quartz, biotite and apatite matrix. The breccia clearly postdates the quartz veins (Fig. 15a). However, its emplacement is considered by Derham and Feely (1988) to be genetically related to the ore-forming processes responsible for Mo-Cu deposit.

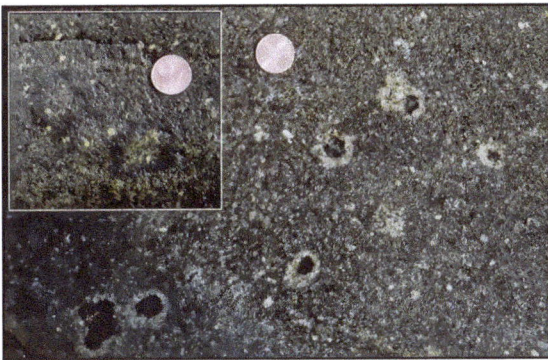

Fig. 16 The two types of orbicular textures in the microgranodiorite plug. The main picture shows the orbs that display a dark core of magnetite and pyrite surrounded by a felsic rim. The inset shows the orbs with a felsic core rimmed by magnetite and pyrite. (From Feely et al. 2006, reproduced with permission from Geological Survey Ireland)

5.2.4 Locality 9.4: Magnetite-Pyrite Orbs (53.318237°, −9.896376°)

The microgranodiorite plug on the west side of the breccia contains centimetric scale orbicular textures defined by a core of microgranodioirite (~1 cm across) surrounded by a rim (~0.5 cm wide) of magnetite and pyrite. Clasts of the microgranodiorite entrained in the breccia display these orbicular textures. Some orbicules, of similar dimensions, occur in the microgranodiorite that have cores defined by magnetite and pyrite surrounded by a rim of plagioclase feldspar and quartz (Fig. 16).

5.2.5 Locality 9.5: Orbicular Granite (53.315531°, −9.886398°)

Access and other information Return to Mace Pier, drive north for 1.3 km to the T-junction, take a right, and continue for 1.4 km then turn right along a narrow road leading to the sea. After 0.3 km the road reaches the head of a small inlet before turning sharply to the right. Park here. There is a lane up to a farmyard on the western side of the inlet. The orbicular granite exposures are reached by walking around the headland to the west for approximately 200 m. From the inlet, the view to the south takes in Mweenish Island and to the west is Saint MacDara's Island. It is believed that he built a wooden church on the island in the sixth century. This was replaced by the present stone church built in the tenth century and is clearly visible from the mainland. Saint MacDara is the principal patron saint of South Connemara's fishermen, and an annual pilgrimage to the island is still held every July.

Dark medium- to fine-grained orbicules occur in a medium grained biotite granite (extending over a 300 × 100 m area) exposed along the shoreline, SE of Mace Pier (Fig. 12). The orbicular granite displays a sharp contact with the Mace–Ards Granite. The orbicules contain a core (4–7 cm in diameter) rich in muscovite and chlorite with subordinate amounts of quartz and feldspar. A rim of biotite-poor granite (1–4 cm wide) usually surrounds each core and on some exposures pairs of orbicules are seen to share a common rim (Fig. 17). Petrographic examination of cores, rims and the host granite reveal a similar texture throughout all three. The modal abundance of biotite, muscovite and chlorite, changes dramatically from the host granite (~5% modal biotite + muscovite) through the rim (trace amounts of biotite) to the core (~50% modal biotite + muscovite + chlorite). Studies of the cores reveal wholesale replacement of the feldspar and cordierite (i.e., pinite) by muscovite and chlorite aggregates. Islands (~2 mm across) of quartz and feldspar surrounded by channels of muscovite and chlorite (1–3 mm wide) occur throughout the cores. The biotite in the cores is texturally similar to that in the host granite and the modal abundance is the same. The geological setting of the orbicular structures strongly suggests a link in space and time with the genesis of the Mo-Cu deposit.

Fig. 17 The Orbicular granite exposed at location 5. Cores contain pinitised cordierite, biotite and chlorite surrounded by quartz and feldspar rims. Some orbicules have shared rims. (From Feely et al. 2006, reproduced with permission from Geological Survey Ireland)

5.2.6 Locality 9.6: The Contact Between the Carna and Cuilleen Granites (53.339008°, −9.870854°)

Access and other information Return to T-junction on the main road, passing Mill Lough on your left, and turn left (a right turn leads back to Carna)—see Fig. 11. Proceed for ~250 m and park in front of a derelict large two-storey grey house (53.340817, −9.870784) on the NE side of the road (Fig. 11). Turn right down a N–S side road and proceed for ~180 m until, just before a sharp right-hand bend in the road outcrops of the Carna Granite cut by aplites, quartz veins and a feldspar porphyry dike, can be inspected. Then cross the road about 50 m further on, the K-feldspar phenocrystic Cuilleen Granite is exposed. Loose blocks of the pink leucocratic Murvey Granite, obtained from a quarry near the northern edge of the Batholith can also be seen.

Exercises

1. Document the different styles of Mo-Cu mineralisation at Mace Head (Locs 9.1, 9.2) and discuss their origin and timing in relation to the emplacement of the Carna Pluton.

2. Describe the different types of orbicular structures encountered in the Mace Head area (Locs 9.4, 9.5). Review the literature dealing with the origin of orbicular structures in granites and discuss the likely origin of the Mace Head orbicular structures.
3. The Mace Head area provides excellent examples of hydrothermal fluid activity in granite, discuss.
4. Document the field evidence that elucidates the timing of emplacement of the K-feldspar breccia at Mace Head (Loc 9.3).

5.3 Excursion 2: The Northwestern Margin of the Carna Pluton, West of Roundstone

Access and other information Park on the east side of the road (R341) from Roundstone to Ballyconneely, at the sharp bend (53.3911339°, −9.978609°), east of Murvey. An abandoned Garnetiferous Murvey Granite quarry is located on the south side of the road (53.391653°, −9.98083°). Although this is not Locality 1 it is nevertheless worth a quick visit as some good samples of this garnet (cherry-red almandine-spessartine) bearing leucogranite can be found.

The major Cleggan–Clifden–Dog's Bay Fault trends SSE across the wood to the east of the main road. The Errisbeg Townland Granite is exposed to the east of this fault and the country rocks belonging to the Metagabbro Gneiss Suite (MGGS) are to the west (Figs. 18, 19). The exposures visited along this traverse (locations 1–4 in Fig. 19) illustrate the relationships between the Errisbeg Townland Granite (ETG), the Murvey Granite, and the Garnetiferous Murvey Granite. The Murvey Granite crystallised above the ETG, whereas the Garnetiferous Murvey Granite crystallised above the Murvey Granite, reflecting an upward directed differentiation trend from the ETG to the Murvey Granites. Field and geochemical evidence supporting this sequence is detailed in Leake (1974). The contacts between the three granites are primary non-faulted contacts. The MGGS country rock dips 57° N.

5.3.1 Locality 9.7: Garnetiferous Murvey Granite (53.386198°, −9.979632°)

Access and other information Walk ~180 m southwards along the R341 then turn right along a small SSW trending narrow road at (53.389593°, −9.977927°). Proceed along this narrow road for ~350 m, (noting the occasional dark grey coloured exposures of amphibolite belonging to the MGGS), to where a N–S line of deciduous trees occurs on the left, and an electricity transformer is on the right. Good exposures of the fine-grained, Garnetiferous Murvey Granite, are on either side of the road. The granite hosts magmatic euhedral-subhedral cherry red almandine-spessartine garnets

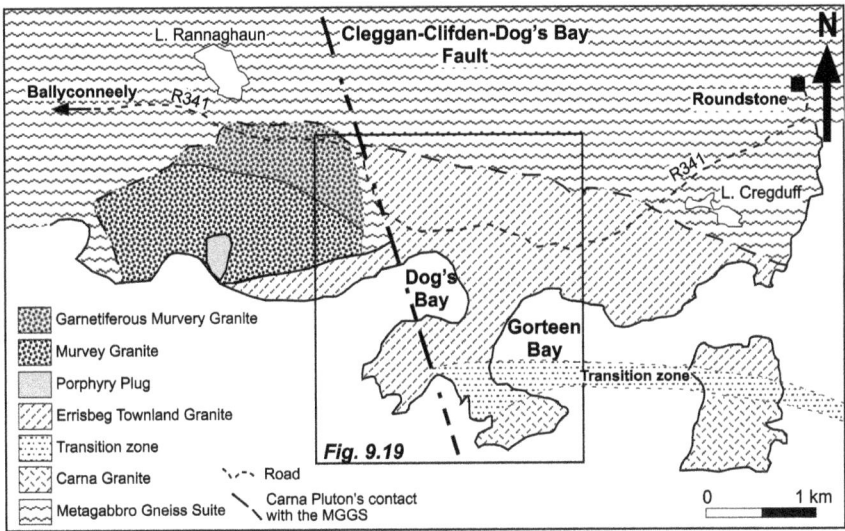

Fig. 18 Geologic map of the northwestern margin of the Galway Batholith adapted from Leake (1974, 2011). The location of Fig. 19 is shown

(<0.5 mm in width)—see Whitworth and Feely (1994). They are visible to the naked eye but warrant examination using a 10× hand lens.

On a clear day, this location affords spectacular views of the landscape. Looking to the NE, dark grey country rocks of the MGGS that form Errisbeg Mountain (elevation 300 m) display a clear ESE-striking intrusive contact with the well exposed light grey coloured ETG (Fig. 20).

The back-to-back arcuate white beaches of Dog's Bay and Gorteen Bay form a tombolo. The narrow beach along the shore, ~200 m west of Dog's Bay, is where the Cleggan–Clifden–Dog's Bay Fault crosses the bay (Fig. 21). On the south side of Gorteen Bay the orthoclase-porphyritic ETG gradually transitions to the Carna Granite. The transition zone is ~200 m wide and is marked by the gradual disappearance of the ETG's K-feldspar phenocrysts, eventually transitioning to the coarse-grained, Carna Granite. This transition zone indicates that both granites were essentially coeval. It is recommended, time permitting after visiting Location 6, to take a traverse from the porphyritic ETG across the transition zone to the aphyric Carna Granite (Fig. 19).

5.3.2 Locality 9.8: Murvey Granite (53.384368°, −9.980015°)

Continue for ~50 m down the road from Location 1 to a gate and pass through; the main Murvey Granite (medium-grained, biotite-bearing but without garnet) is well exposed south of the gate (Fig. 19). Approximately one km west of here, disseminated and quartz vein hosted molybdenite and chalcopyrite mineralisation occur in this

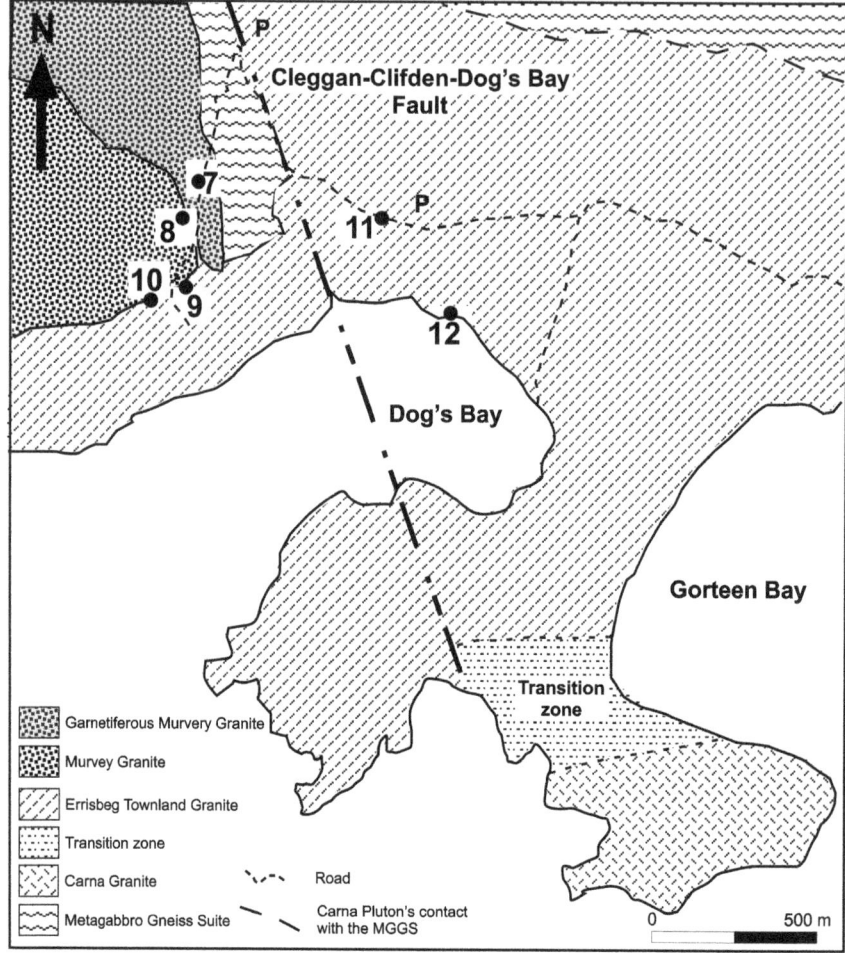

Fig. 19 Geologic map, adapted from Leake (1974, 2011), of the Dog's Bay-Gorteen Bay area showing locations 9.7–9.12 (7–12 respectively) described in the text. P indicates parking area

Murvey Granite. This style of mineralization is similar to that exposed at Mace Head (Excursion 1). Molybdenite from this deposit yields a Re-Os age of 410.5 ± 1.5 Ma, older than the 407.3 ± 1.5 Ma Re-Os age for the molybdenite at Mace Head (Feely et al. 2010).

Fig. 20 Errisbeg Mountain (elevation: 300 m) whose south facing slopes expose the sharp contact between the dark grey country rocks of the MGGS and the light grey ETG. The intrusive contact strikes ESE and dips ~60° NNE. (From Feely et al. 2006, reproduced with permission from Geological Survey Ireland)

5.3.3 Locality 9.9: The Murvey Granite with Rare Mafic Enclaves and the Errisbeg Townland Granite (ETG) (53.383258°, −9.980740°)

Further down the road (~100 m), the well-exposed intrusive contact between the Murvey Granite and ETG, best seen ~30 m south of the road and directly west of a cottage which faces SW (Fig. 19). Here the sharp contact is locally running nearly ENE. Mafic microgranular enclaves (MME), common in the ETG but rare in the Murvey Granite, occur here in outcrops of the Murvey Granite near the contact with the ETG.

5.3.4 Locality 9.10: Sharp Intrusive Contact Between the Murvey Granite and ETG (53.383492°, −9.982382°)

The ENE-striking contact between the ETG and the Murvey Granite (Fig. 22) is clearly seen 40 m to the NW of the road just before the road makes its final bend to the left to run N–S to its termination (Fig. 19). A prominent N–S pale felsic dike is close by. The ETG is finer grained near the contact with the Murvey Granite than it is elsewhere.

Fig. 21 Looking to the southeast from Location 1 onto the back-to-back arcuate beaches of Dog's Bay (facing WNW) and Gorteen Bay (facing ESE) that together form a Tombolo. The narrow beach ~200 m west of Dog's Bay marks the trace of the major Cleggan-Clifden-Dog's Bay Fault. (From Feely et al. 2006, reproduced with permission from Geological Survey Ireland)

Fig. 22 Sharp intrusive contact between the Murvey Granite and the orthoclase feldspar porphyritic ETG. (From Feely et al. 2006, reproduced with permission from Geological Survey Ireland)

5.3.5 Locality 9.11: The Errisbeg Townland Granite (ETG) (53.385465°, −9.971916°)

Return to the parking area and drive ~0.75 km towards Roundstone and park in a layby on the north side of the road opposite a grey brick house on the south side

Fig. 23 Fresh exposure of the orthoclase feldspar porphyritic Errisbeg Townland Granite. (From Feely et al. 2006, reproduced with permission from Geological Survey Ireland)

of the road. A roadside exposure of Errisbeg Townland Granite hosting a mafic microgranular enclave (MME) occurs at this locality (Fig. 23).

5.3.6 Locality 9.12: Layering in the ETG (53.383493°, −9.968807°)

From Loc 9.11 walk east for ~150 m along the R341 (towards Roundstone), then cross over a gate into a field on the south side of the road at (53.385280°, −9.969220°) and walk southward to the shore across abundant horizontal exposures of ETG, keeping to the left along the SSW-oriented wall until the shoreline is reached (Fig. 19). Walk about 50 m to the SE, where there are excellent exposures of the ETG displaying alternating biotite-rich layers and feldspar-rich layers (Fig. 24). Leake (1974) describes right-way-up current bedding and distinct right-way-up graded bedding with biotite rich bottoms grading upwards into a biotite poorer layer which is immediately

Fig. 24 Alternating biotite-rich and K-feldspar-rich layers in the ETG at Dog's Bay, Location 12. The layers dip at ~35° to the north. (From Feely et al. 2006, reproduced with permission from Geological Survey Ireland)

followed by coarser K-feldspar rich layer. According to Leake (1974), the layering is the result of crystal settling under the influence of gravity and sometimes accompanied by magmatic currents. The layers generally dip to the north, typically here at ~35° (Fig. 24). Microgranular mafic enclaves (MME) also occur in this area.

5.3.7 Additional Localities, Time Permitting

Foraminfera Sands, Dog's Bay (53.379344°, −9.962623°)

Go to the parking area at Dog's Bay beach. Good examples of layering in the ETG can be seen in the exposures along the beach. Dog's Bay is famous for its foraminiferal sands that warrant inspection using a hand lens. In some places the foraminifera can account for 60% of the beach sand and some 124 species have been identified. The shells measure <1 mm and are clearly visible under a 10× hand lens (Fig. 25). The foraminifera enter the bay, but as there are no strong currents to transport them back out to sea again, they become trapped and accumulate to form these beach sand deposits. The foraminifera live in shallow and intertidal waters and by far the commonest is *Cibicides lobatulus* which lives attached to seaweeds and hard rocky substrates (Murray 2009).

The Gradual Transition from the ETG to the Carna Granite (53.370622°, −9.960337°)

At the SW end of Gorteen Bay the orthoclase feldspar porphyritic ETG transitions to the Carna Granite. The transition zone is ~250 m wide and is marked by the gradual decrease in abundance of the ETG's K-feldspar phenocrysts, eventually transitioning to the Carna Granite. It is recommended, time permitting after visiting Location 6,

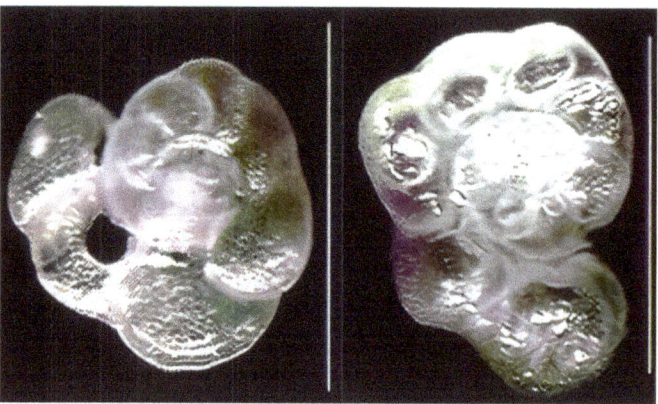

Fig. 25 Foraminifera shells from Dog's Bay beach sand. Scale bar 1 mm

to take a traverse from the porphyritic ETG across the transition zone to the Carna Granite (Fig. 19). This transition is repeated on the south Connemara islands of Dinish, Furnace and Lettermullen where the southern contact of the Carna Pluton is continuously exposed.

Exercises

1. Garnets in granites, magmatic or metamorphic? Discuss the evidence for magmatic garnets in the Murvey Granite (Loc 9.7).
2. Document the mineral layering present in the Errisbeg Townland Granite (ETG) at Dog's Bay. Discuss how seemingly viscous granite magma can accommodate mineral layering that commonly displays graded bedding (Loc 9.12).
3. Discuss how the orientation of the mineral layering in the ETG sheds light on the relative timing of emplacement of the Carna Granite and ETG (Loc 9.12).
4. Mo-Cu mineralisation occurs at the northwestern end of the Carna Pluton and is similar in style to that encountered at Mace Head (Excursion 1). Discuss the efficacy of precise Re-Os geochronometry in the development of a timeframe for granite related Mo-Cu mineralisation in the Galway Granite Complex.
5. Document the field evidence that sheds light on how the Carna Pluton was emplaced into the Connemara Metamorphic Complex.

6 The Kilkieran Pluton (Excursions 3 and 4)

6.1 An Introduction to the Kilkieran Pluton: The Spatial Distribution of Its Major Granite Types

The Galway Batholith's Kilkieran Pluton is the largest pluton in the Galway Granite Complex. It is transected by two major faults, the NNE-trending Shannawona Fault and the NW-trending Barna Fault (Fig. 26) that control the distribution of the granite varieties. The faults cut the pluton into three blocks i.e., the western, central and eastern blocks. Some of the central block granites display features e.g., deformation related fabrics, aligned discoidal mafic microgranular enclaves and magma mingling textures that formed at a deeper level in the pluton. These deeper level foliated and mafic granodiorites e.g., Mafic Megacrystic Granodiorite (MMGr) and Mingling Mixing Zone Granodiorite (MMZGr) are intruded by unfoliated granites e.g. Costelloe Murvey Granite and the Shannapheasteen Granodiorite (Fig. 26). The

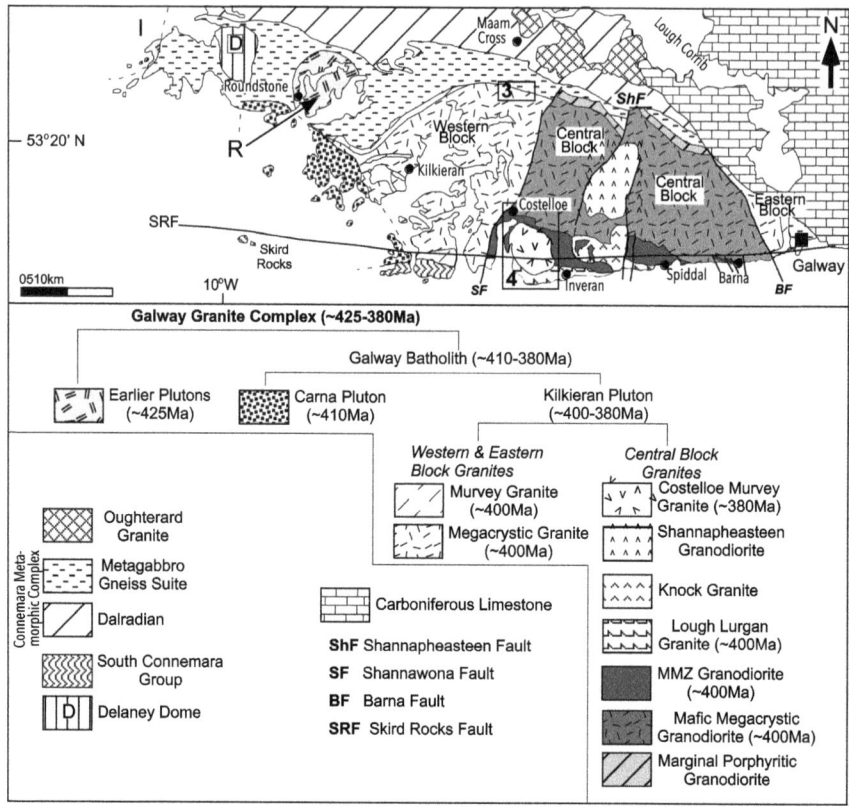

Fig. 26 The Geological setting of the Kilkieran Pluton. Map adapted from Coats and Wilson (1971), Max et al. (1978), Leake et al. (1981), Leake and Tanner (1994), El-Desouky et al. (1996), Crowley and Feely (1997), Baxter and Feely (2002), Baxter et al. (2005), Callaghan (2005), and Leake (2006, 2011). The legend shows how the Galway Granite Complex is classified into its component granite units. See text for details on this classification scheme. Excursion areas 3 and 4 are shown, R and I indicate the location of the Earlier Plutons, Roundstone and Inish

more deeply eroded central block reveals the magmatic processes that were active during the ascent and emplacement of the Kilkieran Pluton. The modal Quartz-Alkali Feldspar-Plagioclase Feldspar (QAP) Streckeisen (1976) ternary diagram (Fig. 27) shows that the rock types encountered in the Kilkieran Pluton range from diorite and quartz diorite (enclaves) to granodiorites, granites and alkali feldspar granite. The QAP diagram shows that the western and eastern blocks expose granite types that plot in the granite field in contrast with the wide spectrum of rock types i.e., from diorite to granodiorite and granite to alkali feldspar granite exposed in the central block. The K-feldspar ($KAlSi_3O_8$) in the Kilkieran Pluton can either be orthoclase or microcline (Leake 2006). Coats and Wilson (1971) reported the presence of megacrysts of orthoclase microperthite in the Megacrystic Granite and microcline megacrysts in

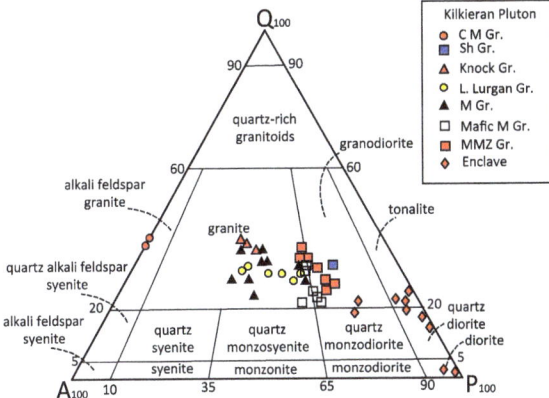

Fig. 27 QAP ternary diagram for plutonic igneous rocks (Streckeisen 1976) showing the range of granite compositions in the Kilkieran Pluton. Legend: Enclave: Mafic Microgranular Enclaves; MMZ Gr: Mingling Mixing Zone Granodiorite; Mafic M Gr: Mafic Megacrystic Granodiorite; M Gr: Megacrystic Granite; L. Lurgan Gr: Lough Lurgan Granite; Knock Gr: Knock Granite; Sh Gr: Shannapheasteen Granodiorite; CM Gr: Costelloe Murvey Granite. Modal data from: El-Desouky et al. (1996), Crowley and Feely (1997) and Leake (2006)

the Mafic Megacrystic Granodiorite. Plagioclase feldspar compositions for the range of granites in the Kilkieran Pluton are presented in Table 2.

Geochronometric studies reveal that the granites of the Kilkieran Pluton are ~400 Ma (see Table 3), in marked contrast, to ~410 Ma age of Carna Pluton (Table 1).

Excursions 3 and 4 were prepared by Martin Feely and Alessandra Costanzo.

Table 2 Plagioclase feldspar compositions for the granites (mafic microgranular enclaves are also included) of the Kilkieran Pluton. Data sources: Wright (1964), Coats and Wilson (1971), Feely 1982, El Desouky et al. (1996), Crowley and Feely (1997) and Leake (2006)

Kilkieran pluton	
Western and eastern blocks	
Rock type	Plagioclase
Megacrystic granite	An_{25-10}
Murvey granite	An_{14-6}
Central block	
Rock type	Plagioclase
Mafic microgranular enclaves	An_{45-35}
Mafic megacrystic granodiorite	An_{33-20}
Mingling mixing zone granodiorite	An_{32-28}
Lough Lurgan granite	An_{30-11}
Knock granite	An_{32-27}
Costelloe Murvey granite	An_{5-2}

Table 3 Geochronology of the Kilkieran Pluton. The Geochronometric data from: [1]Feely et al. (2010), [2] Friedrich and Hodges (2016), [3]Feely et al. (2003) and [4]Feely et al. (2020)

Kilkieran pluton	Zircon single crystal (U-Pb)	Molybdenite in vein quartz (Re-Os)	Biotite in granite Ar/Ar
Western block	[1]Murvey Gr 402.2 ± 1.1 Ma		[2]Megacrystic Gr 402 ± 3 Ma
Central block	[3]Mafic Megacrystic Gr 394.4 ± 2.2 Ma [3]Mafic Megacrystic Gr c.402 Ma [4]Mafic Megacrystic Gr 401.89 ± 0.33 Ma [3]Mafic Microgranular enclave (MME) 397.7 ± 1.1 Ma [3]Mingling mixing zone Gr 399.5 ± 0.8 Ma [3]Costelloe Murvey Gr 380.1 ± 5.5 Ma	[1]Lough Lurgan Gr 399.5 ± 1.7 Ma [4]Mafic Megacrystic Gr 401.0 ± 0.2 Ma	

6.2 The Granites of Western and Eastern Blocks

6.2.1 The Western Block

Two major granite types (i.e., the Megacrystic Granite and a medium to fine-grained leucogranite called the Murvey Granite) occur in the western block (Fig. 26). The latter occurs along the northern and southern margins of the western block (Fig. 26). Two sheet like intrusions of Murvey Granite also occur to the west of the village of Kilkieran (Fig. 26) this yields a Re-Os age of ~402 Ma (Table 3) The Murvey Granite is similar in texture and mineralogy to the older ~410 Ma Murvey Granite of the Carna Pluton (Table 1).

The Megacrystic Granite (MGr) is a coarse-grained porphyritic granite containing salmon pink K-feldspar (~Or_{95}) phenocrysts (up to 8 cm long). The MGr yields a biotite Ar^{39}/Ar^{40} age of ~402 Ma (Friedrich and Hodges 2016)–see Table 3. It is similar in texture and mineralogy to the older (~410 Ma) Errisbeg Townland Granite in the Carna Pluton. The geochronology supports the field observations of Wright (1964), who concluded that the MGr intruded the ~410 Ma Carna Pluton. Furthermore, Wright (1964) noted that the MGr and Murvey Granite contact is frequently sheared. However, locally evidence of chilling in the Murvey Granite indicates that it intruded the MGr.

6.2.2 The Eastern Block

The MGr and the Murvey Granite also occur in the eastern block (Fig. 26). The Murvey Granite outcrops as a strip adjacent to the country rock (~470 Ma Metagabbro Gneiss Suite, MGGS). The Murvey Granite intrudes the MGGS with the contact striking roughly NS, perpendicular to foliation in the gneisses (Coats and Wilson 1971). The poorly exposed contact between the MGr and the Murvey Granite is interpreted to be gradational over ~1 m and the former loses it K-feldspar phenocrysts as the contact is approached (Coats and Wilson 1971).

6.3 The Granites of the Central Block

The central block exposes a wider range of granite types than those exposed in the eastern and western blocks -see the QAP diagram (Fig. 27). Seven major rock types occur in the central block i.e., (1) Marginal Porphyritic Granodiorite (MPGr), (2) Mafic Megacrystic Granodiorite (MMGr), (3) Mingling Mixing Zone Granodiorite (MMZGr), (4) Lough Lurgan Granite (LLGr), (5) Knock Granite (KGr), (6) Shannapheasteen Granodiorite (ShGr) and (7) Costelloe Murvey Granite (CMGr). Granite types 1–4 are foliated and granites 5–7 are unfoliated and intrude the foliated granites. A summary description of each granite type follows.

6.3.1 Marginal Porphyritic Granodiorite (MPGr)

A 500–700 m wide marginal zone of foliated K-feldspar (<2 cm) porphyritic granite called the Marginal Porphyritic Granodiorite (MPGr) occurs along the northern margin of the central block. The edge of the MPGr is conformable with the ESE strike and steep dip (~80° N) of the country rocks and is generally unfaulted. The MPGr displays a gradational contact (~100 m) with the much coarser, foliated Mafic Megacrystic Granodiorite (MMGr).

6.3.2 Mafic Megacrystic Granodiorite (MMGr)

The MMGr has salmon pink K-feldspar (Or_{95} orthoclase or microcline megacrysts, ~5–10 cm long, average ~5 cm). Crowley and Feely (1997) reported that the K-feldspar megacrysts comprise equal proportions of orthoclase and perthite. The MPGr and MMGr contain quartz, zoned plagioclase (andesine and oligoclase), salmon pink orthoclase or microcline, biotite and hornblende (~15% modal abundance), accessories include titanite, zircon and apatite (El Desouky et al. 1996; Crowley and Feely 1997; Leake 2006). The MMGr contains aligned discoidal mafic microgranular enclaves (MME) that are <1 m in longest dimension. Alignments of the enclaves, K-feldspar megacrysts and groundmass minerals (e.g., hornblende and biotite) define a WNW trending fabric dipping steeply to NNE. This fabric becomes more prominent as the conformable intrusive contact with the MMZGr is approached over a distance of ~100 m.

6.3.3 Mingling Mixing Zone Granodiorite (MMZGr)

The MMZGr is 3–4 km wide and extends for ~25 km from the coast at Spiddal to the Carraroe peninsula in the west, where it terminates against the Shannawona Fault (Fig. 26). Near Barna in the east, a separate and narrower (<100 m) WNW trending mingling mixing zone occurs within the MMGr and has a strike length of ~5 km, see details in Baxter (2000).

According to El Desouky et al. (1997) the medium to coarse-grained (1–10 mm) MMZGr comprises incompletely mixed hybrid granodiorite in which mafic microgranular enclaves (MME) are randomly distributed. A mingling spectrum exists from intensely drawn out mafic enclaves to ovoid and even spherical enclaves. The fabric in the MMZGr is defined by alignments of discoidal enclaves and alignments of K-feldspar, plagioclase (An_{45-27}), hornblende and biotite both in the host hybrid granodiorite and the enclaves. The fabric dips to the north-northeast and steepens towards the contact with the MMGr in the north.

World-renowned coastal exposures of the MMZGr displaying abundant discoidal MME can be examined along the shoreline north of the Martello Tower at Rossaveel (Excursion 4).

6.3.4 Lough Lurgan Granite (LLGr)

The LLGr is a medium to coarse (1–7 mm) grained granite with quartz (26%), orthoclase (28%), plagioclase (37%; An_{30}), biotite (~5%) and rare hornblende (<0.5%). Accessory minerals include titanite and apatite. The MMZGr is sharply intruded and its fabric transected by the LLGr east of Spiddal and apophyses of the LLGr are seen within the MMZGr south of Costelloe (El Desouky et al. 1996). Occasional SW-NE alignments of MME are present in the LLGr.

6.3.5 Knock Granite (KGr)

The KGr is a light pink, medium to coarse-grained granite with ~34% quartz, orthoclase (~34%), plagioclase (~25% An_{32-27}), biotite (~5%) and rare hornblende (<1%). Accessories include titanite and apatite. The Knock Granite contains kilometer scale stoped blocks of the MMZGr. Crowley and Feely (1997) reported that in contrast with the granites described above the KGr does not possess a foliation and modal hornblende is <1.0%.

6.3.6 Shannapheasteen Granodiorite (ShGr)

The Shannapheasteen Granodiorite (ShGr), NE of Spiddal, is a fine-grained (0.5–1.5 mm) granodiorite that intrudes the foliated MMGr and contains numerous stoped blocks of the latter while many veins of the ShG have injected the MMGr. It is generally hornblende free, contains chloritised biotite, oligoclase and the K feldspar is orthoclase or microcline (Leake 2006).

6.3.7 Costelloe Murvey Granite (CMGr)

The Costelloe Murvey Granite (CMGr) resembles the other leucogranites (i.e., Murvey Granites) mapped elsewhere in the Carna and Kilkieran plutons. The CMGr displays sharp intrusive contacts with the surrounding MMZGr and the LLGr. These contacts can be inspected during Excursion 4.

The geochronology of the granites that comprise the Kilkieran Pluton reveals emplacement ages for all granites at ~400 Ma which is ~10 million years younger than the Carna Pluton Tables 1 and 3 However, the CMGr yields an age of ~380 Ma which makes it the youngest of all granites recorded to date in the whole of the Galway Granite Complex (Feely et al. 2010).

6.4 The Emplacement of the Kilkieran Pluton

The emplacement of the Kilkieran Pluton was facilitated by sinistral movement on the Skird Rocks Fault (SRF). This created space for both the magma and stoping of the country rock leading to the formation of the Kilkieran Pluton (Leake 2006). Magma crystallisation began at the cooling edges of the Kilkieran Pluton. Magma intrusions into the centre of the pluton continued, causing inflation which compressed the peripheral granite and developed a foliation in the Marginal Porphyritic Granodiorite and in the Mafic Megacrystic Granodiorite and the Mingling Mixing Zone Granodiorite. A schematic showing the proposed structural control of the emplacement of the Kilkieran Pluton adapted from Leake (2006) is presented in Fig. 28.

6.5 Excursion 3: The Northern Contact of the Kilkieran Pluton at Glentrasna

Access and other information Coming from either Clifden or Oughterard along the N59 arrive at Maam Cross junction and turn south onto the R336. Drive for 4.8 km south of Maam Cross (Fig. 26) arriving at the turn off (on the eastern side of the R336; at 53.405927°, −9.542097°) to a minor road leading to Glentrasna. Park at the junction.

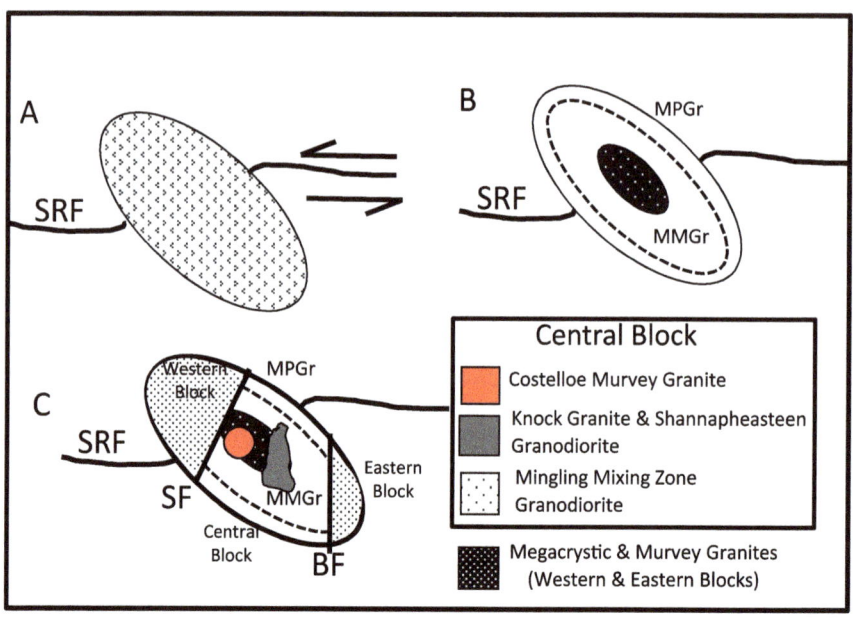

Fig. 28 Schematic showing the proposed structural control of the emplacement of the Kilkieran Pluton by the sinistral Skird Rocks Fault (SRF) after Leake (2006). **A** Sinistral movement on the SRF creates space for magma emplacement. **B** Crystallisation begins at the cooler edges of the pluton. Intrusions, however, continue in the central sector of the pluton causing inflation which results in the development of a magmatic to submagmatic foliation in the MPGr, MMGr and the MMZGr (Baxter et al. 2005) and **C** Depiction of the present exposure after late phase intrusions (i.e., Shannapheasteen Granodiorite, Knock Granite and the ~380 Ma Costelloe Murvey Granite) cut the foliated granodiorites e.g., MMZGr and MMGr. For example, stoped blocks of MMZGr occur in the Knock and Costelloe Murvey Granites and stoped blocks of MMGr occur in the Shannapheasteen Granodiorite. Uplift of the deeper level Central Block, by the Barna and Shannawona Faults, results in the juxtaposition of these granites with the roof zone granites (i.e., MGr and Murvey Granite) exposed in the Western and Eastern Blocks of the Kilkieran Pluton. Abbreviations: MPGr: Marginal Porphyritic Granodiorite, MMGr: Mafic Megacrystic Granodiorite, MMZGr: Mingling Mixing Zone Granodiorite, MGr: Megacrystic Granite, SF: Shannawona Fault, BF: Barna Fault and SRF: Skird Rocks Fault

6.5.1 Locality 9.13: The Intrusive Contact Between the Megacrystic Granite (MGr) and the Country Rocks Near Glentrasna (53.405606°, −9.538851°)

Walk eastwards along this minor road that partly parallels the intrusive contact between the ~400 Ma Kilkieran Pluton and the ~470 Ma Metagabbro Gneiss Suite (MGGS) Fig. 29. To the north are outcrops of the metagabbros i.e., amphibolites (dark grey aphyric hornblende-plagioclase rocks) and orthogneisses (light grey quartz- and feldspar-rich rocks) belonging to the MGGS (Fig. 30a, b). About 200 m from the junction exposures of amphibolite and granite gneiss occur to the north and the granite to the south side of the road. Granite veins cut and surround arrested stoped

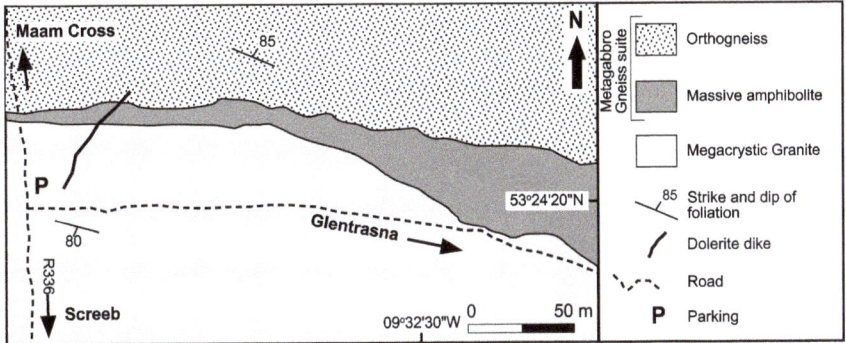

Fig. 29 Geologic map of the contact (west of the Shannawona Fault) between the Megacrystic Granite (Kilkieran Pluton) and the MGGS. P marks parking area

Fig. 30 a Contact between the Megacrystic Granite and the amphibolites and orthogneisses belonging to the MGGS. **b** Intrusive contact between the Megacrystic Granite and the MGGS amphibolite. (From Feely et al. 2006, reproduced with permission from Geological Survey Ireland)

blocks of the MGGS. No contact metamorphism of the MGGS is apparent as granite temperatures were insufficient to trigger metamorphic reactions in the MGGS. Given the bulk composition of the MGGS, temperatures >1000 °C would be required for such reactions to occur, the intruding granite magma was at temperatures <800 °C.

On the way back to the parking area, look to the north of the road and note how the late Devonian (?) NE trending dark grey dolerite dike (Fig. 29) is easily picked out among the granite exposures due to its covering growth of white lichen (53.406155°, −9.541492°).

Outcrops of the Megacrystic Granite (MGr), at the road junction beside the parking area, display a strong quartz ribbon fabric that locally imparts a gneissic appearance to the granite similar to the fabric in the older Grampian gneisses exposed along the R336 ~500 m south of Lough Nahasleam (see Excursion 2, Connemara Metamorphic Complex). The granite exposures here were previously mapped as the older gneisses (Burke 1957, 1963), however, detailed mapping by Leake and Leggo (1963) revealed that the MGr cuts the Grampian gneissic fabrics of the country rocks. The steeply dipping SE trending quartz ribbon fabric was imposed as the granite crystallised

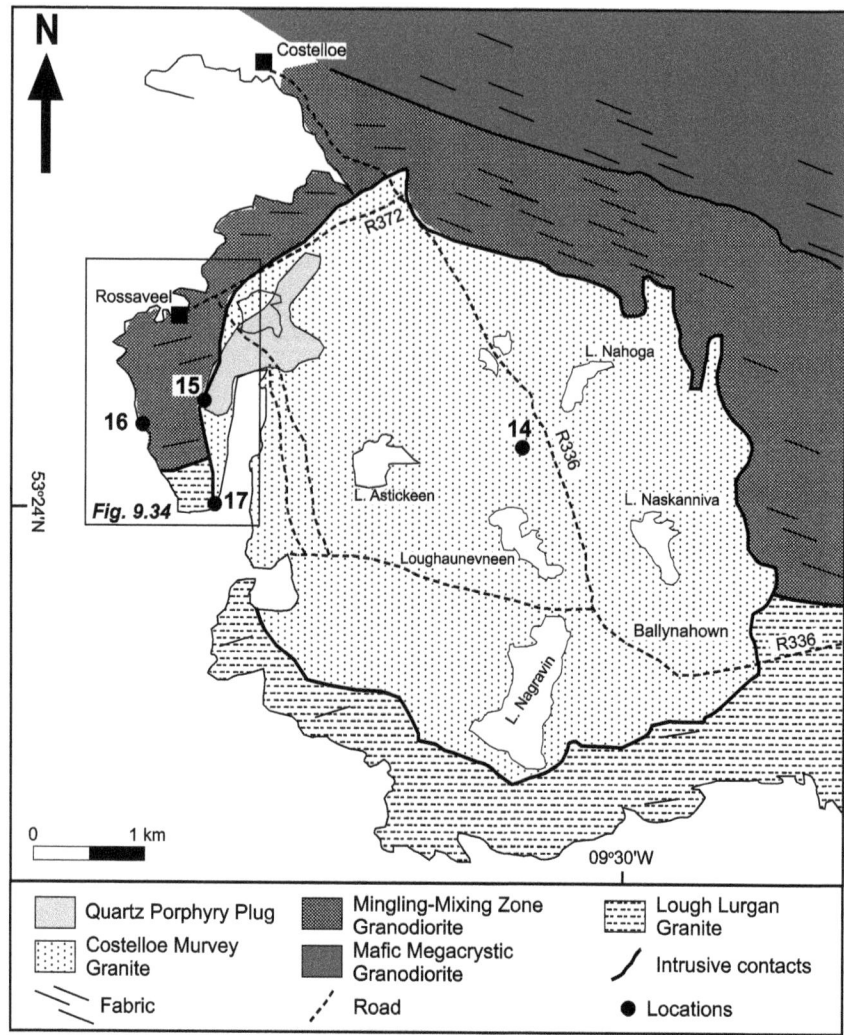

Fig. 31 Geologic map of the Costelloe area showing the spatial distribution of the Costelloe Murvey Granite (CMGr) and its Porphyry Plug, the Mingling Mixing Zone Granodiorite (MMZGr), the Lough Lurgan Granite (LLG) and the foliated Mafic Megacrystic Granodiorite (MMGr). The steeply dipping fabric in the granites is generally defined by the alignment of discoidal mafic microgranular enclaves (MME). Excursion locations 9.14–9.17 are represented by the numbers 14–17 respectively. Map adapted from El-Desouky et al. (1996) and Konig (2003)

(Baxter et al. 2005). The fabric becomes less intense southwards from the contact and is not observable in outcrop at ~500 m south from the junction. The ~400 Ma Megacrystic Granite is a pink coarse-grained granite with large (~3 cm long) salmon-pink crystals of orthoclase or microcline set in a groundmass of quartz, orthoclase, plagioclase and biotite. It is similar in appearance to the ~410 Ma Errisbeg Townland Granite of the Carna Pluton (Excursion 2 above).

Exercises

1. Assuming that there is no geochronology available to you, document the field evidence that sheds light on the relative ages of the Kilkieran Pluton and the rocks of the Connemara Metamorphic Complex.
2. Discuss the reasons for the lack of thermal metamorphism in the country rocks exposed along the Glentrasna road.
3. Describe the fabric in the Megacrystic Granite and discuss the granite emplacement mechanism(s) that can be deduced from the granite and country rock (Metagabbro Gneiss Suite) exposures at this locality.

6.6 Excursion 4: The Costelloe to Rossaveel Sector of the Kilkieran Pluton: Costelloe Murvey Granite and the Mingling Mixing Zone Granodiorite

6.6.1 Locality 9.14: Costelloe Murvey Granite (CMGr) Quarry (53.252130°, −9.507791°)

Access and other information This abandoned dimension-stone quarry is located in the leucocratic CMGr on the west side of the R336, 1.2 km north of the crossroads at Ballynahown, and 3.2 km south of the crossroads at Costelloe bridge, where the R336 and R372 meet (Fig. 31). Park at the entrance to the quarry.

The CMGr has a roughly circular surface exposure with an area of approximately 30 km^2. Gravity modeling indicates that the CMGr intrusion has a maximum depth of ~3 km (Feely 1982). The CMGr intruded at ~380 ± 5.0 Ma into the ~400 Ma Mingling Mixing Zone Granodiorite (MMZGr) and the Lough Lurgan Granite (LLG)-(Feely et al. 2010). It is a pink, medium- to coarse-grained granite containing, approximately 38% quartz, 33% perthitic orthoclase, 27% plagioclase (albite) and 2% biotite. The granite displays both vertical and horizontal joints. and has some notable geochemical characteristics that include relatively high radioelement abundances (U ~ 15 ppm; Th ~ 40 ppm and K ~ 4%). The radioelement rich nature of the CMGr makes it a High Heat Production (HHP) granite (~7 μW m^{-3}). Its suitability as a Hot Dry Rock (HDR) source of energy was assessed during the late 1980s and more recently in the 2000s (Feely and Madden 1986, 1987, and Farrell et al. 2014, 2015). A vertical, continuously cored borehole with a depth of 137.8 m was drilled into the CMGr in 1982 to obtain temperature and heat flow measurements. A thermal gradient of 21.3 °C km^{-1}, with a heat flow of 77 mW m^{-2} was determined from the borehole indicating that the CMGr can only be considered as a very low-grade heat source.

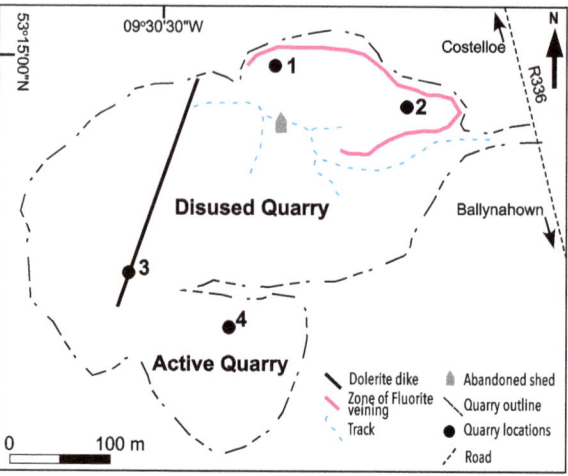

Fig. 32 Map of the disused CMGr quarry at Locality 9.14 (Fig. 31). Quarry locations 1, 2, 3 and 4 indicate where the images (**a**, **b**, **c** and **d**) in Fig. 33 were captured. The map also shows the main zone of fluorite veining in the quarry. The late Devonian dolerite dike is cut by fluorite veins (location 3). Clusters of cherry red spessartine garnet occur in rare pegmatite veins at location 4

A horizontal pegmatite containing clusters (~2 cm across) of cherry red spessartine garnets and a garnetiferous aplite were recorded in the oldest section of the quarry (Whitworth and Feely 1994; Figs. 32, 33d). However, ongoing granite extraction in this part of the quarry means that these exposures may not survive, however, loose blocks of the pegmatite may still be available for inspection in the vicinity of location 53.251390°, −9.510712°.

NNE-trending vertical fluorite veins contain green, purple and colourless fluorite, along with other vein minerals that can include quartz (smoky in places), calcite, epidote, chlorite, barite, galena, chalcopyrite, sphalerite and pyrite. Crystals of purple to green fluorite (Fig. 33a, b), smoky and clear quartz, galena and chalcopyrite are worth searching for although over the years, the best material has been removed. Dark red dendritic haematite is also visible on many joint surfaces and is the result of precipitation of the iron oxide from surface peaty waters percolating down along the granite joints. A NNE trending dolerite dike is also exposed in the quarry and is cut by the hydrothermal fluorite veins (Figs. 32, 33c). Extensive alteration of both granite and dolerite is genetically related to emplacement of these veins and can be readily observed looking towards the NNW and NNE quarry walls (Fig. 32, areas 1 and 2). The fluorite veins form part of a regional suite of fluorite + Pb, Cu and Fe sulphide veins that transect the Galway Batholith and are considered to represent late-Triassic hydrothermal mineralisation. The ~230 Ma (Ar–Ar age; pyroxene and whole rock) age obtained for the dolerite probably reflects the age of the hydrothermal mineralization rather than the dolerite's emplacement age (Feely et al. 2020). The dolerite is a member of the mid-Palaeozoic (Late Devonian?) dolerite dike suite that transects the Galway Batholith (Mohr et al. 2018). These fluorite and sulphide bearing ~230 Ma veins form part of the N Atlantic-European Triassic-Jurassic hydrothermal mineralization province. The rifting of the N Atlantic facilitated crustal thinning and subsidence of continental crust and initiated hydrothermal activity at the margins of

Fig. 33 a Examples of green and **b** purple fluorite veins cutting the CMGr (loc.1 and 2 in Fig. 32) **c** the fluorite veins cut the CMGr and the late Devonian dolerite dike (loc.3 in Fig. 32). **d** Clusters of cherry red spessartine garnet hosted by a rare pegmatite (loc.4 in Fig. 32). (From Feely et al. 2006, reproduced with permission from Geological Survey Ireland)

Mesozoic basins, triggering fluorite vein emplacement—see Feely et al. (2020) for a review of this topic.

6.6.2 The Rossaveel Peninsula

Access and other information From the CMGr quarry (locality 9.14) drive northwest on the R336 for ~3 km to the crossroads and turn left onto the R372, signposted for the Port of Rossaveel (Fig. 31). After 1.9 km turn left onto a narrower road, and then proceed to the junction with the second minor road on the right (Fig. 34). Parking is available in the vicinity of this junction (P: Fig. 34; at 53.262596°, −9.545838°). Proceed on foot along this minor road that leads across the northern tip of the sea inlet, to locality 9.15 and from there to the Martello Tower to reach locality 9.16 (Fig. 34).

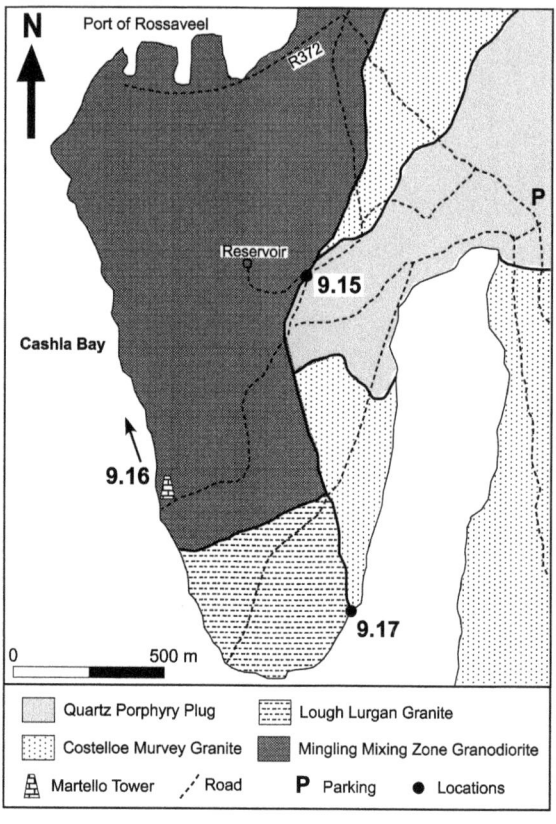

Fig. 34 A geologic map of the Rossaveel Peninsula showing the distribution of the main granite types. Excursion locations 9.15, 9.16 and 9.17 are shown as 15, 16 and 17 respectively

6.6.3 Locality 9.15: Quartz Porphyry's Intrusive Contact with the Mingling Mixing Zone Granodiorite (MMZGr) (53.261114°, −9.556202°)

The quartz porphyry plug that intrudes the CMGr (Konig 2003) can be examined (locality 2, Fig. 34) on the way to the coastal exposures of the Mingling Mixing Zone Granodiorite (MMZGr). Having reached the T-junction (at 53.259866°, −9.556202°) turn right onto the road leading up to the reservoir where the intrusive contact between the MMZGr and the porphyry is well exposed at locality 9.15 (Fig. 34). Here angular xenoliths of the MMZGr occur within the porphyry. Return to the T-junction (at 53.259866°, −9.556202°) and note the exposures of the porphyry present along the roadside. The porphyritic texture comprises rounded millimetric scale quartz phenocrysts (i.e., bird's eye quartz texture), orthoclase euhedra and books of biotite set in a fine-grained groundmass (Fig. 35).

Elsewhere, the intrusive contact between the porphyry and the CMGr is both sharp and locally gradational the latter indicates the coeval relationship between the two magmas. The porphyry represents a high-level differentiate of the CMGr magma.

Fig. 35 The bird's eye quartz texture displayed by the Quartz Porphyry. The arrows indicate examples of the rounded quartz phenocrysts

The porphyritic texture reflects early phenocryst formation followed by rapid magma cooling to form a sub-volcanic rhyolite. Proceed from here to the Martello Tower and locality 9.16 (Fig. 34).

6.6.4 Locality 9.16: Martello Tower Coastal Section: Mingling Mixing Zone Granodiorite (MMZGr) (53.255045°, −9.562116° to 53.257970°, −9.563495°)

Access and other information From location 9.15 return (<100 m) to the T-junction and proceed southwestwards along the track—an old military road—leading to the early nineteenth century Martello tower on the seashore (at 53.255045°, −9.562116°). The Martello tower was built by the British Ordnance around 1804 to defend the British Empire against a possible invasion by Napoleon Bonaparte's troops (Fig. 36). However, this invasion never materialised. Forty were built around the Irish coast. This example here at Rossaveel was built using cut blocks of limestone probably sourced from the Carboniferous limestone quarries on the shores of Lough Corrib, Co. Galway.

The Central Block (i.e., between the Shannawona and Barna Faults) of the Kilkieran Pluton is notable for its relative abundance of variably foliated diorite to granodiorite exposures that display field and textural characteristics indicative of interaction between coeval diorite and granite magmas. The MMZGr occurs between Costelloe in the west and Spiddal in the east. It provides spectacular examples of coeval diorite and granite interactions on continuous, wave-polished coastal exposures. In general, the MMZGr is a highly variable hybrid granodiorite that contains an abundance of mafic microgranular enclaves (MME). To the north, the MMZGr has a concordant intrusive contact with the foliated Mafic Megacrystic Granodiorite

Fig. 36 The early nineteenth century Martello tower, located on the Rossaveel peninsula ~1.3 km south of the Port of Rossaveel (see Fig. 34)

(MMGr in Fig. 26, Kilkieran Pluton) and was intruded by the CMGr, the Quartz Porphyry and the LLGr.

Outcrops displaying evidence for coeval diorite and granite magma interactions occur for ~750 m, along the coastline north of the Martello tower. Three main lithologies can be recognised in these exposures, that together make up the MMZGr: dark grey mafic microgranular enclaves (MME), a highly variable medium- to fine-grained hybrid granodiorite, and a grey medium-grained granite (see Fig. 37).

The light grey granite typically contains quartz (~28%), orthoclase (~43%) plagioclase (~27%) and biotite (~3%). The hybrid is very variable, but an average modal abundance indicates the presence of quartz (~20%), orthoclase (~17%), plagioclase (An_{32-28} ~ 50%), biotite (~10%), hornblende (~3%) and accessory titanite, apatite and magnetite. The enclaves contain phenocrysts of plagioclase (An_{30-40}) set in a fine grained (<1 mm) groundmass of plagioclase (46–54%; An_{30-40}), hornblende and biotite (22–25%), quartz (15–18%), orthoclase (4–14%) and accessory titanite, acicular apatite and zircon. The random orientation of groundmass grains in the MME contrasts with a fabric defined by local alignments of biotite and hornblende especially along MME margins. Plagioclase phenocrysts in the MME are commonly enveloped by this fabric. The exposures of MMZG display highly irregular local variations in mafic mineral (hornblende and biotite) over distances of <1 m. The hybrid invariably contains MME that range from <5 cm to 0.5 m in longest dimension. Sharp chilled margins are displayed by some enclaves; however, others display diffuse and gradational margins with the enclosing hybrid. All three lithologies display intricate intermingling patterns with each other (Figs. 37, 38). The MME display a wide range of morphologies. Ovoid, rounded and subangular examples occur and some display serrated to cuspate margins. Lenticular apophyses (<3 m long) of the hybrid occur

Fig. 37 A typical coastal exposure of the Mingling Mixing Zone Granodiorite (MMZGr) north of the Martello Tower (location 9.16) showing three main lithologies: **A** dark grey mafic microgranular enclaves (MME), **B** highly variable hybrid granodiorite and **C** light grey coloured granite. Note the upper section of the image where a transition from the dark grey MME (**A**) to the enveloping hybrid (**B**) and marginal lighter grey granite layer (**C**). The lobate margins between the granite and hybrid are testament to the coeval nature of the magmas. This is repeated in the lower half of the image where flow aligned layers of hybrid (**B**) and granite (**C**) can be distinguished

Fig. 38 Coastal exposure of the MMZGr north of the Martello Tower (location 9.16). A steeply dipping planar fabric defined by the alignment of mafic microgranular enclaves (MME) hosted by hybrid granite. (From Feely et al. 2006, reproduced with permission from Geological Survey Ireland)

Fig. 39 Titanite (aka sphene)–plagioclase Ocelli. This magma-mingling texture consists of a core of titanite typically in an ophitic intergrowth with calcic plagioclase (An_{30-40}) laths, surrounded by a zone of calcic plagioclase ± quartz ± orthoclase

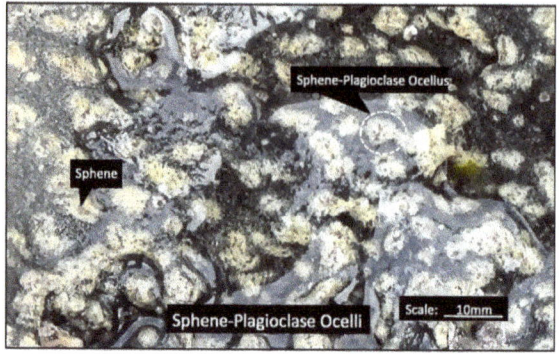

within the granite while irregular patches, stringers and tendrils of the granite occur within the hybrid (Fig. 37).

Layers of the granite alternate with hybrid layers on many exposures. Individual layers vary in width (30–30 cm) and commonly expose aligned lenticular enclaves (<0.5 m long). In general, both the layering and enclave alignment are vertical and trend in a NE direction. Locally, however, this trend abruptly changes (due in part to local faults) to NS and SE directions. Some of the enclaves display an abundance of titanite (also known as sphene)—plagioclase ocelli (~2–10 mm across—see Fig. 39), a texture indicative of magma mixing and mingling (Baxter and Feely 2003).

6.6.5 Locality 9.17 Contact Between the Lough Lurgan and Costelloe Murvey Granites (53.253612°, −9.553118°)

If time and tide permit, a visit to this locality is recommended, as it exposes the sub-horizontal intrusive contact between the ~380 Ma Costelloe Murvey Granite (CMGr) and the ~400 Ma Lough Lurgan Granite (LLGr). Access to this locality requires a walk from the Martello Tower around the southern tip of the peninsula (Fig. 34). The contact is exposed on the seaward side of a prominent sub-vertical slab of rock at the top of the shore. The contact is sharp, strikes 140° and dips 18°S. The younger (~380 Ma) CMGr displays a chilled margin of ~10 cm wide against the overlying ~400 Ma medium-grained, unchilled Lough Lurgan Granite (Fig. 40).

Exercises

1. Document the mineralogical evidence that supports the classification of the Costelloe Murvey Granite (CMG) as an Alkali Feldspar Leucogranite (Loc 9.14).

2. Construct a timeframe to document the sequence of geological events and processes that are exposed in the CMG Quarry (Loc 9.14). This timeframe should focus on granite emplacement, hydrothermal mineralisation, aplites and pegmatites, mafic dike(s) and granite jointing.
3. Document the field and mineralogical evidence for magma mingling and mixing exposed along the Martello Tower coastal section (Locs 9.15, 9.16).
4. Document the field and textural evidence present in this excursion area for subvolcanic activity at ~380 Ma (Middle Devonian), discuss its significance or otherwise for the regional tectonics at that time.

Fig. 40 The intrusive contact between the Costelloe Murvey Granite and the older Lough Lurgan Granite. The CMGr displays a well-defined chilled margin of ~10 cm against the LLGr. Inset is a wider-angle view of this contact (adapted from Feely et al. 2006)

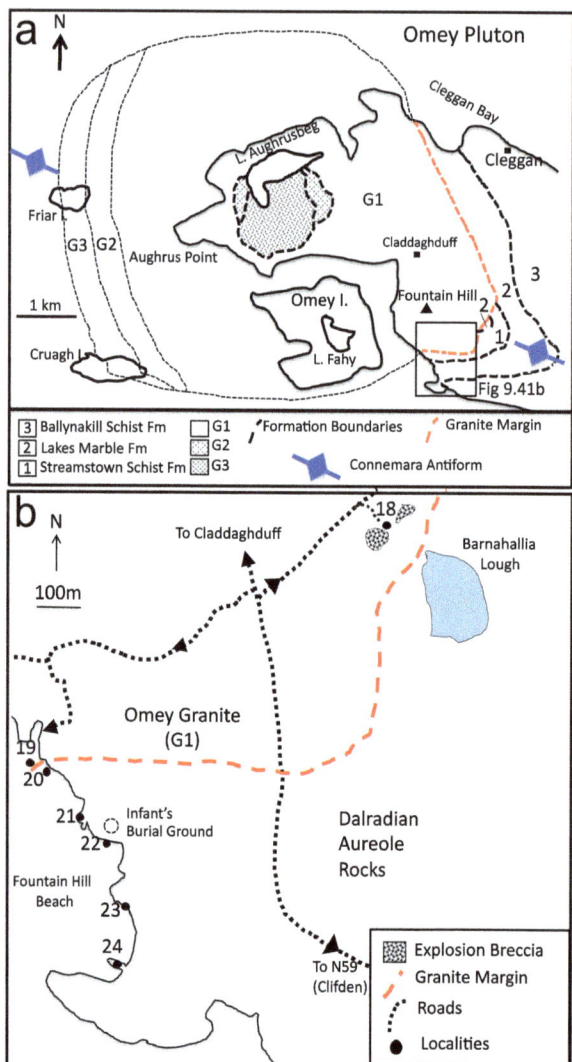

Fig. 41 **a** Geologic map of the Omey Pluton and its country rocks (after Townend 1966; Leake et al. 1981; Leake and Tanner 1994; McCarthy et al. 2015b). **b** Geologic map of the Fountain Hill beach area showing the spatial distribution of the field localities (9.18–9.24) described in the text

Table 4 Geochronology of the Earlier Plutons. Chronometry from: [1]Feely et al. (2018), [2] Feely et al. (2007), [3] McCarthy et al. (2015a) and [4]Friedrich and Hodges (2016)

Earlier plutons			
Pluton	U-Pb single crystal Zircon	Re-Os Molybdenite	Ar/Ar Biotite
Inish pluton	[1]Main Granite: 422.72 ± 0.19 Ma [1]Marginal Granite: 423.77 ± 0.26 Ma		
Omey pluton		[2]G2 Granite 422.5 ± 1.7 Ma	
Roundstone pluton	[3]GI Granite c.420 Ma		[4]G1Granite 420 ± 4 Ma

7 The Omey Pluton and Its Thermal Aureole at Fountain Hill Beach, South of Claddaghduff.

7.1 Introduction

The Omey Pluton is a member of the ~422 Ma Earlier Pluton suite (Omey, Inish, Roundstone and Letterfrack)—see Table 4 The southern margin of the Omey Pluton and its Dalradian aureole are continuously exposed at Fountain Hill beach, south of Claddaghduff (Fig. 41a, b). This location provides an excellent opportunity to study granite emplacement, an explosion breccia and granite-related contact metamorphism. The Dalradian country rocks were subjected to regional metamorphism during the Grampian orogeny (~475–462 Ma, chapter "The Ordovician Arc Roots of Connemara") and were then thermally overprinted during the emplacement of the Omey Pluton. This excursion visits five locations; their geological setting and spatial distribution are shown in Fig. 41a, b.

This excursion was prepared by Bruce WD Yardley, William McCarthy and Martin Feely.

7.2 Geological Setting

The Omey Pluton intruded the Dalradian metasedimentary rocks of the Streamstown Schist Fm. and the Lakes Marble Fm (see Chapter "The Ordovician Arc Roots of Connemara"). The pluton displays a circular outcrop pattern that is ~6 km in diameter. It is a composite intrusion consisting of three main granite types, G1, G2 and G3. The G1 granite makes up ~75% of the surface area of the pluton (Fig. 41a, b). G1 is a medium-grained (~8 mm) equigranular, biotite + hornblende granite. G2 occurs between G1 and G3 and is a medium-grained biotite + hornblende granite similar to G1, and locally contains K-feldspar megacrysts (~4 cm in longest dimension). G3 is a small circular intrusion (diameter ~1.2 km) exposed on the Aughrus Peninsula.

It is a medium to fine grained (~2–6 mm) pink granite and displays a porphyritic texture defined by ~15 mm plagioclase and K-feldspar phenocrysts and ~12 mm quartz grains. The pluton was emplaced into the south-easterly plunging (~25°) hinge region of the D4 Connemara antiform by a combination of centrally focused magma ascent followed by lateral sequential emplacement of G1, G2 and finally G3. The emplacement was controlled by the symmetry of the antiform, forming a convex upward lens-shaped igneous intrusion (mushroom shaped) called a phacolith (McCarthy 2015b).

The geology and contact metamorphic history of the Omey Pluton's aureole rocks are documented by Cobbing (1969), Ferguson and Harvey (1979) and Ferguson and Al-Ameen (1985). Here, thermally overprinted pelites and carbonates of the Middle Dalradian, Streamstown Schist and Lakes Marble Formations display notable contact metamorphic lithologies including andalusite hornfelses and spectacular wollastonite -grossular -vesuvianite (idocrase)-diopside skarns. Andalusite and cordierite occur up to 1.4 km and metamorphic biotite up to 2 km from the granite margin (Leake and Tanner 1994). Ferguson and Al-Ameen (1985) calculated the inner aureole P–T conditions at $615 \pm 25°$ C and 2.5 ± 0.25 kbar.

7.3 Excursion 5: The Omey Pluton and Its Aureole

Access and other information To get there, travel to Clifden then head north on the N59 (Westport Road) past the catholic church on the right and continue for 8.5 km then turn left onto the Claddaghduff road and proceed along this road, with Streamstown Bay on your left, for approximately 8 km, Barnahallia Lough is to the east (Fig. 41b). To access locality 9.18 adjacent to Barnahallia Lough, turn right and proceed on foot along the country lane shown in Fig. 41b. To access Fountain Hill beach and localities 9.19–9.24 turn left, find parking along this country road and proceed on foot. A left turn into a narrow sandy laneway leads to the beach and locality 9.19 (Fig. 41b). The beach section ends by the Clifden Eco Campsite, and parties may be able to make prior arrangements to use it to access the southern end of the beach.

NB: The traverse of Fountain Hill beach should only be attempted at low tide, and neap tides may still cause wet feet in the middle of the traverse. Time your trip carefully using Clifden tide times.

7.3.1 Locality 9.18: Barnahallia Explosion Breccia and the Omey Pluton Contact (53.533564°, −10.125674°)

Access and other information Walk ~500 m along the country lane and cut across the fields on your right just as the track begins to veer to the right and walk ~100 m across the field to the southeast, towards Barnahallia Lough (Fig. 42). Several outcrops of granite and country rock can be observed in this area.

Fig. 42 The explosion breccia composed of angular clasts of Dalradian country rock and Omey granite blocks hosted in a relatively fine-grained matrix

Multiple sets of 10–50 cm thick, granitic to dioritic dikes transect the G1 facies of the pluton at this locality. Dikes display unchilled sharp linear to lobate contacts. In addition, diffuse contacts with the granite are locally developed. Quartz ocelli and reaction rims around feldspar phenocrysts in some of the relatively mafic dykes can be observed using a hand lens. All these features indicate that multiple batches of magma coexisted and interacted as the pluton inflated.

Outcrops of explosion breccia occur closer to the country rock contact. Three distinct pods of breccia with diameters ranging from 10 to 30 m are present. Spectacular angular clasts, some ranging up to ~50 cm in size occur in a fine to medium grained matrix (Fig. 42). Breccia clasts appear to be derived from facies of the Omey Pluton and the Dalradian country rock. The breccia formed during the emplacement of the Omey Pluton.

The pluton—country rock contact from Barnahallia Lough towards Fountain Hill, exposes multiple sheets of granite that intrude along the steeply inclined laminations in the country rocks. Abrupt undulations in the granite-country rock contact in this area occur parallel to regional joint sets (NNW-SSE, WNW-ESE). These features are interpreted to be broken bridges that formed when intruding sheets of magma coalesced to form progressively thicker, longer sheets (Fig. 43).

The pluton—country rock contact from Barnahallia Lough towards Fountain Hill, exposes multiple sheets of granite that intrude along the steeply inclined laminations in the country rocks. Abrupt undulations in the granite-country rock contact in this area occur parallel to regional joint sets (NNW-SSE, WNW-ESE). These features are interpreted to be broken bridges that formed when intruding sheets of magma coalesced to form progressively thicker, longer sheets (Fig. 43).

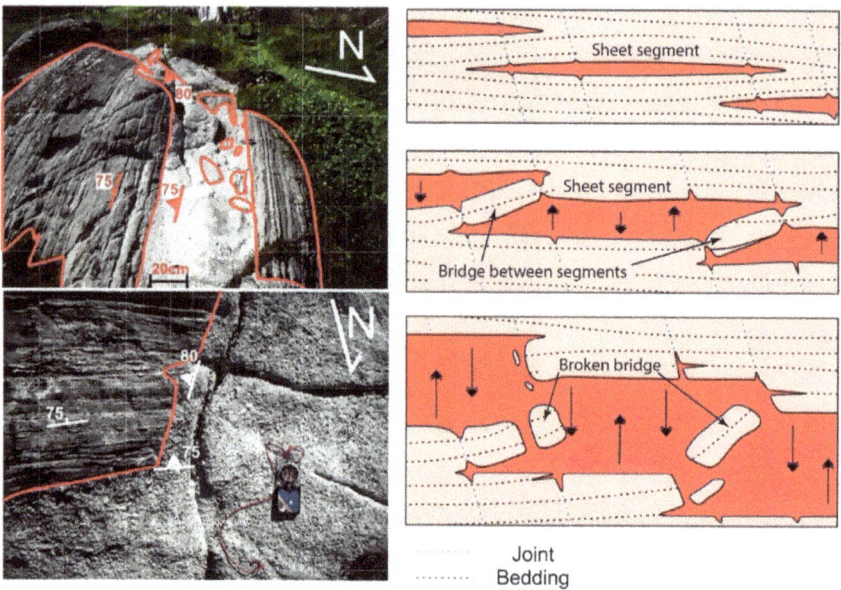

Fig. 43 Examples of steeply inclined granite sheets that define stepped contacts and broken bridge structures that formed during the piece meal injection and inflation of granite magma into the deformed Dalradian country rocks

7.3.2 Locality 9.19: Fountain Hill Beach: Omey Granite (G1) Exposures (53.529029°, −10.138880°)

Enter the beach via the narrow sandy path, below a conspicuous holiday home (at 53.530821°, −10.138096°). The medium-grained G1 Omey granite is exposed in numerous outcrops along the coast to the right but the tidal island in front of you is ideal for inspection of the granite. Here, wave-polished outcrops facilitate recognition of the granite's essential mineralogy comprising approximately equal amounts of pink orthoclase and white plagioclase (oligoclase-andesine), slightly less glassy quartz and about 5% dark minerals -biotite and a little hornblende. Large (<2 cm) crystals of orthoclase are set in a finer groundmass and it is sometimes possible to observe tiny white inclusions of plagioclase in the orthoclase crystals marking growth zones.

7.3.3 Locality 9.20: Fountain Hill Beach: Contact Zone Between the Omey Pluton and the Streamstown Schist Formation (53.528494°, −10.137942°)

The granite's contact, ~50 m around the corner to the left, is marked by thin sheets of finer grained granite intruded into the psammitic and pelitic hornfels (Fig. 44a). The

Fig. 44 a Intrusive contact zone of the Omey G1 granite with hornfelsed schists. Note how the pink fine-grained granite sheets parallel the bedding in the country rocks. **b** Detail of granite sheet at the intrusive contact. The granite is cut by a composite vein with very fine saccharoidal aplite at the edges, but very coarse pegmatite in the core, products of coexisting melt and aqueous fluid respectively

granite sheets are concordant with bedding in the hornfels and appear to preserve the process of levering of the country rock to form xenoliths, a process known as stoping, The contact is also cut by discordant pale sheets of very fine grained saccharoidal aplite, representing the final, water-saturated melt which froze because of the pressure drop as it flowed into fractures. Locally, coexisting coarse grained pegmatite is also present (Fig. 44b). The hornfels look to have a relict schistosity, but are very tough with no preferential splitting direction. Hand lens inspection of the pelite bands, preferably in sunshine, reveals the presence of small andalusite prisms formed by the contact metamorphism—note their size and aspect ratio for comparison with more distal outcrops. There are also folds present, formed in the regional second phase of folding, and these are cut by aplite sheets. Note that these folds are very tight and almost vertical; further away from the granite they become more open and plunge south.

7.3.4 Locality 9.21 Fountain Hill Beach: The Skarns of the Lakes Marble Formation (53.527476°, −10.136552°)

Continue south across the beach, crossing more hornfelsed psammite and pelite. At the end of the beach cross the landward side of a rock ridge but before continuing into the fault gully ahead, turn right to face the sea and explore the gully ahead of you. The rocks here are skarn, but were originally the calcite marbles (whose protoliths were siliceous dolomitic limestones) of the Lakes Marble Fm (Chapter "The Ordovician Arc Roots of Connemara", Excursion 1) but have been extensively modified by fluids and contact metamorphism from the Omey Pluton. Near the head of the gully a grey outcrop is all that remains of the marble. Here, there are chalky white rosettes of wollastonite and centimeter-size pink grossular-rich garnet (see Fig. 46b from Locality 9.22). Elsewhere, much of the original marble is replaced by skarn beds consisting of pink grossular, with minor brown vesuvianite (idocrase), (Fig. 45). Other interbeds are green and rich in diopside, plagioclase and locally amphibole. They appear to represent original psammite bands.

Fig. 9.45 Vesuvianite (idocrase) in skarn—Lakes Marble Formation

Fig. 46 a Andalusite prisms, developed during contact metamorphism of pelitic hornfels. Crystal size increases away from the granite contact. **b** Skarn developed by metasomatism of a Lakes Marble Fm calcite marble. Dark grey calcite in the centre of the original bed is partly replaced by grossular-rich garnet and minor white wollastonite. Figures 46a, b (© Bruce Yardley) appeared in the supplementary material to Yardley and Warren (2021)

NB: if you wish to take samples of skarn, do so here from the many loose boulders. There are other outcrops you will visit shortly, but they are rightly protected and should not be hammered under any circumstances.

7.3.5 Locality 9.22: Fountain Hill Beach: Large Andalusite Porphyroblasts and Adjacent Marble Skarn (53.527022°, −10.135220°)

Go back to the head of the skarn gully and head south through a fault gully. Where it emerges on the beach (*this is where the tide can catch you out*), turn left and examine the surfaces of steeply dipping bedded hornfels. (Fig. 46a). The pelitic beds are crowded with andalusite prisms that are much easier to see than the ones at the contact. This is because contact metamorphism commonly produces finer grain sizes in the more rapidly heated rocks at a contact, an effect analogous to the chilled margin of an igneous intrusion. Further up the beach there are superb outcrops of skarn in which dark grey calcite is studded with pink garnet, white wollastonite, sporadic green diopside and brown vesuvianite porphyroblasts (Fig. 46b). Some garnets are zoned with the darker pink reflecting a higher content of andradite garnet component.

7.3.6 Locality 9.23: Regional Structures in the Aureole, Part 1 (53.525631°, −10.134782°)

Cross the beach and make your way across a rocky point or proceed inshore until you come to the last outcrops before a large beach. Throughout this N–S traverse of the thermal aureole regional, second phase folds abound and away from the granite they are seen to be tight and overturned to the north, so that one limb is almost vertical, and the other dips south at around 50°. They strike approximately east–west. On this side of the bay, gently dipping fold limbs predominate, so that the folds verge towards an antiformal structure to the north. The pelitic lithologies here locally preserve regional metamorphic almandine garnet, but there are also skarn interbeds and andalusite in the pelites.

7.3.7 Locality 9.24: Regional Structures in the Aureole, Part 2 (53.524343°, −10.135058°)

Cross the sandy beach to the south side where a prominent rib of rock runs out into the sea. This exposure contains a similar interbedded pelitic and psammitic hornfels with minor skarn to the previous locality, but the bedding is nearly vertical. Walk up along the rib of rock, then turn round: you should now see easterly plunging small-scale folds, of the second fold phase. The orientation of the limbs and hinges are the same as at Locality 9.23, but the outcrop looks different because the rocks mostly belong to the steep limb and the folds verge south. Beneath the sand of the bay is the core of a major regional second phase synformal fold.

Returning to the beach you may notice a pale dike crossing the outcrop on the landward side. This dike is of broadly granitic composition and could be related to the Omey pluton, but such dikes are widespread in Connemara and so its origin is uncertain.

Exercise

1. Document the field evidence that indicates multiple batches of magma coexisted and interacted during the emplacement of the Omey Pluton.
2. The granite at locality 9.19 has some particularly clean surfaces below the high tide line, and the four major minerals can be seen clearly on these. (a) Estimate the relative proportions of quartz, plagioclase, K-feldspar and biotite (assume all the dark grains are biotite) and measure the typical size and shape of each. Recalculate the proportions of quartz and the feldspars to 100% without biotite and plot your answer on Fig. 10. (b) Find a good example of a large K-feldspar grain with inclusion. Which minerals occur as inclusions in K-feldspar and which do not? What can you deduce from this about the sequence in which the different mineral species appeared as the melt solidified? Discuss why the included grains are much smaller than the K-feldspar host. Is it a reflection of the trapping mechanism or representative of the crystal size population in the partially solidified magma at the time?
3. (a) Examine the andalusite prisms at locations 9.20, 9.22 and 9.23 or 9.24. In each case choose 5 crystals at random and measure their width and length. Also find the largest crystal you can and measure that also. Is the change in size away from the contact what you expected? Discuss possible reasons for your observations and draw comparisons with the variation in size of crystals in an igneous rock away from its contact. (b) Although not apparent in outcrop, Ferguson and Al-Ameen (1985) reported that K-feldspar occurs with andalusite in much of the aureole, whereas sillimanite is restricted to a few isolated occurrences at the granite contact and evidence of partial melting is lacking. What constraints do these observations place on the likely depth of emplacement of this part of the Omey granite?

References

Archibald DB, Macquarrie LMG, Murphy JB, Strachan RA, McFarlane CRM, Button M, Larson KP, Dunlop J (2021) The construction of the Donegal composite batholith, Irish Caledonides: Temporal constraints from U-Pb dating of zircon and titanite: GSA Bull. https://doi.org/10.1130/B35856.1

Archibald D B, Murphy J B (2021) A slab failure origin for the Donegal composite batholith, Ireland as indicated by trace-element geochemistry. In: Murphy B, Strachan RA, Quesada C (Eds) Pannotia to Pangaea: neoproterozoic and Paleozoic orogenic cycles in the circum-Atlantic region. Geol Soc Lond Spec Pub 503:347–370

Atherton MP, Ghani AA (2002) Slab breakoff: a model for Caledonian, Late Granite syn-collisional magmatism in the orthotectonic (metamorphic) zone of Scotland and Donegal, Ireland. Lithos 62:65–85

Baxter S (2000) The geology of part of the Galway Granite near Barna, Co. Galway, Ireland. Unpublished PhD thesis, National University of Ireland, Galway

Baxter S, Feely M (2002) Magma mixing and mingling textures in granitoids: examples from the Galway Granite, Connemara, Ireland. Miner Petrol 76:63–67

Baxter S, Graham NT, Feely M, Reavy RJ, Dewey JF (2005) Fabric studies of the Galway Granite, Connemara, Ireland. Geol Mag 142:81–95

Brown PE, Ryan PD, Soper NJ, Woodcock NH (2008) The Newer Granite problem revisited: a transtensional origin for the Early Devonian Trans-Suture Suite. Geol Mag 145:235–256

Buchwaldt R, Kroner A, Toulkeredes T, Todt W, Feely M (2001) Geochronology and Nd-Sr systematics of Late Caledonian granites in western Ireland: new implications for the Caledonian orogeny. Geol Soc Am Abstracts with Programs 33(1):A32

Burke KC (1957) An outline of the structure of the Galway Granite. Geol Mag 94:452–464

Burke KC (1963) Age relations of Connemara Migmatites and Galway Granite. Geol Mag 100:470–471

Callaghan B (2005) Locating the Shannwona Fault: field and barometric studies from the Galway Batholith, western Ireland. Irish J Earth Sci 23:85–100

Chew DM, Strachan RA (2014) The Laurentian Caledonides of Scotland and Ireland. In New Perspectives on the Caledonides of Scandinavia and Related Areas. Edited by F Corfu, D Gasser, DM Chew, Geol Soc Lond Spec Pub 390:49–91

Coats JS, Wilson JR (1971) The eastern end of the Galway Granite. Min Mag 38:138–151

Cobbing EJ (1969) The geology of the district north-west of Clifden, Co Galway. P Roy Irish Acad B 67:303–325

Crowley Q, Feely M (1997) New perspectives on the order and style of granite emplacement in the Galway Batholith, western Ireland. Geol Mag 134:539–548

Derham JM, Feely M (1988) A K-feldspar breccia from the Mo-Cu stockwork deposit in the Galway Granite, west of Ireland. J Geol Soc London 145:661–667

Dewey JF, Holdsworth RE, Strachan RA (1998) Transpression and transtension zones. In: Holdsworth RE, Strachan RA, Dewey JF (Eds) 1998. Continental Transpressional and Transtensional Tectonics. Geol Soc Lond Spec Pub, 135:1–14

Dewey JF, Strachan RA (2003) Changing Silurian-Devonian relative plate motion in the Caledonides: sinistral transpression to sinistral transtension. J Geol Soc London 160:219–229

Domeier M (2016) A plate tectonic scenario for the Iapetus and Rheic oceans. Gondwana Res 36:275–295

El Desouky M, Feely M, Mohr P (1996) Diorite-granite magma mingling and mixing along the axis of the Galway Granite batholith, Ireland. J Geol Soc London 153:361–374

Fairhead JD, Walker P (1977) The geological interpretation of gravity and magnetic surveys over the exposed southern margin of the Galway Granite, Ireland. Geol J 12:17–24

Farrell T, Jones A, Muller M, Feely M, Brock A, Long M, Waters T (2014) IRETHERM: the geothermal energy potential of Irish radiothermal granites. EGU Gen. Assem. 2014, Austria, Vol. 16, EGU2014–6749:16

Farrell T, Muller M, Rath V, Feely M, Jones A, Brock A, The Iretherm Team (2015) IRETHERM: The Geothermal Energy Potential of Radiothermal Granites in a Low-Enthalpy Setting in Ireland from Magnetotelluric Data. Proceed World Geothermal Congress 2015, Melbourne, Australia:1–10

Feely M (1982) Geological, Geochemical and Geophysical studies on the Galway Granite in the Costelloe-Inveran Sector, Western Ireland. Unpublished PhD thesis, National University of Ireland

Feely M, Coleman D, Baxter S, Miller B (2003) U-Pb zircon geochronology of the Galway Granite, Connemara, Ireland: Implications for the timing of late Caledonian tectonic and magmatic events and for correlations with Acadian plutonism in New England. Atlantic Geol 39:175–184

Feely M, Costanzo A, Gaynor SP, Selby D, McNulty E (2020) A review of molybdenite, and fluorite mineralisation in Caledonian granite basement, western Ireland, incorporating new field and fluid inclusion studies, and Re-Os and U-Pb geochronology. Lithos 354–355:1–12

Feely M, Gaynor S, Venugopal N, Hunt J, Coleman DS (2018) New U-Pb zircon ages for the Inish Granite Pluton, Galway Granite Complex, Connemara, Western Ireland. Irish J Earth Sci 36:1–7

Feely M, Leake B, Baxter S, Hunt J, Mohr P (2006) A geological guide to the granites of the Galway Batholith, Connemara, western Ireland. Geological Survey Ireland, Field guides Series. Geological Survey Ireland, Dublin:70pp. ISBN 1-899702-56-3

Feely M, Madden J (1986) A quantitative regional gamma-ray survey on the Main Galway Granite, western Ireland. In: Andrew CJ, Crowe RWA, Finlay S, Pennell WM, Pyne JF (Eds), Geology and Genesis of Mineral Deposits in Ireland. Irish Association for Economic Geology, Dublin:195–199

Feely M, Madden J (1987) The spatial distribution of K, U, Th and surface heat production in the Galway Granite, Connemara, western Ireland. Irish J Earth Sci 8:155–164

Feely M, Madden J (1988) Trace element variation in the leucogranites within the main Galway Granite, Connemara, Ireland. Min Mag 52:139–146

Feely M, McCabe E, Kunzendorf H (1991) The evolution of REE profiles in the Galway Granite, western Ireland. Irish J Earth Sci 11:71–89

Feely M, McCabe E, Williams CT (1989) U, Th and REE bearing accessory minerals in a high heat production (HHP) leucogranite within the Galway Granite, western Ireland. T I Min Metall B 98:27–32

Feely M, Selby D, Conliffe J, Judge M (2007) Re-Os geochronology and fluid inclusion microthermometry of molybdenite mineralisation in the late-Caledonian Omey Granite, western Ireland. Appl Earth Sci (Trans. IMM. B) 116(3):143–149

Feely M, Selby D, Hunt J, Conliffe J (2010) Long-lived granite-related molybdenite mineralization at Connemara, western Irish Caledonides. Geol Mag 147:886–894

Ferguson CC, Al-Ameen SI (1985) Muscovite breakdown and corundum growth at anomalously low f H2O: a study of contact metamorphism and convective fluid movement around the Omey granite, Connemara, western Ireland. Min Mag 49:505–515

Ferguson CC, Harvey PK (1979) Thermally overprinted Dalradian rocks near Cleggan, Connemara, western Ireland', P Geologist Assoc, 90:43–50

Friedrich AM, Hodges KV (2016) Geological significance of $^{40}Ar/^{39}Ar$ mica dates across a mid-crustal continental plate margin, Connemara (Grampian orogeny, Irish Caledonides), and implications for the evolution of lithospheric collisions. Can J Earth Sci 53:1258–1278

Konig M (2003) A detailed study of Feldspar Porphyries spatially related to the Costelloe Murvey Granite, Connemara, Ireland, Unpublished Diploma Mapping Project Thesis, Rheinische Friedrich-Wilhelms-Universität Bonn

Leake BE (1974) The crystallisation history and mechanism of emplacement of the western part of the Galway Granite, Connemara, western Ireland. Min Mag 39:498–513

Leake BE (1978) Granite emplacement: the granites of Ireland and their origin. In Bowes, DR Leake BE (Eds), Crustal Evolution in Northwest Britain and Adjacent Regions. Geol J, 10: 221–248

Leake BE (1986) The geology of SW Connemara, Ireland: a fold and thrust Dalradian metagabbroic-gneiss complex. J Geol Soc London 143:221–236

Leake BE (2006) Mechanism of emplacement and crystallisation history of the northern margin and centre of the Galway Granite, western Ireland. T Roy Soc Edin-Earth 97:1–23

Leake BE (2011) Stoping and the mechanisms of emplacement of the granites in the Western Ring Complex of the Galway granite batholith, western Ireland. T Roy Soc Edin-Earth 102:1–16

Leake BE, Leggo PJ (1963) On the age relations of the Connemara migmatites and the Galway Granite, West of Ireland. Geol Mag 100:193–204

Leake BE, Tanner PWG, Senior A (1981) The geology of Connemara: geological map (1: 63,360) with cross-sections, fold traces and metamorphic isograd map. Glasgow University

Leake BE, Tanner P WG (1994) The Geology of the Dalradian and Associated Rocks of Connemara, Western Ireland. Royal Irish Academy, Dublin, 96pp

Lees A, Feely M (2016) The Connemara Eastern Boundary Fault: a review and assessment using new evidence. Irish J Earth Sci 34:1–25

Lees A, Feely M (2017) The Connemara Eastern Boundary Fault: a correction. Irish J Earth Sci 35:55–56

Max MD, Geoghegan LCB, M, (1978) The Galway Granite and its setting. Geol Sur Ireland Bull 2:223–233

Max MD, Long CB, Keary R, Ryan PD, Geoghegan M, O'Grady M, Inamdar DD, McIntyre T, Williams CE (1975) Preliminary report on the geology of the northwestern approaches to Galway Bay and part of its landward area. Geol Sur Ireland, Report Series RS75/3

Max MD, Ryan PD, Inamdar DD (1983) magnetic deep structural geology interpretation of Ireland. Tectonics 2:431–451

McCarthy W (2013) An evaluation of orogenic kinematic evolution utilizing crystalline and magnetic anisotropy in granitoids. Unpublished PhD thesis, University College Cork, National University of Ireland

McCarthy W, Petronis MS, Reavy RJ, Stevenson CT (2015a) Distinguishing diapirs from inflated plutons: an integrated rock magnetic fabric and structural study on the Roundstone Pluton, western Ireland. J Geol Soc Lond 172:550–565

McCarthy W, Reavy RJ, Stevenson CT, Petronis S (2015b) Late Caledonian transpression and the structural controls on pluton construction; new insights from the Omey Pluton, western Ireland. T Roy Soc Edin-Earth 106:11–28

Mohr P, Hunt J, Riekstins H, Kennan PS (2018) Distinguishing dolerite dike populations in post-Grampian Connemara. Irish J Earth Sci 36:1–16

Murphy T (1952) Measurements of gravity in Ireland: gravity survey of central Ireland. Dublin Inst Adv Stud, Geophys Mem 2(3):31

Murray JW (2009) Wind transport of foraminiferal tests into subaerial dunes: an example from western Ireland. J Micropalaeo 28:185–187

Neilson JC, Kokelaar BP, Crowley QG (2009) Timing, relations and cause of plutonic and volcanic activity of the Siluro-Devonian post-collision magmatic episode in the Grampian Terrane, Scotland. J Geol Soc London 166:545–561

Pidgeon SJ (1969) Zircon U-Pb ages from the Galway Granite and the Dalradian, Connemara, Ireland. Scott J Geol 5:375–392

Pracht M, Lees A, Leake B, Feely M, Long B, Morris J, McConnell B (2004) Geology of Galway Bay: A geological description to accompany the Bedrock Geology 1:100,000 Scale Map Series, Sheet 14, Galway Bay. Geol Sur Ireland, Dublin, pp76

Ryan PD, Dewey JF (2004) The South Connemara Group reinterpreted: a subduction-accretion complex in the Caledonides of Galway Bay, western Ireland. J Geodyn 37:513–529

Selby D, Creaser RA, Feely M (2004) Accurate Re-Os molybdenite dates from the Galway Granite, Ireland. A critical comment to: Disturbance of the Re-Os chronometer of molybdenites from the late-Caledonian Galway Granite, Ireland, by hydrothermal fluid circulation. Geochem J 38:291–294

Streckeisen A (1976) To each plutonic rock its proper name. Earth Sci Rev 12:1–33

Townend R (1966) The geology of some granite plutons from western Connemara, Co. Galway, P Roy Irish Acad B 65:157–202

Whitworth MP, Feely M (1994) The compositional range of magmatic Mn-garnets in the Galway Granite, Connemara, Ireland. Min Mag 58:163–168

Wright P C 1964) The petrology, chemistry and structure of the Galway Granite of the Carna area, Co. Galway. P Roy Irish Acad B 63:239–264

Yardley BWD, Warren CE (2021) An introduction to metamorphic petrology, 2nd edn. Cambridge University Press, Cambridge, p 333

Devonian and Carboniferous

John R. Graham

Abstract Devonian rocks record parts of the history of oblique collision that produced sinistral shear along the Fair Head–Clew Bay Line. This is evidenced by the Ox Mountains granodiorite and small basins formed under local transtensional conditions associated with a transfer zone between the North Ox Mountains fault and the southern Clew Bay fault zone. These basins contain Lower Devonian (probably Emsian) and Middle Devonian (Eifelian) conglomerate dominated successions deposited in small basins. Provenance was mainly from the southeast but locally from the northwest. Well-rounded quartzite clasts hint at recycling and possible cannibalisation but other first cycle clasts give information on successions that are no longer extant. Consolidation of the Laurentian margin ceased at some stage during the later Devonian. Renewed extension in the Mississippian was manifested by reactivation of many earlier faults such that facies patterns and sediment thicknesses are markedly controlled by the underlying Caledonian template. Substantial non-marine to marginal marine successions are seen beneath the extensive Mississippian carbonates that characterise much of Ireland; their location being strongly controlled by the positive topography of western Ireland following the Caledonian orogenic cycle.

1 Introduction

The Upper Palaeozoic history of western Ireland spans the period of final consolidation of the Laurentian margin to that of reactivation of several of its main structures during the subsequent Variscan orogenic cycle. This chapter examines the sedimentary record of both the final stages of sediment accumulation of the 'Caledonian' cycle and the initial stages of the Variscan cycle where the effects of the 'Caledonian' cycle are most evident.

J. R. Graham (✉)
Department of Geology, Trinity College Dublin, Museum Building, Dublin D2, Ireland
e-mail: jrgraham@tcd.ie

© The Author(s), under exclusive license to Springer Nature Switzerland AG 2022
P. D. Ryan (ed.), *A Field Guide to the Geology of Western Ireland*, Springer Geology Field Guides https://doi.org/10.1007/978-3-030-97479-4_10

2 Devonian

The pattern of oblique collision with associated sinistral shear in this region that is seen in the northern two Silurian basins (Chapter "The Silurian of North Galway and South Mayo") continued through into the Devonian (Graham 2021; Graham et al. 2020; Riggs et al. 2021). A reliable Lower Devonian age of 412.3 ± 0.8 Ma for the Ox Mountains Plutonic Complex (Chew and Schaltegger 2005) provides clear demonstration that sinistral shear was present at this time. Continued uplift of the Laurentian margin produced a series of small basins whose location was controlled by the local geometry of major faults (Graham 1981a, 2009). The currently preserved remnants are present in the area between Clew Bay and Lough Conn (Fig. 1) and the salient features of these successions can be examined at several accessible localities. Exposure is very variable in quality but is sufficient to make a reasonably detailed field map of the successions (Graham 1981a).

Dating of exclusively non-marine sections, particularly in the Paleozoic, relies mainly on palynology and locally on fish faunas. In this area two significant productive palynological samples provide the basis for age assignation (Graham et al. 1983) (Fig. 1). The older assemblage from the Derryharriff Formation indicates an imprecise but definitively Lower Devonian age, whereas the younger assemblage

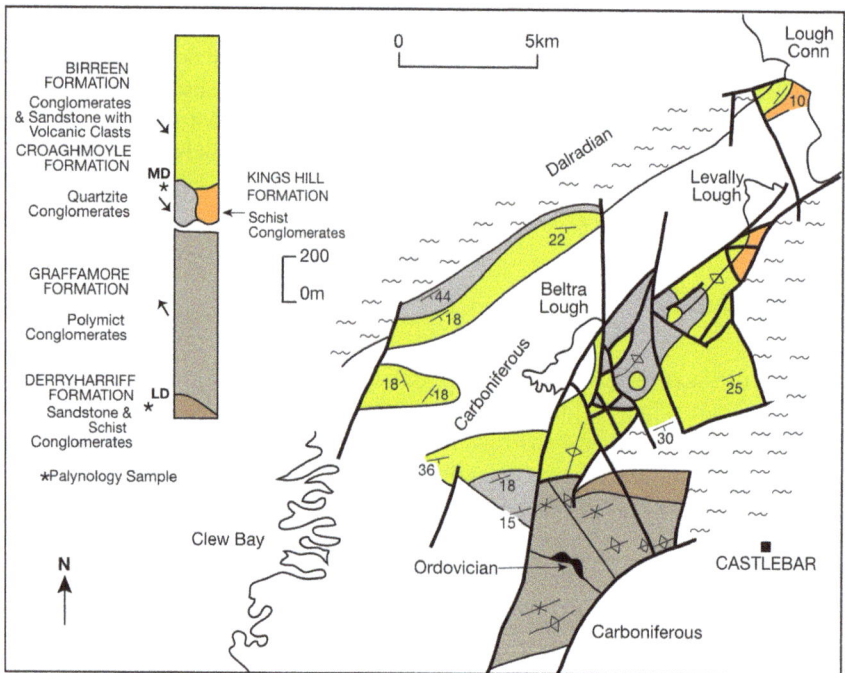

Fig. 1 Outcrop of the Devonian rocks of western Ireland (after Graham 1981a) with field localities indicated. Abbreviations used: LD = Lower Devonian; MD = Middle Devonian

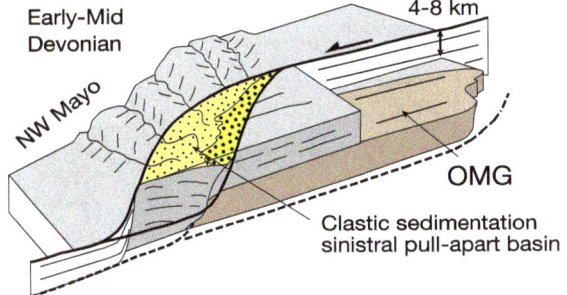

Fig. 2 Cartoon diagram illustrating the development of the Devonian basins under sinistral transtension (after McCaffrey 1997). OMG = Ox Mountains granite; grey colour = Dalradian

from the Croaghmoyle Formation indicates a Middle Devonian (early-mid Eifelian) age. These ages are consistent with the field observations that fold profiles in the Islandeady Group are tighter than those in the Beltra Group (Graham 1981a). The contact between the Islandeady and Beltra groups occurs in sparsely exposed ground but appears to be a fault.

Although only these two groups of Devonian rocks are preserved there may well have been other small basins that formed in a similar way but were subsequently uplifted and cannibalised into younger basins. There is some suggestion of this from the well-rounded nature of more resistant clasts such as quartzites that is suggestive of recycling (Locs. 10.4 and 10.5).

The area in which these rocks accumulated occupies a major fault transfer zone between the North Ox Mountains Fault and the southern Clew Bay Fault zone (McCaffrey 1997). McCaffrey (ibid.) suggested that basin size and geometry were controlled by the North Ox Mountains and Knockaskibbole faults and that the Ox Mountains block was being pulled north-eastwards (Fig. 2). Conversely the pattern of open folds and faults in these rocks indicates formation during dextral transpression that signifies a major change in kinematics in this part of the orogen (McCaffrey 1997; Fig. 3). Locally the timing of this kinematic switch can only be constrained as post-Eifelian and pre-Visean and is loosely ascribed to the Upper Devonian.

Whilst it is difficult to reliably constrain the original basin sizes, the Beltra Group (Fig. 1) does demonstrate sediment input from both NW and SE sources in its early history suggesting cross-basin dimensions of just a few km. The common presence of alluvial fan facies (Graham 1981a) would also be consistent with steep slopes and small basin sizes. Poor grain size segregation within all the conglomeratic successions is also consistent with very limited basin sizes.

The Devonian successions contain valuable provenance information which has only been superficially investigated, in part because many modern techniques were not available at the time of the main fieldwork. Particularly intriguing are the abundant red sandstone clasts and the numerous volcanic clasts in the Beltra Group (Locs. 10.4–10.6). Dating of the volcanic clasts and detrital zircon profiles from the sandstones would produce valuable additional data in unravelling the later history of the orogen. The thickest and most widespread formations (Lower Devonian Graffa More Formation and Middle Devonian Birreen Formation) both indicate a south-easterly

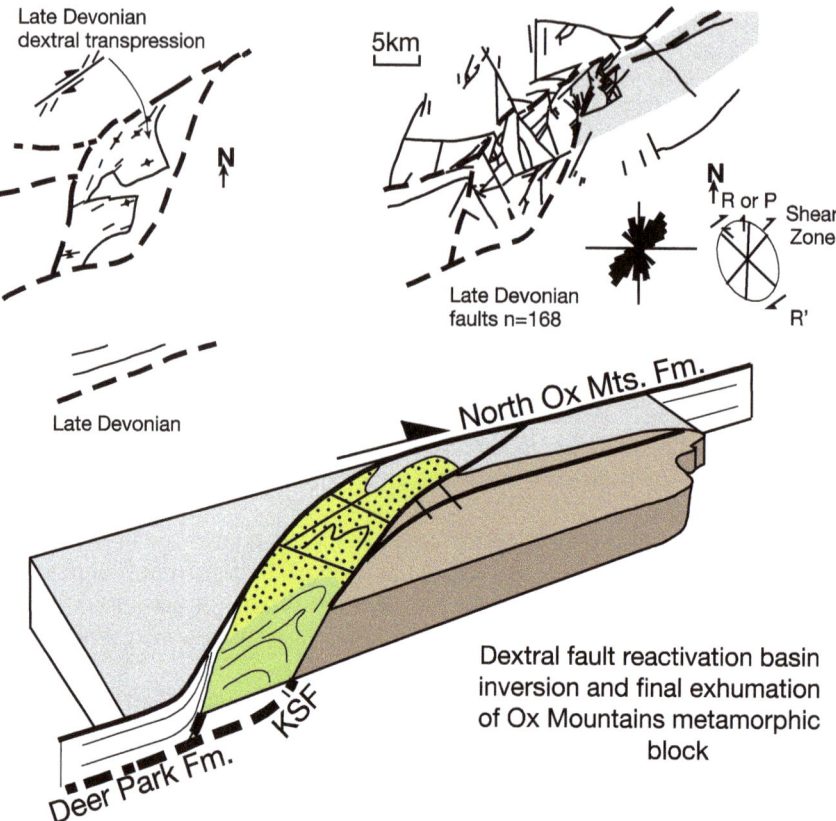

Fig. 3 Summary of tectonic structures in the Devonian rocks (the dotted region of the block diagram) with and their interpretation (after Graham 1981a; McCaffrey 1997). The Dalradian is shown in grey and the Ox Mountains granite in brown

derivation but have significantly different clast assemblages (Graham 1981a). Immature sandstone clasts in the Graffa More Formation may well represent erosion of the South Mayo Ordovician succession but this would need to be tested with some sandstone geochemistry and heavy mineral analysis. The presence of quartzose red sandstones, presumably of younger age, in the Birreen Formation may well be an indication of sampling different source areas during continued sinistral displacement.

2.1 Locality 10.1: Letter (53.853261°, −9.423138°)

Access and Additional Information From Castlebar take the R311 for Newport and after c. 3 km fork left for Islandeady on the L1810. After c. 2.5 km turn right at a crossroads (53.839452°, −9.402822°) on an unmarked road that heads uphill to the

Devonian and Carboniferous

hamlet of Letter. Pass the new school on the left and the old, abandoned school on the right proceeding to a T junction and then turn left. Proceed along this road, rising slightly until an obvious metalled track leaves on the left and park at the junction (53.853261°, −9.423138°) (Fig. 4).

This area is noteworthy for the small inlier of Lower Palaeozoic rocks, Letter Formation, that may be part of the South Mayo succession (Graham and Smith 1981).

The unconformable contact of the Graffa More Formation on the Letter Formation is no longer exposed although small parts of the Letter Formation can be seen at the side of the current road at (53.852195°, −9.421796°) and at the entrance to the house just to the east.

From the parking walk up the track until just after the first sharp right hand bend where there is easy access to open boggy moorland on the right (N). These are the typical glaciated rock surface exposures seen in the Graffa More Formation (Fig. 5). Return to the track and continue west. A small old quarry on the left (S)

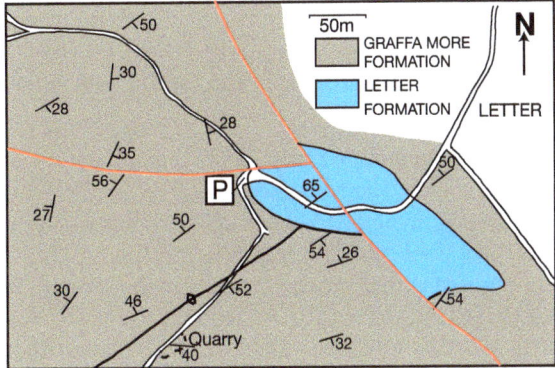

Fig. 4 Geological map of the Letter area. See Fig. 1 and caption for further details

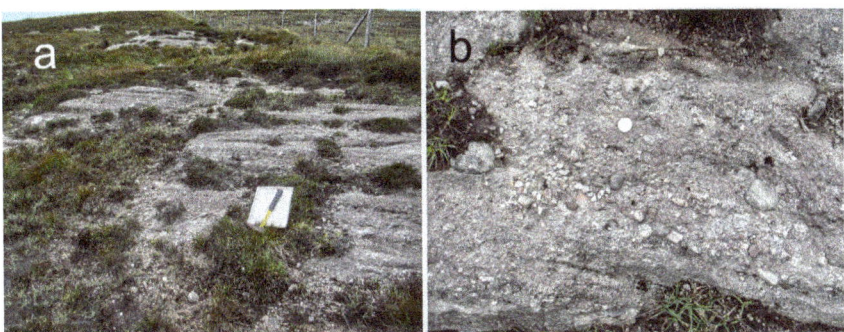

Fig. 5 Field images of Graffa More Formation **a** Typical exposure style of polymict conglomerates **b** Thin sandy lenses in flat bedded conglomerates

side of the track at (53.851191°, −9.423703°) shows typical flat bedded polymict conglomerates of the Graffa More Formation and the typical exposure style can be seen in the moorland to the south. Return to the track and access exposures on the north side of the track entering via a gate. The variable dips seen in these localities demonstrate that the rocks are folded about NE trending axes (Fig. 4) although folds are mapped out rather than seen due to their wavelength of tens to hundreds of metres. Good exposures in this area show the typical polymict nature of the conglomerates (see Graham 1981a, Fig. 5) and the dominance of flat bedding; some beds display imbrication. Thin graded beds of conglomerates probably represent individual floods. The lack of evidence for any large barforms suggests flows were shallow and poorly channelised (Graham 1981a, b). Return to the parking and then walk a c. 200 m west along the road to some large roadside exposures of cobble and pebble conglomerates (53.853600°, −9.424884°). These are particularly rich in clasts of the underlying Letter Formation including many of green slate and are inferred to lie just above the unconformity.

From here return to the parking.

2.2 Locality 10.2: Graffa More (53.867401°, −9.360925°)

Access and Additional Information From Castlebar take the R311 towards Newport and after c. 3 km turn right onto an unmarked road at 53.853297°, −9.350380°. Just after passing the highest point of the road park at the entrance to a small, gated track to the left at (53.867401°, −9.360925°) (Fig. 6).

Sandstones are exposed at the entrance to the track but overall, these are a small proportion of the Graffa More Formation. About 50 m along the track a small quarry on the left is partly in bedrock and partly in Quaternary cover. The bedrock shows

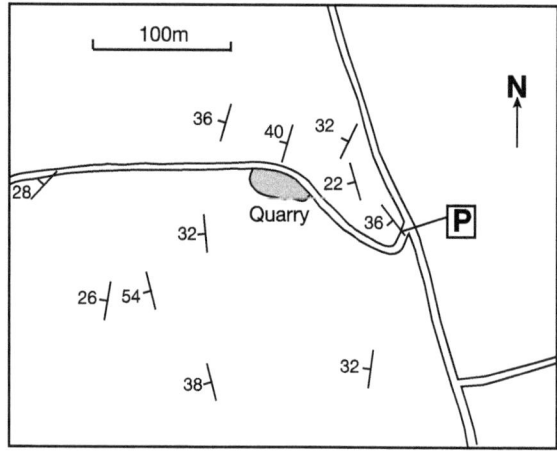

Fig. 6 Sketch map of Graffa More locality

Fig. 7 a Typical exposure style of polymict conglomerates **b** Thin sandy lenses in flat bedded conglomerates

the typical features of the Graffa More Formation conglomerates which are poorly sorted, flat bedded polymict pebble conglomerates directly comparable to those seen at locality 10.1 and similar interpretations apply.

Exit the quarry on the SW side and examine typical exposures of the formation that emerge from beneath the bog. Everywhere in this area the exposure style is of glaciated rock surfaces with very limited vertical relief (Fig. 7a). The polymict conglomerates are relatively invariant with just a few sandy lenses and occasional sandstone beds (Fig. 7b). The facies compares closely to that seen at Letter (Loc. 10.1) some 4 km to the SW. Use one of the gates to regain access to the track (Figs. 6 and 8).

Just before regaining the road use a gate on the north side of the track to access a series of low crags that parallel the road and offer some limited vertical sections through the conglomerates and rare sandstone beds. The thin sandstones are interpreted to represent the waning flows of individual floods.

2.3   Locality 10.3: Croaghmoyle (53.925941°, −9.373238°)

Access and Additional Information From Castlebar take the R311 for Newport and after 4 km fork right on the R312 for Crossmolina. At Glenisland church turn right on an unlabelled road for c. 2 km past an area of forest and at a staggered crossroads turn left on a metalled road which heads north towards the communications mast visible at the top of Croaghmoyle hill. The road uphill is steep but there is abundant parking at the top near the mast at 53.925941°, −9.373238° (Fig. 8).

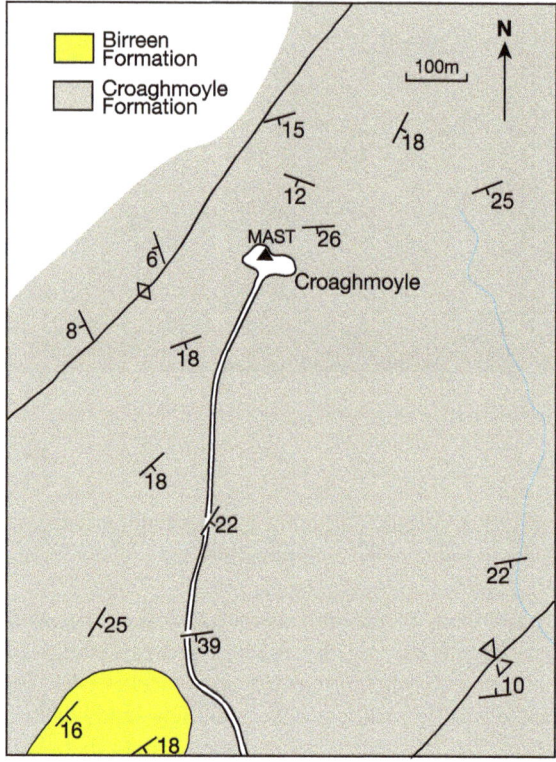

Fig. 8 Geological map of the Croaghmoyle area see Fig. 1 and caption for further details

This area is dominated by conglomerates of the Croaghmoyle Formation. Exposures to the east and north on small crags give a good overall view of facies present in the Croaghmoyle Formation (Fig. 9). Several beds of unsorted conglomerate are interpreted as debris flow deposits (Fig. 9a) that are typical of the proximal parts of alluvial fans. Other crags display well bedded conglomerates deposited by fluidal

Fig. 9 a Unsorted texture of debris flow deposit **b** Interbedding of conglomerates dominated by quartzite clasts and sandstones

flows and interbedded sandstones (Fig. 9b). Exposures by the east side of the road below the mast show interbedded conglomerates and sandstones; sandstones from here have yielded macroplant fragments suggesting the surfaces of these alluvial fans were at least partly vegetated.

The large crags to the west of the road are part of the overlying Birreen Formation that lies above the quartzite conglomerates of the Croaghmoyle Formation. This relationship is better seen at Burren Mountain (Locality 10.4).

2.4 Locality 10.4: Burren Mountain (53.924538°, −9.339399°)

Access and Additional Information From Castlebar take the R130 towards Pontoon and after c. 1 km fork left on L1721 towards Burren. After crossing a marked hump-back bridge take the next left at the abandoned school which is an unmarked road giving access to the wind farms. After passing a small piece of woodland on the left there is a slanted T junction with parking at the roadside at 53.924538°, −9.339399° (Fig. 10). From here walk right (E) as far as a marked bend in the road where the road crosses a stream. Access the hillside via a gate and track on the east side of the stream at 53.929181°, −9.332183°. Head diagonally NE across the fields and then cross the fence at the NW corner of the fields to the open hillside. Close to this (53.931254°, −9.334289°) are exposures of quartzite conglomerates and sandstones of the upper part of the Croaghmoyle Formation (Fig. 11).

From here bear NW rising slightly and passing scattered exposures of Croagh-moyle Formation; large blocks of Birreen Formation are present, but these rocks will be examined in situ later. From the obvious shoulder at the western end of the hill walk NE following the base of the spectacular cliff sections (Fig. 12a). The cliffs are formed by rocks of the Birreen Formation. These consist of flat bedded pebble and locally cobble conglomerates and interbedded sandstones (Fig. 12b–e). A few beds show good imbrication, but cross-beds are rare, where present they indicate transport from the S and SE (Graham 1981a). There are numerous small scours (Fig. 12d) but no significant channels suggesting shallow flows. Compositionally the conglomerates contain three major clast types, metamorphic clasts (mainly quartzites), red sandstones (Fig. 12e) and volcanic clasts, mainly pink/purple quartz bearing porphyries (Fig. 12f), (Graham 1981a, Fig. 5).

Continue along the base of the cliffs to the col and then descend slightly to the lowest exposures on the SE side of the valley at 53.941558°, −9.330030°. These are formed of quartzite conglomerates and green weathering sandstones of the Croagh-moyle Formation. Thin silty seams from this locality have yielded palynomorphs that indicate an early to middle Eifelian age (Graham et al. 1983). Some macroplant fragments are also present. These can be seen to underlie and pass conformably upward into the typical conglomerates and sandstones of the Birreen Formation which are spectacularly exposed in the cliff faces above. It is possible to examine these with

Fig. 10 Geological map of Burren Mountain

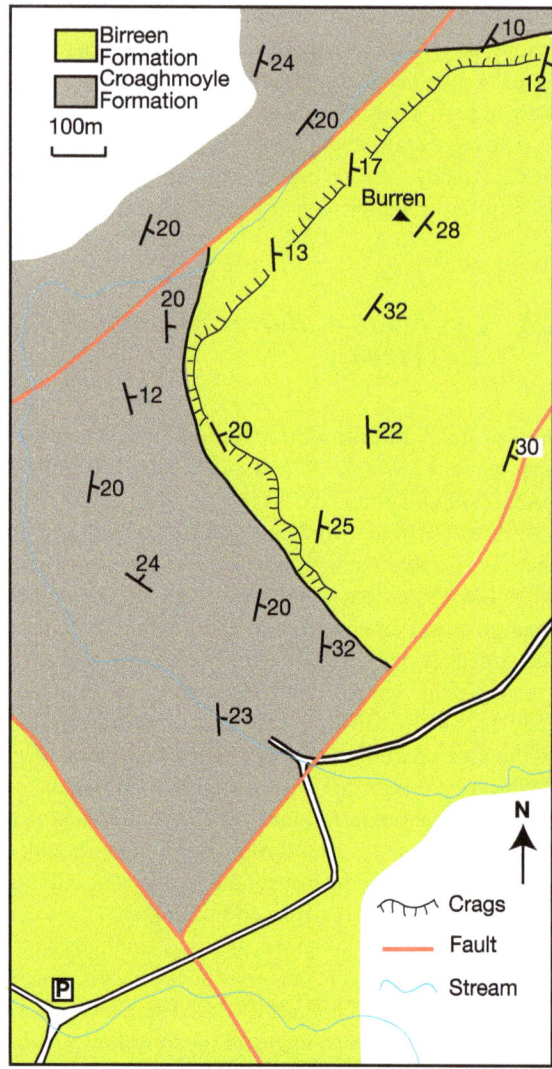

care as the slopes are very steep and this should not be attempted in wet weather or if the conditions underfoot are very wet.

Return to the parking is a reverse of the outward journey but keeping much lower than the base of the cliffs will speed up the journey.

Fig. 11 Croaghmoyle Formation **a** Conglomerates cutting into coarse sandstones **b** Quartzite clast dominant conglomerates and sandstones

2.5 Locality 10.5: Kings Hill (53.954232°, −9.315546°)

Access and Additional Information From Castlebar take the R310 for Pontoon; after c. 1 km fork left on the L1721 towards Burren, proceed to the highest point of the road, called Windy Gap, and park in the large car park at 53.954232°, −9.315546° with fine views over Glen Nephin to the north (Fig. 13a). The car park is built on Dalradian rocks which are exposed just to the south. Access the hillside to the east via an obvious gate and then walk northwards along a ridge of Dalradian rocks towards the col on the skyline (Fig. 13a). At the col walk left (west) to access the well exposed crags, which are made from the Middle Devonian Kings Hill Fm. The crags on the east side of the hill offer the safest access. If wished, these crags can be followed round the southern ridge of the hill and the facies can also be examined on the spectacular crags on the western side of the hill.

At this locality the Kings Hill formation is at its most proximal and coarsest grained. The conglomerates are unsorted in many places and may represent debris flows. Some beds do show imbrication (Fig. 13c) representing normal fluidal flows showing palaeocurrents from the east (Graham 1981a). The clasts are almost exclusively metamorphic and match well with the adjacent metamorphic rocks of the Raheen Barr succession (Long and Max 1977) suggesting limited lateral transport after deposition. Approximately 180 m of these conglomerates are present and the overall interpretation is of an alluvial fan setting (Graham 1981a). To the north of this locality some small stream exposures suggest that the Kings Hill Formation is conformably overlain by the Birreen Formation. This is better seen at Tonacrock (Locality 10.6).

Having examined the Kings Hill Formation walk carefully downslope to an obvious lower set of crags that are formed by the Birreen Formation (Fig. 13a, 14). This traverse has crossed a NNW-SSE trending fault that makes a small feature in the landscape. The Birreen Formation differs from the Kings Hill Formation both in bedding style (Fig. 14a–c) and in composition (Fig. 14d). The conglomerates

Fig. 12 Field images of the Birreen Formation, Burren Mountain **a** Spectacular cliff exposures (person in centre for scale) **b** Poorly sorted flat bedded conglomerates **c** Flat bedded conglomerates with discontinuous sandy lenses **d** Small erosional scour in near flat bedded conglomerates **e** Typical clast assemblage; the pencil points to a red sandstone clast **f** Flat bedded conglomerates; some of the pink weathering porphyry clasts and well-rounded quartzite clasts are indicated

are dominated by near horizontal bedding with some shallow scours often lined with coarser clasts. Cross-beds are not observed at this locality but on a regional basis define, along with imbrication measurements, a sediment derivation from the south to south-east. The formation is characterised by three main clast types namely metamorphic, fine-grained red sandstones, and volcanic clasts, mainly quartz-bearing pink/purple porphyries (Fig. 14d) that are of rhyolitic or dacitic composition (Graham 1981a, b, Fig. 5); there are some rare granitic clasts. Interestingly the quartzites, which

Devonian and Carboniferous

Fig. 13 a Overview of the Kings Hill locality **b** Close up view of well imbricated schist clast conglomerate, Kings Hill Formation **c** Poorly sorted, imbricated schist clast conglomerates, Kings Hill Formation

dominate the metamorphic clasts, are often the best rounded of all (Fig. 14b and c) suggesting that they may be recycled from earlier conglomerates.

Descend from these crags directly towards the road where a gate allows exit to the roadside. A large roadside exposure just uphill from here at 53.955750°, −9.316367° is of typical Birreen Formation. Return uphill on the road to the car park.

Exercises

1. Produce a geological map of the small area around Kings Hill.
2. How important are the clast assemblages in distinguishing different conglomerates?
3. From which directions were the clasts derived and how far away were the source areas?
4. Why in some of the conglomerates are the most resistant clasts the best rounded?
5. What is the likely tectonic setting for these deposits?

2.6 Locality 10.6: Tonacrock (54.022820°, −9.292056°)

Access and Additional Information This locality can be accessed from the R315 which runs from Pontoon to Crossmolina. Approaching from the north pass a minor road turn to the left and then, after c. 850 m, park at the roadside by the first roadside

Fig. 14 **a** Crag exposures of the Birreen Formation with Glen Nephin (Carboniferous rocks) in the mid-distance and Nephin (Dalradian quartzite) with well-developed north face corries in the distance **b** Typical bedding style of cobble and pebble conglomerates and pebbly sandstones **c** Close up of bedding; individual beds are defined by variations in grain size, often subtle **d** Large porphyry clast

Devonian and Carboniferous

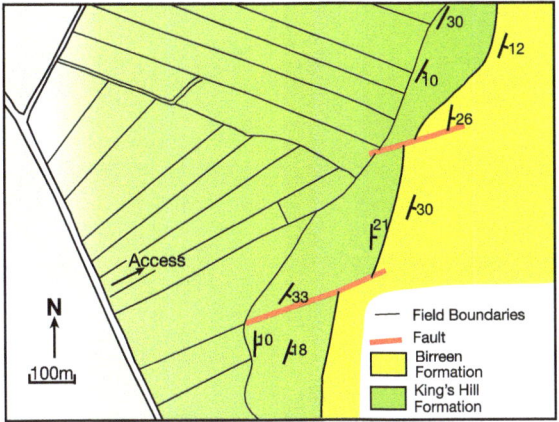

Fig. 15 Simplified map of the Tonacrock locality

houses. Walk back northwards until passing the last buildings and a track to a set of barns. Enter the next field via a gate at 54.020722°, −9.298126° and walk upslope through this long field close to and parallel with the southern field boundary. Walk to the NE corner of this field and then cross a low stone wall on the right into a smaller field with exposures of red sandstone in the upper part (54.022820°, −9.292056°) (Fig. 15) that dip beneath the conglomerate crags to the east.

Cross the fence onto the open hillside where the nearest exposures are of relatively coarse-grained Kings Hill Formation (Fig. 16a). As well as being finer-grained than the exposures at Kings Hill, probably due to a more distal setting, the composition of the metamorphic clasts varies. Although schist clasts are common, clasts of psammite and semi-pelite are also common here (Fig. 16b). This is interpreted as reflecting variation in the local basement rocks which in such localised deposits would be expected to manifest in the conglomerates. Interbedded finer-grained conglomerates are common here and reflect the much more distal alluvial fan setting. About 100 m of Kings Hill Formation are present at this locality. The superposition of the conglomerates over red sandstones is taken to mark the progradation of the fan into this area.

From these first crags gradually walk northwards, contouring to get the best ground. There are some cross faults (Fig. 15) but despite these a gradational change to conglomerates of a different composition can be discerned. This is the conformable upward passage into the Birreen Formation which is characterised by metamorphic clasts, mainly quartzite, red sandstone clasts, and a variety of volcanic clasts (Fig. 16c and d). As the Birreen Formation also gradationally overlies the Croaghmoyle Formation (Localities 10.4 and 10.5) this demonstrates the lateral equivalence of the Kings Hill and Croaghmoyle formations. The composition and facies are closely comparable to those seen at Kings Hill some 7 km to the SSW. From the northern exposures cut diagonally towards the boundary fence of the open moorland, walk south to regain

Fig. 16 a Metamorphic clast cobble conglomerates of the Kings Hill Formation **b** Close up of the metamorphic clast assemblage **c** Conglomerates and minor sandstones of the Birreen Formation **d** Large porphyry clast in Birreen Formation

the original crossing point of this fence and then return to the road by the incoming route.

3 Carboniferous of the Clew Bay Area

Renewed subsidence during the early Carboniferous was focussed on the reactivation of major structures developed during the 'Caledonian' cycle. There is evidence that the area considered by this field guide remained topographically positive throughout the Tournaisian and Visean (Sevastopulo and Wyse Jackson 2009). The localities used in this guide serve to demonstrate these two aspects. The differential subsidence produced during fault reactivation is best demonstrated by the successions developed in the Clew Bay area [Fig. 17 (Locs 10.7–10.10)]. The style and variation of nearshore facies are best seen in the spectacular coastal exposures of North Mayo (Locs. 10.11–10.15).

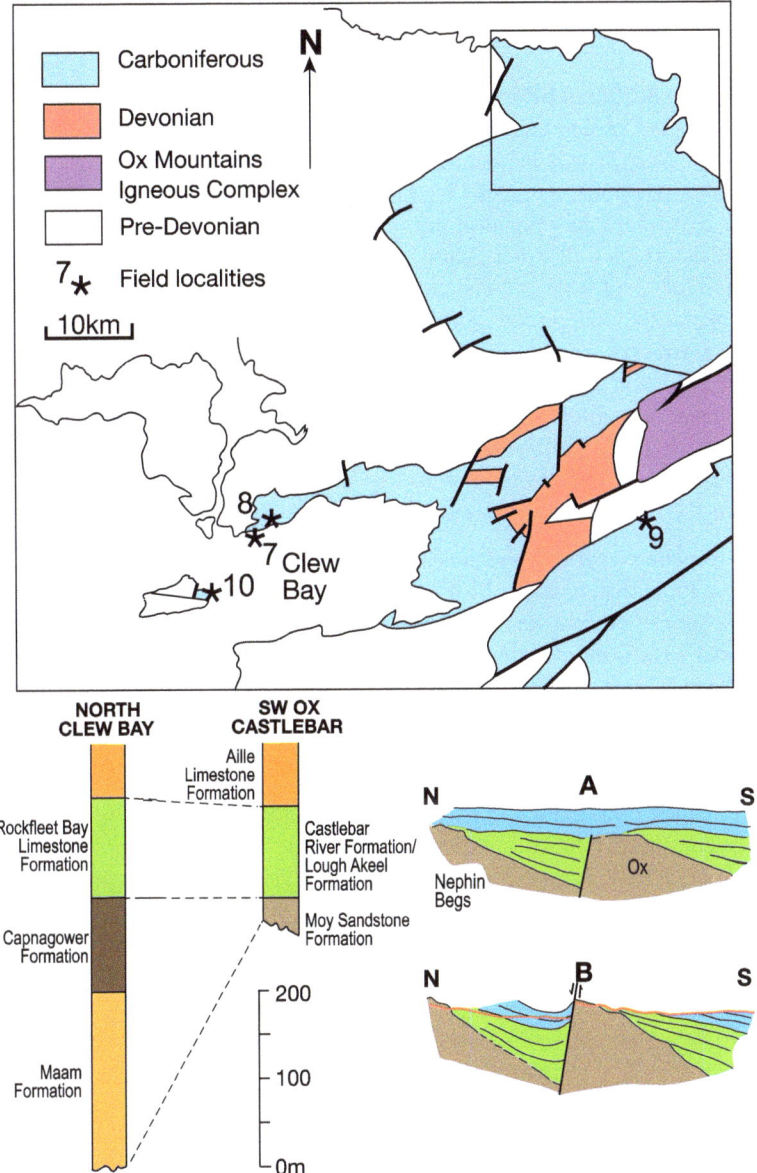

Fig. 17 The map shows the location of field localities in the Clew Bay area; the inset is the location of Fig. 29 in which the North Mayo localities are shown (after Graham 2010). The stratigraphic columns show the thickness variations across the SW Ox Mountains. The cross sections show the schematic development (A, upper) and current erosion level (red line) (B, lower) of the Carboniferous successions. The numbers 7, 8, 9 and 10 on the map refer to localities 10.7, 10.8, 10.9 and 10.10 respectively

3.1 Locality 10.7: Bolinglanna (53.864850°, −9.899847°)

Access and Additional Information From Mulrany take the coast road west (L1404) signposted for Corraun and, in the small settlement of Bolinglanna, take a small road left at a crossroads leading directly south towards the coast; park at the end (Fig. 18) (53.864850°, −9.899847°). Go through the small gate and follow the path into the bay where the unconformity of the Carboniferous Maam Formation with the Birreencorragh Schist Formation of the Dalradian is exposed in the tidal rocks at (53.863708°, −9.900323°) and at (53.862804°, −9.900172°). Larger clasts up to 35 cm match the underlying Dalradian. Only the lowest few metres of the Maam Formation are exposed in this bay. Much of this foreshore is formed of boulders and great care must be taken if it is wet.

Walk eastward around the headland of Dalradian rocks to the next bay where the lower part of the Maam Formation is faulted against the Dalradian but from here eastwards forms a near continuous section through 195 m of the Maam Formation (Graham 1981b) (Fig. 19a). This section is composed of three main facies:

- Erosively based pebbly sandstones with internal structures dominated by flat lamination. Individual beds are 20–120 cm thick, each representing a flood (Fig. 19b). These sandstones may be amalgamated into multistorey packages up to 20 m thick without significant mud deposition, although thin mud drapes between floods are common. The dominance of flat lamination over cross-bedding is probably due to relatively shallow flows. Paleocurrents from the subordinate cross beds and from parting lineation in the flat bedded sandstones indicate flow from the north-west (Graham 1981b). These are interpreted as the deposits of wide bedload channels.
- Interbedded sandstones and mudrocks with sandstones 20–60 cm thick forming over half of the facies. These sandstones are lenticular over 20–50 m and commonly have demonstrably erosive bases. The sandstones are very poorly sorted, containing both scattered pebbles and high silt contents. Evidence for desiccation is common in the muddier beds. These are interpreted as proximal overbank deposits where large floods overtopped the main bedload dominant channel belt.

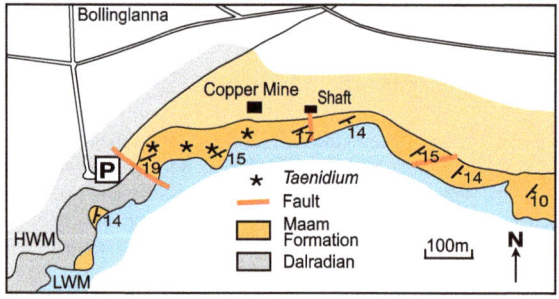

Fig. 18 Simplified map of the Bollinglanna coastal section

Fig. 19 a General view of Maam Formation exposed in the lower part of the cliff beneath the Quaternary deposits east of the Copper Mine **b** Erosively based pebbly sandstone in the upper part of the cliff; the rucksack rests on desiccated mudrocks **c** *Taenidium barreti* burrow in coarse grained sandstone **d** Concentration of Taenidium barreti burrows in coarse grey sandstone

- Mudrock dominant intervals with abundant evidence of polygonal desiccation cracks and occasional carbonate concretions that are interpreted as soil forming. Subordinate sandstones are lenticular and very poorly sorted. This facies is interpreted as deposits a little more distant from the main channel belt.

A notable feature of some of the facies 1 sandstones is the presence of large burrow systems of *Taenidium barreti* (formerly *Beaconites barreti*) (Graham and Pollard 1982). These are encountered in the first pale weathering pebbly sandstone near the back of the first beach (Figs. 18 and 19c, d) but there are numerous other examples in this lower part of the section. Locally the density of these burrows is so great that the bed tops display an irregular hummocky surface. The builders of these structures are unknown, but they are known from early Devonian to Permo-Triassic; arthropods seem the most likely.

A small fault interrupts the section at a shaft for an old copper mine (Fig. 18; 53.866398°, −9.892116°) and a reasonably low tide is needed to round the headland here and access the rest of the section. Much of this foreshore is formed of boulders and great care must be taken if it is wet.

3.2 Locality 10.8: Lough Ard (53.877675°, −9.878718°)

Access and Additional Information From Bolinglanna take the bog road from 53.870613°, −9.901843° to the end by Lough Ard at (53.877675°, −9.878718°. This road is only suitable for four-wheel drive vehicles and those with a reasonable clearance.

This locality provides an interesting exercise in delimiting palaeotopography and in demonstrating that the majority of the sediments are not derived from the immediately underlying Dalradian (Fig. 20). There is good exposure of both the lower part of the Maam Formation and the underlying quartzites of the Annafrin Formation of the Dalradian succession.

Immediately adjacent to the remnant topographic high the lowest beds are very coarse grained breccias and conglomerates dominated by clasts of the local quartzites (Fig. 21a and b). However, bed textures and imbrication indicate there has been some local transport. A few tens of metres away from this the Maam Formation is dominated by red pebbly sandstones with abundant vein quartz clasts. Concentration of vein quartz requires transport of at least a few km indicating that most of these strata are derived from a little further to the north-west (Graham 1981a, b).

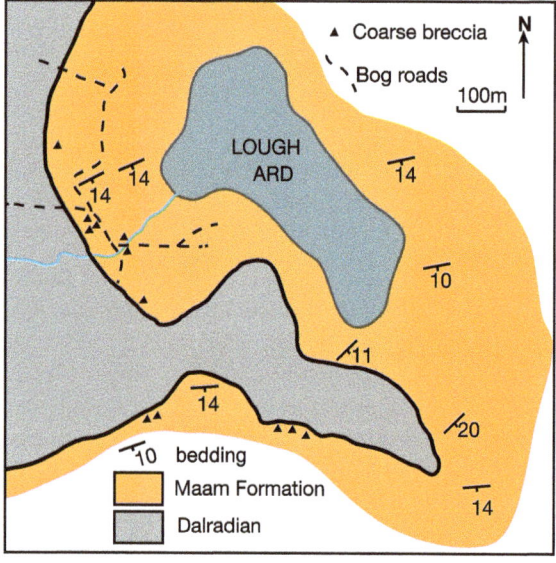

Fig. 20 Geological map of the Lough Ard locality

Fig. 21 **a** Crudely bedded quartzite breccias **b** Close up of quartzite breccias showing imbrication

3.3 Locality 10.9: Castlebar (53.882225°, −9.273217°)

Access and Additional Information This locality demonstrates the very thin succession of sub-carbonate clastics seen in the Carboniferous successions on the south-east side of the Ox Mountains. This should be visited along with locality Bolinglanna and Clare Island to appreciate the marked contrast in the sub-carbonate successions.

From Castlebar take the R310 towards Pontoon and after c. 2 km park on the left in a wide gateway at (53.882225°, −9.273217°) (Fig. 22). There is abundant exposure in roches moutonnées adjacent to the gateway of the local Dalradian rocks. These have a very strongly developed vertical schistosity that strikes north-east (Fig. 23a). In the field these are quartz-muscovite-feldspar rocks and the high proportion of both quartz and muscovite suggests they are metasediments. These have been assigned to Lower Lismoran Formation.

Walk north-eastwards top the next road junction where these rocks are exposed at the junction. Exposure is very poor in this area but the gently dipping Lower Carboniferous limestones can be seen from here on the south-east side of Lough Akeel. Walk down the small road towards these and just before the river on the left hand side of the road numerous large loose blocks of grey quartzose sandstone are visible at (53.881431°, −9.269147°). Although these are not in situ similar sandstones assigned to the Moy Sandstone Formation are seen in situ along strike. Continue to the next road junction and access the old quarry sections immediately opposite (53.878673°, −9.267874°). The lower part of the quarry face is formed of coarse grained sandy, oolitic, bioclastic limestone (Lough Akeel Formation) (Fig. 23b) and the gentle south-easterly dip is evident. It appears that only c. 10 m of clastics underlie the limestones in this area which contrasts strongly with the 300 m seen on the north-western side of the Ox Mountains in the western Clew Bay area. This implies that there was synsedimentary accommodation focussed on earlier major faults.

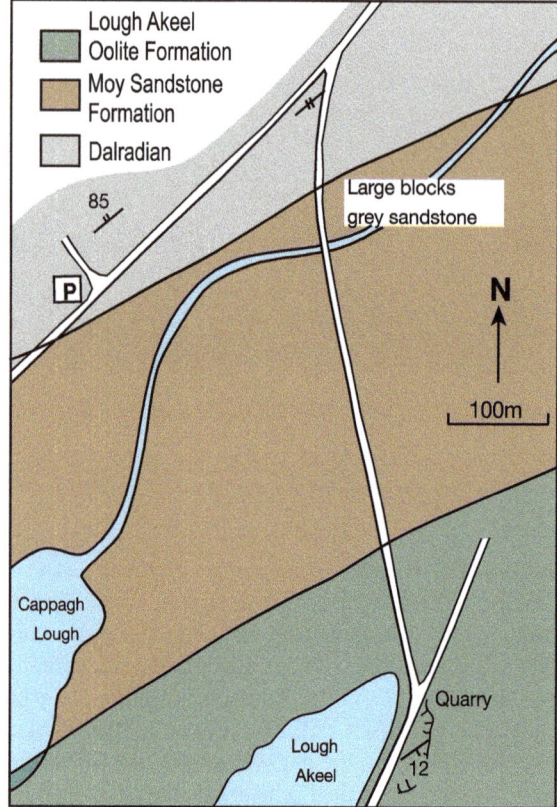

Fig. 22 Simplified map for locality 10.9

3.4 Locality 10.10: Capnagower, Clare Island (53.812515°, −9.949958°)

Access and Additional Information This locality displays both inaccessible cliff sections and accessible coastal sections through the 100 m thick Capnagower Formation that lies conformably above the Maam Formation. This locality should be visited in conjunction with Bolinglanna to appreciate the thick early Carboniferous clastic succession deposited north of the Ox Mountains and with Castlebar to see the contrasting thin clastic succession on the south-east side of this fault block.

From Clare Island harbour walk northwards along the road to the lighthouse to the Mill (Fig. 24; 53.811225°, −9.965449°) and take the grassy track to Portlea, bearing right parallel to the coast after the last gate. Walk to the eastern end of Portlea and examine the section just northeast of an obvious waterfall.

These rocks belong to the lower part of the Capnagower Formation and expose a spectacular section through a complex paleosol horizon (Fig. 25;

Fig. 23 a Schists of the Lismoran Formation **b** Pebbly oolitic limestones of the Lough Akeel Formation

Fig. 24 Location map for locality 10.10

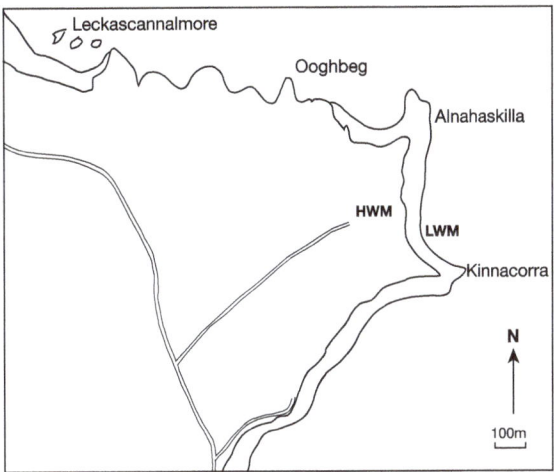

53.812566°, −9.959487°). These paleosols provide both palaeoenvironmental and palaeoclimatic information. The key features are abundant desiccation cracks, colour mottling of red and drab patches, calcareous and ferruginous concretions, rootlets, clastic dykes and mukarra (pseudoanticline) structures (Fig. 26). Together these features are characteristic of vertisols that form in seasonally dry climates. It has been estimated that this example indicates a monsoonal climate with a mean annual precipitation of 800–1000 mm (Graham 2001). To the north, this paleosol horizon is cut out by the erosive base of a small channel. Another key feature of this part of the

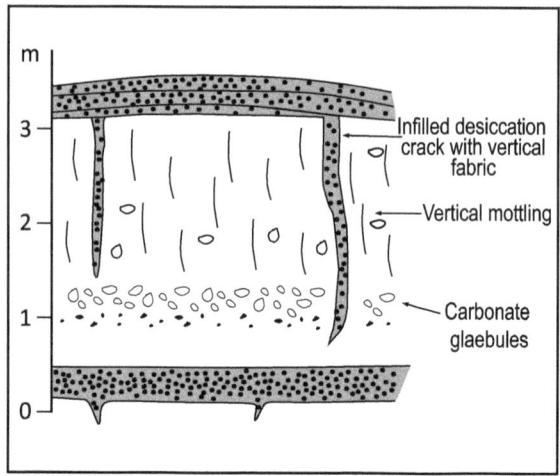

Fig. 25 Diagrammatic summary of paleosol horizon

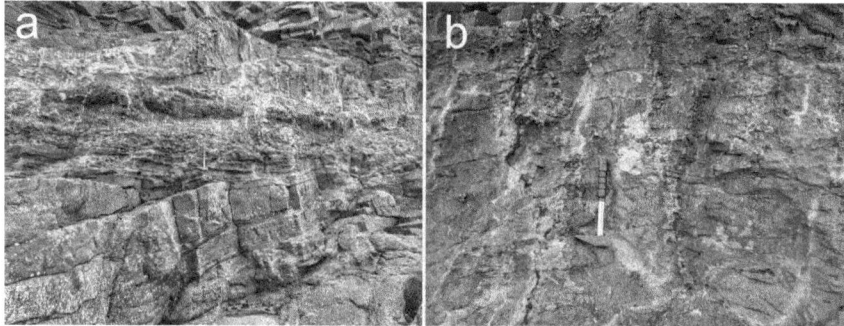

Fig. 26 a Overview of main paleosol horizon (above hammer) **b** Detail of palaeosol horizon showing desiccation crack fills, colour mottling and concretions

succession is the common small channels that are interpreted as crevasse channels, formed where the main channels overflowed onto the adjacent flood plains.

Access the cliff top just south of the waterfall and carefully walk east along the edge, keeping on the seaward side of the fence. Although not accessible, the cliffs expose spectacular sections through the Capnagower Formation, including channels of varying sizes (Fig. 27). The best examples of these are seen just east of Leckascannalmore (Fig. 27a and b, 53.812379°, −9.950141°) and on the west side of Ardal (Fig. 27c; 53.816°, −9.950°). Just east of Ardal it is necessary to track inland to a gate to cross a beautifully built stone wall.

Fig. 27 **a** Cliff section east of Leckascannalmore with channel cut bank in left part of cliff **b** Channel cross section with fine grained fill east of Leckascannalmore **c** Overview of typical Capnagower Formation Ardal

Where the coast bends southwards there are excellent views of the impressive storm beach at Kinnacorra Point (53.811289°, −9.943376°). The youngest Carboniferous beds exposed on Clare Island are seen near this point, in which bivalves, ostracods and gastropods indicate brackish-water conditions. These beds are interpreted to lie just below the level of the limestones that characterize much of the regional Carboniferous succession. An easily accessible section can be examined south of Kinnacorra (Fig. 28).

4 Carboniferous of North Mayo

The North Mayo coast displays some spectacular exposures through the marginal marine Visean succession that developed here as a result of the continued positive relief of the western Irish Caledonides. The localities to be visited are shown in Fig. 29 along with an outline of the local Carboniferous geology and stratigraphy.

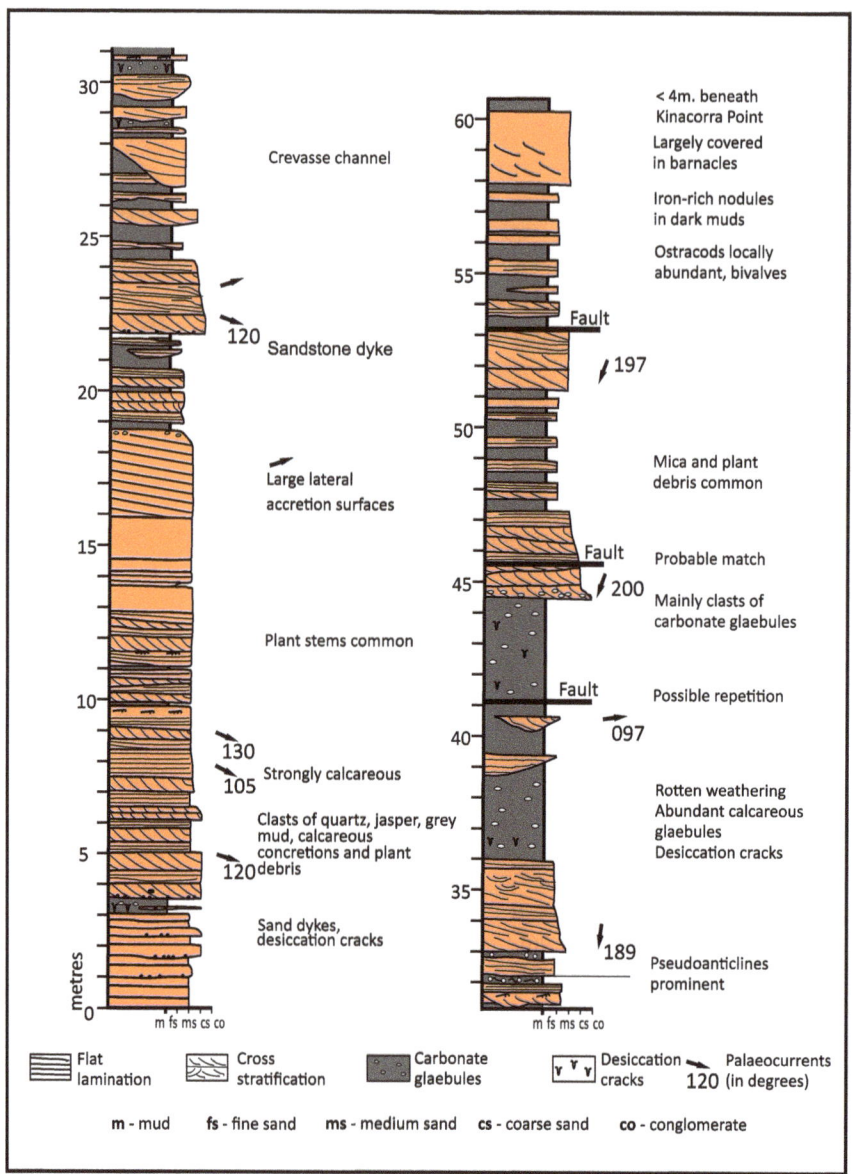

Fig. 28 Sedimentological log of the Capnagower Formation, Kinnacorra (after Graham 2021)

Devonian and Carboniferous

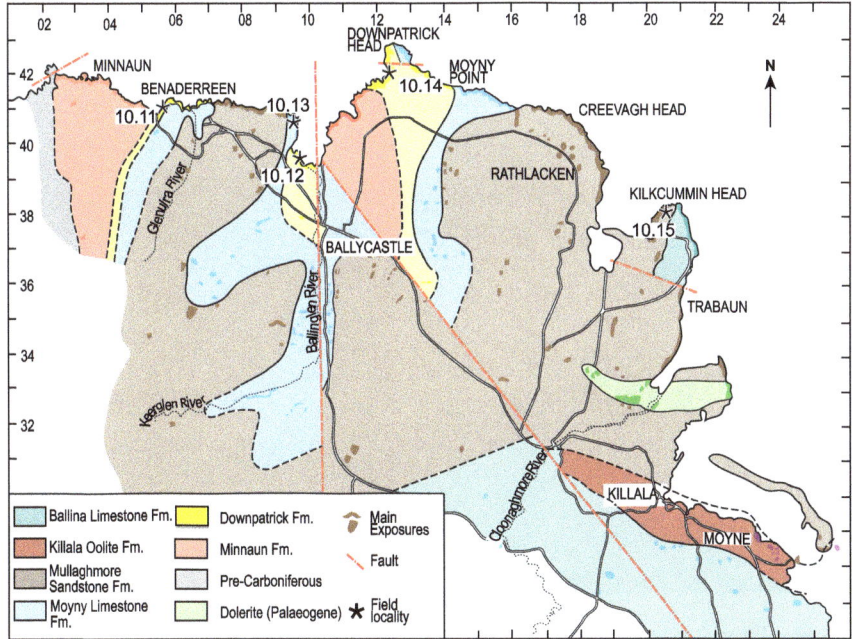

Fig. 29 Outline geology of the Carboniferous of North Mayo (after Graham 2010) with field localities indicated. Numbers in the margin relate to the Irish National Grid

4.1 Locality 10.11: Ceide (54.308565°, −9.455869°)

Access and Additional Information Take the R314 west of Ballycastle to the well signposted Ceide Fields Visitor Centre and park in the large car park. With care cross the road to the seaward side and a safe viewing platform (54.308565°, −9.455869°).

Spectacular cliff sections of Benadereen headland (Fig. 30) to the east show good two-dimensional sections of the Downpatrick Formation with several channel cross sections visible. The topmost part of the cliff is formed by tabular limestone beds of the base of the Moyny Limestone Formation. Downpatrick headland is visible to the east under most weather conditions.

The base of the cliff can be shown to be near the base of the Downpatrick Formation by tracing through inaccessible cliff sections to the west where it rests conformably on the fluvially dominated Minnaun Formation. The whole succession can be shown to be thinner here than in the Ballycastle area to the east (Graham 2010) and this is interpreted as onlap onto the persistently high areas formed by the Dalradian strata of North Mayo.

Fig. 30 Cliff section at Beenaderreen; the tabular grey beds in the upper part of the cliff section are the base of the Moyny Limestone Formation

4.2 Locality 10.12: Ballycastle Beach (54.294872°, −9.386396°)

Access and Additional Information Excellent exposures at the western end of Ballycastle beach demonstrate the typical nature of the Downpatrick Formation. The section cannot be accessed from the main Ballycastle beach carpark as the Ballinglen river forms an inaccessible barrier part way along the beach. Instead take the R314 west from Ballycastle and then taking a small, unmarked road at a crossroads c1km from Ballycastle (54.289948°, −9.386431°). At a sharp left hand bend after 250 m a gate provides access to short track to the back of the beach where there is limited parking (54.294872°, −9.386396°; Fig. 31). The track is only suitable for vehicles with reasonable clearance and its condition varies.

These safely accessible rocks require only half-tide and are excellent for student projects that involve constructing sedimentological logs and interpreting palaeoenvironments (Fig. 32). The first visible exposures are of micritic limestones that contain an ostracod/gastropod fauna but lack open marine fossils. These are interpreted as near coastal lagoonal deposits. The succeeding rocks are mainly desiccated red mudrocks with abundant soil forming carbonate concretions and subordinate sandstones (Fig. 33). Continue until encountering an obvious limestone horizon with abundant *Syringopora* that marks the first influx of truly marine deposits into the

Fig. 31 Location map for localities 10.12 and 10.13

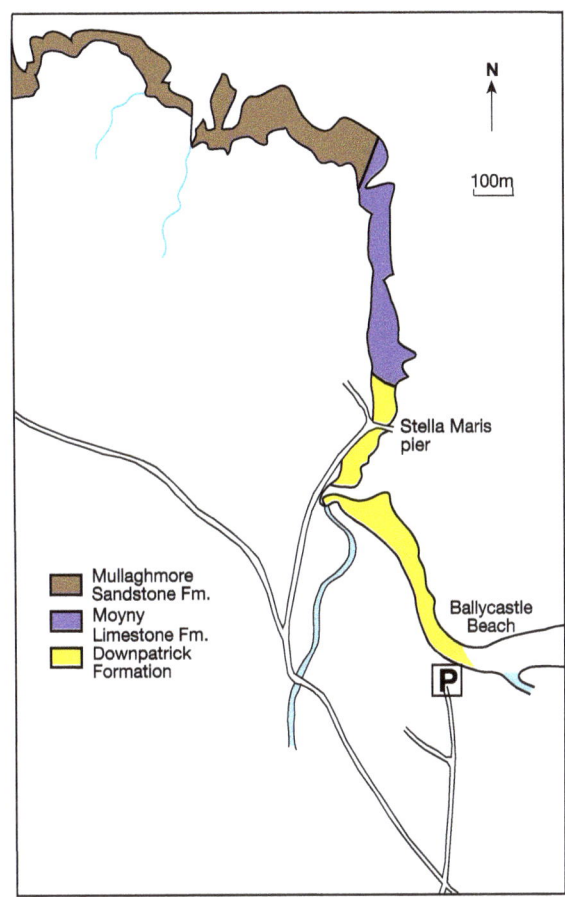

succession; beyond this there is a return to the mainly non-marine deposits that are interpreted to have formed on a low lying coastal plain.

Exercises

1. Produce a sedimentological log of the rocks exssposed on the foreshore.
2. What environments are represented here and on what evidence is their assignation made?

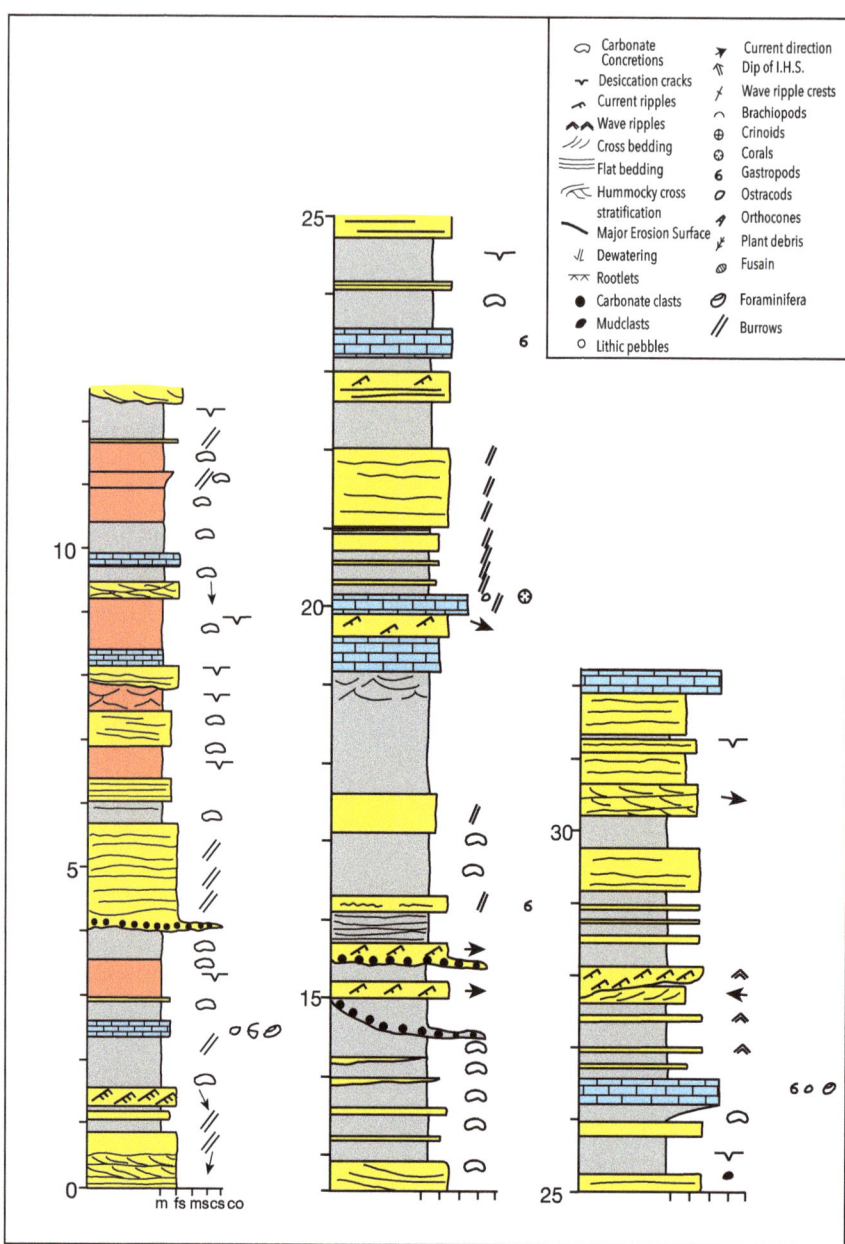

Fig. 32 Sedimentological log of the Ballycastle beach section

Fig. 33 a Micritic ostracod rich limestone overlain by variegated desiccated red mudrocks **b** Desiccated red mudrocks with abundant carbonate concretions

4.3 Locality 10.13 Stella Maris (54.301173°, −9.389916°)

Access and Additional Information Take the R314 west from Ballycastle and after c. 2 km and immediately after crossing the bridge over the Bellaminnaun river fork right to Stella Maris pier (54.301173°, −9.389916°) where there is abundant parking (Fig. 31). This section requires a reasonably low tide for ease of access and best examination of the geology. Cross the wall to the north of the pier onto the uppermost part of the Downpatrick Formation.

A thick tabular limestone marks the base of the Moyny Formation and the next part of the succession shows the typical mixed carbonate/clastic nature of this formation which is interpreted to have been deposited in very marginal shallow marine conditions. At 54.306265°, −9.390069°) there is an obvious change in the cliff section with a sharp contact at the base of a thick feldspathic sandstone that marks the base of the Mullaghmore Sandstone Formation (Fig. 34). This erosive base marks a local sea level drop estimated to be at least 9 m (Graham 1996, 2010, 2017). However, there is evidence for shallowing and local emergence seen in the topmost part of the Moyny Limestone Formation where some rootlet horizons are seen (Fig. 35).

The lowest sandstone horizon is here 11 m thick and extensively cross bedded (Fig. 36). Follow the upper part of this sandstone to a marker micrite band that occurs just above it that can then be traced into the next cove to the west. Walk down towards the low tide area (54.308945°, −9.394672°) and look back and observe that the sandstones and thin mudrocks are here deposited on large, inclined surfaces taking the micrite as the local horizontal (Fig. 37). These are interpreted as lateral accretion surfaces to the main channel area represented by the thick cross bedded sandstones and indicate that these fluvial deposits were deposited in sinuous coastal plain fluvial channels.

Fig. 34 Sharp base to the Mullaghmore Sandstone Formation

Fig. 35 a, b Rootlet horizons in the upper part of the Moyny Limestone Formation

From here walk across the cliff tops to where a small stream enters and regain the foreshore (54.309219°, −9.399745°). Excellent examples of trough cross bedding seen in three dimensions illustrate the palaeocurrent direction (102 m on Fig. 36). Walk north on the tidal rocks and from the lowest rocks available look back at the low cliff section. Gently left dipping surfaces represent lateral accretion surfaces related to the channel centre deposits represented by the trough cross bedded sandstones (Fig. 38). The flooding of this channel by open marine deposits is seen in the

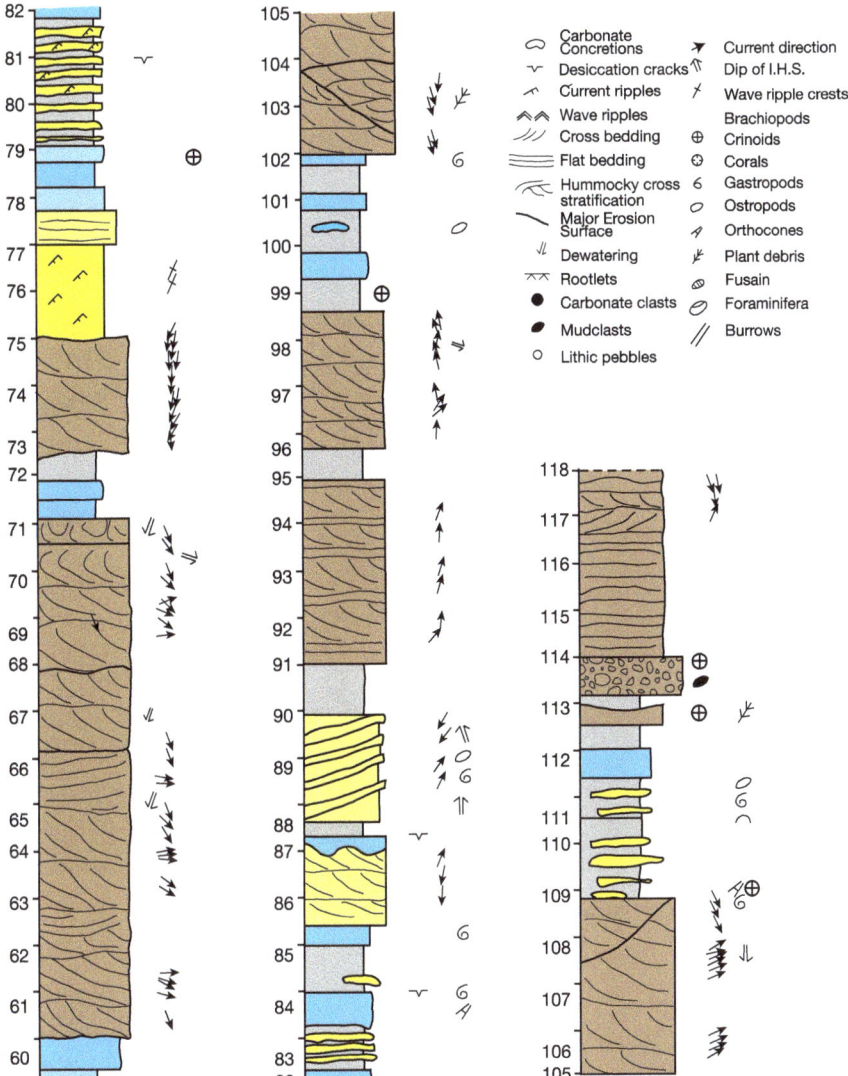

Fig. 36 Sedimentological log of the Mullaghmore Sandstone Formation Stella Maris. blue colours are predominantly carbonate (after Graham 2010)

low cliff section where muddy deposits preserve isolated lenses of limestone with orthocones and other open marine fauna thus demonstrating the coastal nature of the environment. From here return to the pier either by the coastal rocks or though the fields on the low cliff top.

Fig. 37 Lateral accretion surfaces dipping left; the pale micrite at the base of the low cliff provides the original horizontal

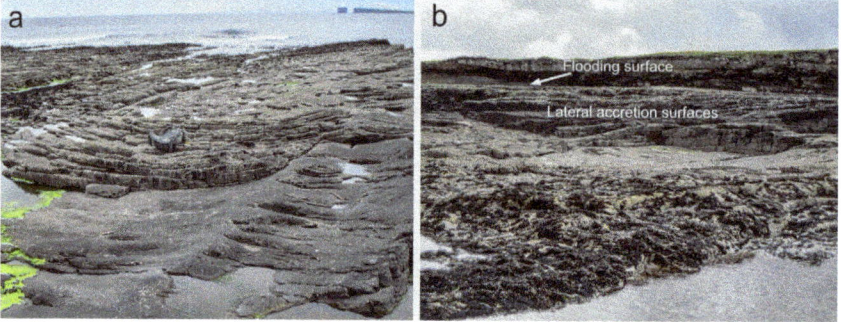

Fig. 38 **a** Large trough cross beds **b** Large lateral accretion surfaces; rucksack and map board on the top flooding surface for scale

4.4 Locality 10.14: Downpatrick Head (54.318659°, −9.345733°)

Access and Additional Information This spectacular locality demonstrates the ubiquitous presence of sinuous channels in both the coastal plain sediments of the Downpatrick Formation and in the subtidal sediments of the overlying Moyny Formation. From Ballycastle follow the brown tourist signs for Downpatrick Head. Approaching the headland, stop at the first large parking area on the left (54.318659°, −9.345733°) and access the tidal rocks immediately adjacent to this (Fig. 39). This requires at least half tide to appreciate the structures.

The succession here in the Downpatrick Formation is dominated by intersecting sets of lateral accretion surfaces formed in fluvial to tidal channels (Fig. 40); locally developed limestone horizons can be used to approximate the original horizontal.

From here proceed to the second larger car park a few hundred metres further north (54.322815°, −9.345878°). In this area marine erosion has exploited the weaker breccias formed along a series of small normal faults and produced some spectacular blow holes. Walk to the northern part of the headland to view the sea stack of Doonbriste

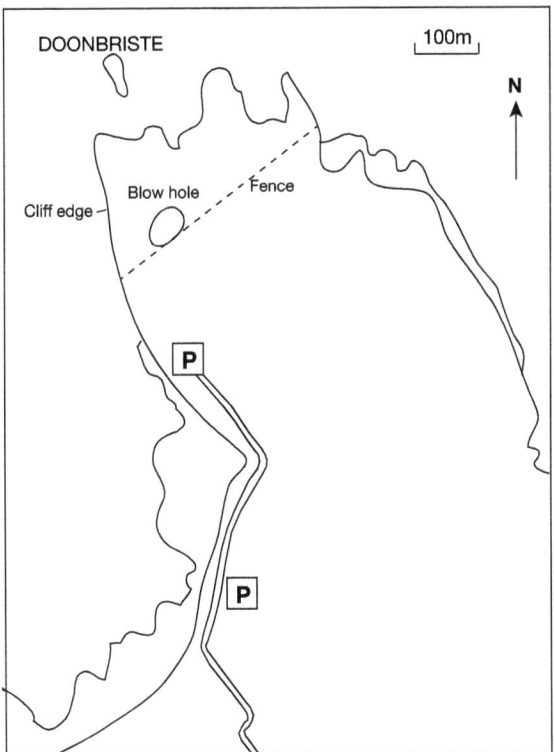

Fig. 39 Location map for Downpatrick Head

Fig. 40 Lateral accretion surfaces of small tidal channels **a** The micrite bed just below the pebble beach provides the original horizontal **b** Intersecting sets of lateral accretion surfaces in the intertidal rocks

Fig. 41 a East side of Doonbriste; The tabular limestones in the upper part mark the base of the Moyny Limestone Formation; a prominent pale weathering sandstone in the middle of the cliff shows lateral accretion surfaces dipping left and fining upwards **b** West side of Doonbriste; small mud filled channel sections are visible beneath the main pale-weathering sandstone horizon

which displays two excellent vertical sections through the succession. The upper part of the stack is formed of the tabular limestone beds of the Moyny Limestone Formation and the lower part is formed of the upper part of the Downpatrick Formation. The latter shows two excellent sections through a thick sandstone with lateral accretion surfaces (Fig. 41). Doonbriste was attached to the mainland in historical times but became separated by enlargement of and connection of blowholes formed along small faults. The current headland will no doubt suffer the same fate in the near future.

Fig. 42 Lateral accretion surfaces in muddy Moyny Limestone Formation

From the headland walk south-east to the first prominent fence and then follow this eastward to the other coast of the headland. The fence can be easily rounded at the seaward end to access a small amphitheatre in the Moyny Formation (Fig. 42) (54.326495°, −9.342039°). Here gently dipping lateral accretion surfaces formed in muddy subtidal channels are evident from the clifftop. These muddy sediments contain an open marine fauna including some well-preserved crinoids. The associated overlying bioclastic limestones have a varied fauna with particularly prominent gastropods. From here recross the fence and follow this to return to the car park.

4.5 Locality 10.15: Kilcummin Head (54.287490°, −9.214534°)

Access and Additional Information The purpose of this locality is to examine excellent cliff exposures in the Mullaghmore Sandstone Formation with a view to interpreting palaeoenvironments and evidence for sea level changes. It is an excellent locality for students to attempt this and to construct sedimentological logs.

Fig. 43 Location map for Kilcummin Head

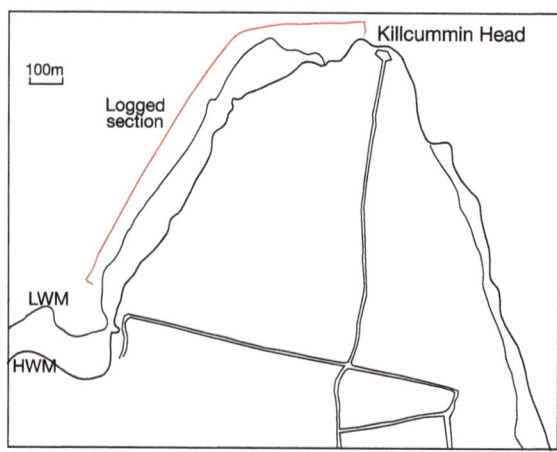

From the small village of Kilcummin head west and then turn right at the first crossroads to Humbert Landing (54.287490°, −9.214534°) where there is a large car park and turning area (Fig. 43). It is best to walk west and then south from here along the grassy cliff top to study the section in ascending order. If time is limited the upper part of the section from where the coastline turns southwards can be used. A log of the whole section is shown in Fig. 44. The lower part of the sections (54.281780°, −9.226022°) is dominated by thin bedded sandstones with abundant wave ripples that show both subparallel lamination and hummocky cross stratification (HCS) in cross sections. Dips on the HCS are very low but some domes and hollows are exposed on bedding planes.

A thick sandstone unit (20.3–30.3 m on Fig. 44) displays a bipolar current pattern demonstrating that these were tidally influenced fluvial channels in a coastal setting. The sandstone occurs just above a prominent surface with large wave ripples (Fig. 45a). Where the coastline turns to face northwards there is an excellent example of large scale hummocky cross-stratification (Fig. 45b; 51.4–54.8 m on Fig. 44). The top part of the section (60–70 m in Fig. 44) has been described in detail by Graham (2017). There is evidence for solution of carbonate that is interpreted to have occurred during emergence followed by a major flooding surface at the base of the Ballina Limestone Formation (Figs. 45c, d and 46). Care must be taken on this part of the section if it is wet and/or windy as there is a vertical cliff on the northern side of the section.

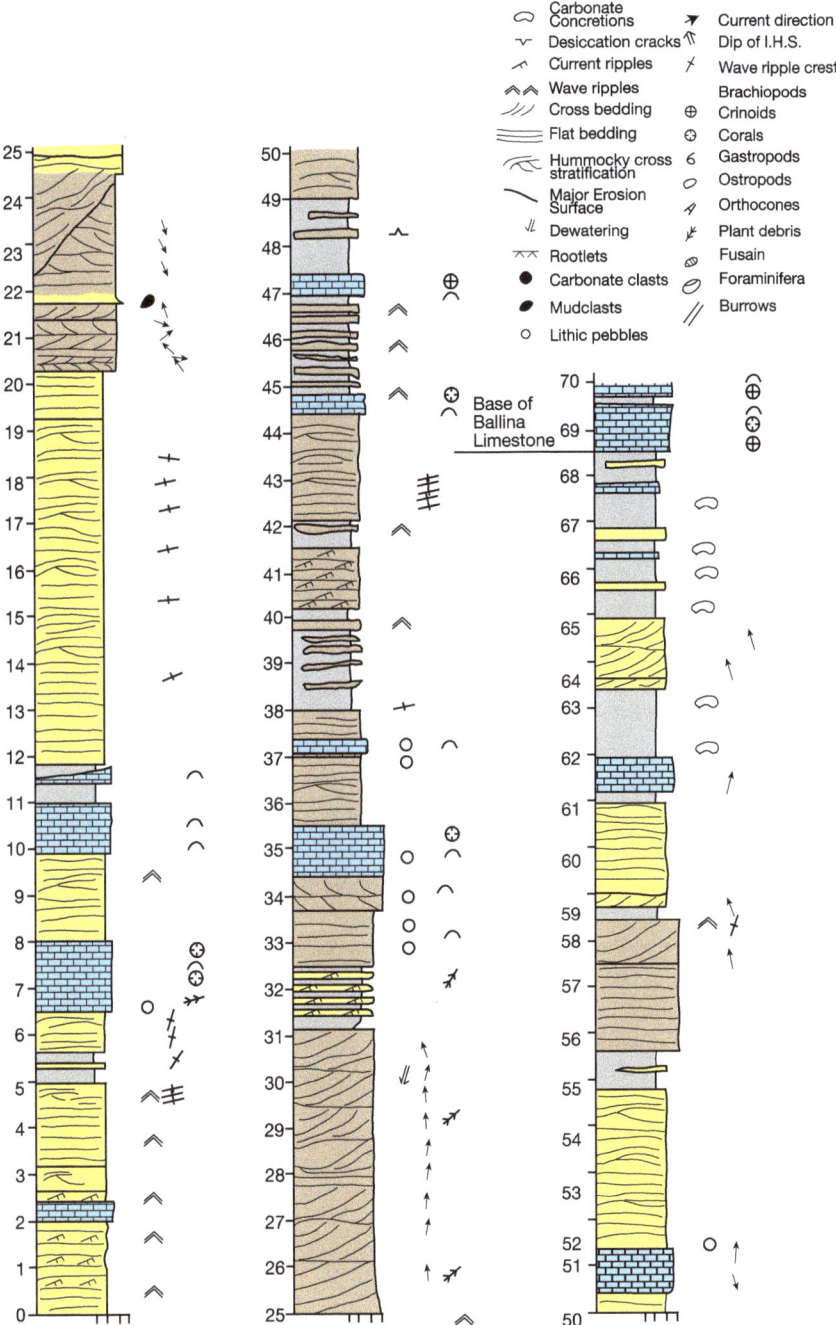

Fig. 44 Sedimentological log of the Kilcummin Head section; blue colours are predominantly carbonate (after Graham 2010)

Fig. 45 **a** Large wave ripples, Mullaghmore Sandstone Formation **b** Hummocky cross stratification, Mullaghmore Sandstone Formation **c** Flooding surface at the sharp base of the Ballina Limestone Formation **d** Solution collapse feature in the upper part of the Mullaghmore Sandstone Formation

Exercises

1. Make a sedimentological log of the uppermost 20 m exposed on the headland.
2. Make a list of the structures that you observe and interpret the processes that produced them.
3. What evidence is there for sea level changes and what are the key surfaces that need explanation?
4. What could be responsible for these sea level changes? How would you test your hypotheses?

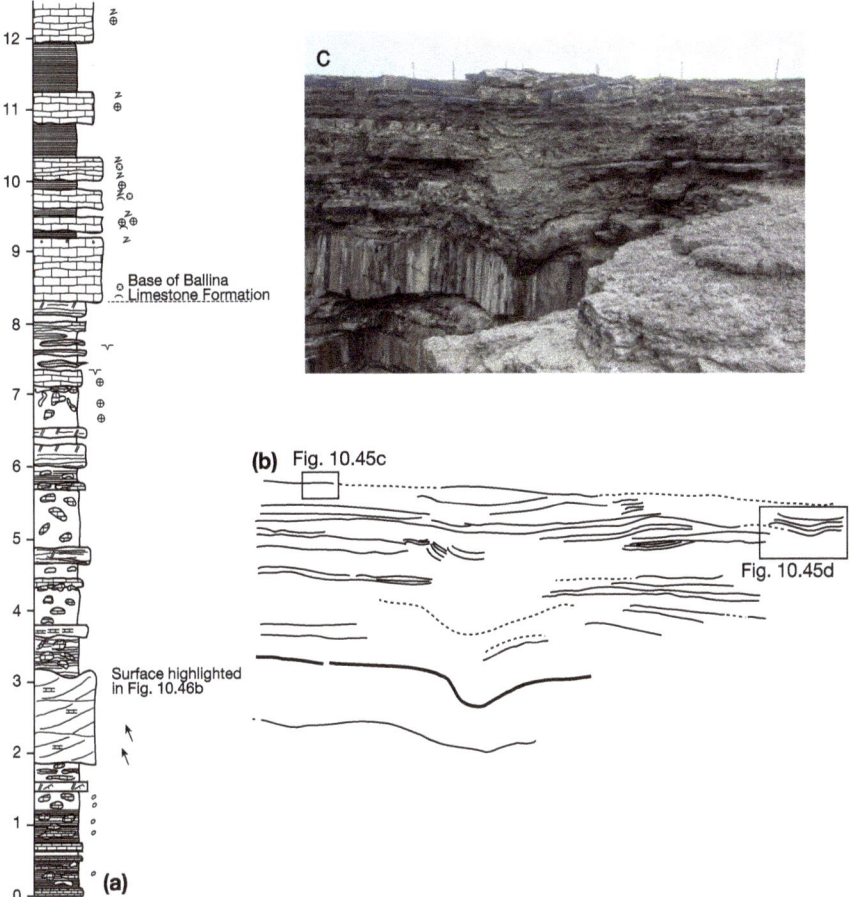

Fig. 46 a Detailed log of the upper part of the accessible section at Kilcummin Head **b** Simplified line drawing of the area shown in (**c**) **c** Inaccessible cliff section at Kilcummin Head, after Graham (2017)

References

Chew DM, Schaltegger U (2005) Constraining sinistral shearing in NW Ireland: a precise U-Pb zircon crystallisation age for the Ox Mountains Granodiorite. Irish J Earth Sci 23:55–63

Graham JR (1981a) The 'Old Red Sandstone' of County Mayo, Northwest Ireland. Geol J 16:157–173

Graham JR (1981b) Fluvial sedimentation in the Lower Carboniferous of Clew Bay, County Mayo, northwest Ireland. Sediment Geol 30:195–211

Graham JR (1996) Dinantian river systems and coastal zone sedimentation in northwest Ireland. In: Strogen P, Somerville ID, Jones, GL (eds) Recent advances in Lower Carboniferous geology. Geol Soc Lond Spec Pub 107:183–206

Graham JR (2001) The Carboniferous rocks of Clare Island. In: Graham JR (ed) New survey of Clare Island, Volume 2 Geology, Royal Irish Academy, Dublin, pp 75–86

Graham JR (2009) Devonian. In: Holland CH, Sanders IS (eds) The geology of Ireland, 2nd edn. Dunedin Academic Press, Edinburgh, pp 175–214

Graham JR (2010) The Carboniferous geology of North Mayo. Irish J Earth Sci 28:25–45

Graham JR (2017) The Mullaghmore sandstone formation of north-west Ireland: a regional Mississippian lowstand deposit. Irish J Earth Sci 35:1–16

Graham JR (2021). Geology of South Mayo: a field guide. Geological Survey of Ireland, Dublin, 158pp

Graham JR, Pollard JE (1982) Occurrence of the trace fossil Beaconites antarcticus in the Lower Carboniferous fluviatile rocks of County Mayo, Ireland. Palaeogeogr Palaeocl 38:257–268

Graham JR, Smith DG (1981) The age and significance of a small Lower Palaeozoic inlier in County Mayo. J Earth Sci-Dublin 4:1–6

Graham JR, Richardson JB, Clayton G (1983) Age and significance of the Old Red Sandstone around Clew Bay, NW Ireland. Trans R Soc Edin-Earth 73:245–249

Graham JR, Riggs NR, McConnell B (2020) A reliable age for the Louisburgh-Clare Island succession and its significance in the stabilisation of the Laurentian margin in Ireland. Irish J Earth Sci 38:1–10

Long CB, Max MD (1977) Metamorphic rocks in the SW Ox Mountains inlier, Ireland: their structural compartmentation and place in the Caledonian orogeny. J Geol Soc Lond 133:413–432

McCaffrey KJW (1997) Controls on reactivation of a major fault zone: the Fair Head-Clew Bay line in Ireland. J Geol Soc Lond 154:129–133

Riggs N, McConnell B, Graham J (2021) Sedimentary provenance of Silurian basins in western Ireland during Iapetus closure. In: Kuiper YD, Murphy JB, Nance RD, Strachan RA, Thompson MD (eds) New developments in the Appalachian-Caledonian-Variscan Orogen; Geol Soc Am Spec Paper 554. https://doi.org/10.1130/2021.2554(16)

Sevastopulo GD, Wyse Jackson PN (2009) Carboniferous: Mississippian (Tournaisian and Visean) In: Holland CH, Sanders IS (eds) The geology of Ireland, 2nd edn. Dunedin Academic Press, Edinburgh, pp 215–268

The Geology of Western Ireland: A Record of the 'Birth' and 'Death' of the Iapetus Ocean

Paul D. Ryan, David M. Chew, Robert A. Cliff, Alessandra Costanzo, J. Stephen Daly, John F. Dewey, Martin Feely, Rose Fitzgerald, Michael J. Flowerdew, John R. Graham, Bernard E. Leake, Barry Long, Claire A. McAteer, Ken McCaffrey, William McCarthy, Julian F. Menuge, Ian S. Sanders, Ray Scanlon, and Bruce W. D. Yardley

P. D. Ryan (✉) · A. Costanzo · M. Feely
School of Natural Sciences, Earth and Ocean Sciences, National University of Ireland Galway, University Road, Galway, Ireland
e-mail: paul.ryan@nuigalway.ie

D. M. Chew · J. R. Graham · I. S. Sanders
Department of Geology, Trinity College Dublin, Museum Building, Dublin Dublin 2, Ireland

R. A. Cliff · B. W. D. Yardley
School of Earth & Environment, University of Leeds, Leeds LS2 PJT, UK

J. S. Daly · R. Fitzgerald · C. A. McAteer · J. F. Menuge
UCD School of Earth Science, Science Centre West, University College Dublin, Belfield, Dublin Dublin 4, Ireland

J. F. Dewey
University College, High Street, Oxford OX1 4BH, UK

M. J. Flowerdew
CASP, West Building, Madingley Rise, Madingley Road, Cambridge CB3 0UD, UK

B. E. Leake
School of Earth and Ocean Sciences, Cardiff University, Main Building, Cardiff C10 3AT, UK

B. Long · R. Scanlon
Geological Survey Ireland, Beggars Bush, Haddington Road, Dublin 04 K7X4, Ireland

K. McCaffrey
Department of Earth Sciences, Durham University, Stockton Road, Durham D1 3LE, UK

W. McCarthy
School of Earth and Environmental Science, Bute Building, Queen's Terrace, St Andrews K16 9AL, UK

© The Author(s), under exclusive license to Springer Nature Switzerland AG 2022
P. D. Ryan (ed.), *A Field Guide to the Geology of Western Ireland*, Springer Geology Field Guides https://doi.org/10.1007/978-3-030-97479-4_11

1 Introduction

This chapter reviews the evidence found in the remarkable and varied geology of western Ireland (Fig. 1) for the opening and the closing of the Iapetus Ocean adjacent to the Laurentian margin. The geology visited is briefly reviewed along with a short summary of its likely tectonic significance. Ages are given either as absolute ages followed by the name of the appropriate stratigraphic interval where the determination was isotopic, or by the stratigraphic interval name followed by the absolute age range for that interval where the determination was stratigraphic. Ages are cited using the timescale of Walker et al. (2018). The reader is referred to the individual chapters for more detail. An overview of the changing tectonic regimes that shaped the geology studied is provided in Fig. 2. The geology of western Ireland and its

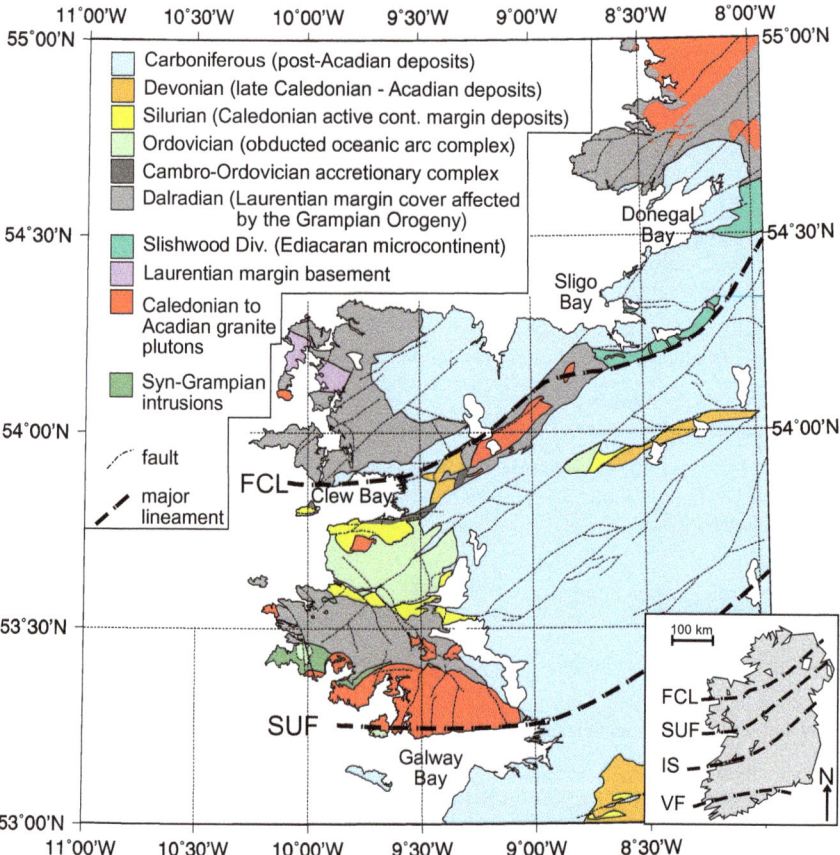

Fig. 1 Outline geological map of western Ireland showing the rock units studied and their likely tectonic setting. Abbreviations: cont. = continental; FCL = Fair Head—Clew Bay Line; IS = Iapetus Suture; SUF = Southern Uplands Fault; VF = Variscan Thrust Front

Time	Process	Cartoon
Late Devonian 380 Ma to Early Carboniferous ~330 Ma	Marine transgression. Dextral transpression related to Variscan docking of Armorica in the south	García-Navarro & Fernández (2004)
Late Silurian 424 Ma to Late Devonian 380 Ma	Tear in down-going slab, mafic underplating and granite genesis. Acadian sinistral transpression then transtension	Atherton & Ghani (2002)
Late Silurian ~424 Ma	Highly oblique sinistral Laurentia - Baltica 'Caledonian' continent-continent collision	Chew & Stillman (2009)
Late Ordovician Early Silurian 463 - 430 Ma	Subduction flip & formation of magmatic arc at Laurentian Margin. Convergence now sinistrally oblique	Dewey & Mange (1999)
Middle Ordovician 475 - 463 Ma	Dextrally oblique arc-continent collision causing Grampian Orogeny and subsequent collapse	Dewey & Mange (1999)
Mid-Cambrian 510 Ma to Middle Ordovician 475 Ma	Subduction initiates forming intra-oceanic arc & accretionary complex. Onset Iapetus closure. Late Dalradian deposition	Dewey & Mange (1999)
Cryogenian >700 Ma to Cambrian ~520 Ma	Breakout of Laurentia, rifting hyper-extension, mafic volcanism, Dalradian deposition opening of Iapetus	Chew et al. (2008)
Early Neoproterozoic 1015±4 Ma - 958+17/-19 Ma	Grenville Orogen, Laurentia collides with Amazonia? assembly of Rodinia and subsequent slow collapse	Hynes & Rivers (2010)
Late Mesoproterozoic 1177±4 Ma	Magmatism on Laurentian margin, ocean closure and collision with Amazonia?	Hynes & Rivers (2010)
Palaeoproterozoic 1753±3 Ma	Continental growth during Gothian Orogeny along accretionary margin of 'Columbia'	Wang et al. (2021)

Fig. 2 A summary of the tectonic history of western Ireland from the Palaeoproterozoic to the Lower Carboniferous. Highly stylised cartoons, based upon tectonic models proposed in the literature, are used to illustrate the changing tectonic setting throughout time. The Carboniferous continental mass, Laurussia, is as defined by Torsvik et al. (2002). Continental lithosphere is hachured, oceanic lithosphere is black, granites are red, faults are dashed lines, dykes are vertical black lines, mafic continental underplate is black, vulcanism is portrayed as a grey ash cloud, asthenospheric upwelling by curved arrows, thrusts by half arrows, collapse by divergent hollow arrows and sense of strike slip by arrow heads and tails

setting in the wider context of the geology of Ireland and beyond is reviewed in Holland and Sanders (2009).

2 Palaeoproterozoic and Mesoproterozoic

The Annagh Gneiss complex is examined in chapter "The Basement and Dalradian Rocks of the North Mayo Inlier". It comprises Palaeoproterozoic and Mesoproterozoic orthogneisses deformed and metamorphosed during the polyphase Grenvillian orogeny (Daly 1996) which marked the assembly of the supercontinent Rodinia. The oldest rocks, the Mullet gneiss (1753 ± 3 Ma, Statherian, Palaeoproterozoic), were formed from juvenile magma during a major Palaeoproterozoic crust forming event. (Locs 2.1 and 2.2). The Cross Point gneiss (c. 1280 Ma, Ectasian, Mesoproterozoic) represents a phase of anorogenic magmatism (Loc. 2.5). Early Grenville deformation and metamorphism took place between 1177 Ma (Stenian, Mesoproterozoic), the date of the formation of the Doolough gneiss (Loc. 2.6), and 1015 Ma (Stenian, Mesoproterozoic). This was followed by migmatisation and pegmatite formation (Loc. 2.1) between 995 and 960 Ma (Tonian, Neoproterozoic). This sequence of events correlate with those found in the Laurentian and Baltic shields (Daly 1996). Evidence for vulcanism associated with subsequent rifting at around 600 Ma (Ediacaran, Neoproterozoic) is provided by meta-dolerites that crosscut the Grenvillian fabrics (Locs 2.1, 2.3, 2.4 and 2.5). These data establish that NW Mayo comprised a part of the Laurentian continent that was subject to rifting during the Neoproterozoic.

3 Neoproterozoic Rock Units

3.1 The Slishwood Division

The Slishwood Division examined in chapter "The Slishwood Division and Its Relationship with the Dalradian Rocks of the Ox Mountains" is a testament to the complexity of the Laurentian Margin during Neoproterozoic rifting. The paragneiss protolith (Locs 3.1 and 3.3) was deposited on the Laurentian margin between 926 Ma (Tonian, Neoproterozoic), the age of the youngest detrital zircon, and 596 Ma (Ediacaran, Neoproterozoic), the age of the emplacement of mafic intrusions (Loc. 3.1) and may correlate with the Grampian Group of the Dalradian (Flowerdew et al. 2005; Sanders 1994). Mafic and ultramafic bodies (Locs 3.1 and 3.2) were emplaced as these rocks became detached from the Laurentian margin. At some time before or during the Grampian Orogeny the rocks were buried to at least 45 km and underwent eclogite facies metamorphism. Decompression to granulite facies occurred during the Grampian Orogeny, discussed below, at around 470 Ma (Floian to Dapingian; Flowerdew and Daly 2005). Subduction related tonalites were emplaced (Locs 3.4 and 3.5) during this time and the Slishwood Division was tectonically interleaved with the Dalradian (Loc. 3.7) along the North Ox Mountains Slide (NOMS) (Flowerdew et al 2000). The sheared structural contact with the Dalradian along the NOMS entrains mantle material (Loc. 3.7). Late pegmatites (Loc. 3.6) date post-Grampian orogenic collapse at c. 455 Ma (Sandbian; Flowerdew et al. 2000).

3.2 The Dalradian

The Dalradian Supergroup comprises a mid-Neoproterozoic to Early Palaeozoic sequence which crops out on the islands of the Shetlands and the Inner Hebrides, the

Highlands of Scotland south of the Great Glen Fault, in northwestern and western Ireland and may continue to Newfoundland (see review of Stephenson et al. 2013). It is interpreted as a southeasterly prograding clastic wedge (Soper and England 1995) deposited during and immediately after the break-out of Laurentia from the core of Rodinia (Van Staal et al. 1998). Four main groups are named in order of decreasing age (Harris et al. 1994): the Grampian Group; Appin Group; Argyll Group; and Southern Highland Group.

The Dalradian Supergroup metasedimentary and volcanic rocks of NW Mayo and Connemara, County Galway are studied in chapters "The Basement and Dalradian Rocks of the North Mayo Inlier", "The Central Ox Mountains" and "The Ordovician Arc Roots of Connemara" of this guide. The Scotchport Schists, attributed to the Grampian Group, rest uncomformably upon Laurentian basement comprising the Annagh Gneiss Complex (Loc. 2.2). The psammitic Grampian Group (Loc. 2.4) marked the initial phase of rifting followed by thermal subsidence during which the compositionally more varied Appin Group (Locs 2.10–2.12) was deposited (Soper and England 1995). A key stratigraphic horizon recognised throughout the Dalradian and used for regional lithostratigraphic correlation is the Cryogenian Boulder Bed (C. 700 Ma) that marks the base of the Argyll Group (Locs 2.9 and 5.s2). The Argyll Group (Locs 2.13–2.17, 4.1, 4.3, 4.5, 4.6 and 5.1–5.5) recorded major rifting associated with mafic vulcanism (Locs 2.14, 2.15a, 2.17, 3.7a, 4.3, 4.6 and 5.1) and a serpentinite mélange (Loc. 2.15f) produced by serpentinisation of sub-continental lithospheric mantle at the ocean-continent transition during hyper-extension (Chew 2001; Chew and Van Staal 2014). The Ediacaran to Lower Palaeozoic Southern Highland Group was deposited during rifting and subsequent onset of drift associated with the opening of the Iapetus Ocean and may have prograded onto oceanic crust (Soper and England 1995). Rocks attributed to the Southern Highland Group crop out in the Ox Mountains (Cloonygowan Formation, Loc. 4.4) but these may belong to the Clew Bay Complex (see below). In summary, these rocks demonstrate rifting and subsequent hyper-extension of the Laurentian margin at around 600 Ma (Ediacaran, Neoproterozoic) associated with the subsequent opening of the Iapetus Ocean and the formation of the Slishwood micro-continent.

4 The Cambro–Ordovician Rock Units

4.1 Clew Bay Complex

The low metamorphic grade, highly sheared, turbidites of the Clew Bay Complex, studied in chapters "The Basement and Dalradian Rocks of the North Mayo Inlier" and "The Ordovician Fore-Arc and Arc Complex" are poorly dated palaeontologically to mid-late(?) Cambrian to mid-late(?) Ordovician (c. 510–460 Ma). They are of Laurentian provenance (Chew 2003) and are attributed to a north facing accretionary complex that was obducted onto the Laurentian margin during Grampian arc-continent collision (Chew 2003; Dewey and Mange 1999; Dewey and Ryan 1990). They are separated from Dalradian rocks to the north by the Achill Beg Fault (Loc. 2.24). However, Chew (2003) concluded that the rock units on either side of this fault share the same 460 Ma (late Darriwilian), Grampian, S2 fabric. Compositionally the Clew Bay Complex comprises quartz-wackes (Locs 2.18 and 2.21) often displaying syn-sedimentary and dewatering structures, thinly interbedded graphitic schists (Loc. 2.19) and limestones (Loc. 2.20). Blocks of sandy or conglomeratic material often have no internal fabric but occur as phacoids surrounded by sheared and quartz-veined mudrocks (Loc. 6.1). These rocks share the same deformation history as the Dalradian to their north (Locs 2.18 and 2.21). The Cloonygowan Formation of the Central Ox Mountains (Loc. 4.4) is compositionally similar to the quartz-wackes of the Clew Bay Complex: however, it is not clear whether it formed part of the accretionary prism.

4.2 The Deer Park Complex

The Deer Park Complex is an ophilitic mélange (Ryan et al. 1983) tectonically interleaved with the Clew Bay Complex (Locs 2.23, 6.2 and 8.17). Blocks of mafic and ultramafic igneous rocks (Loc. 6.3) and sediment are surrounded by a sheared matrix of serpentinite or talc-carbonate (Loc. 6.2). The metamorphic facies of the blocks varies from greenschist to amphibolite. Hornblendes from a metabasite yielded a $^{40}Ar-^{39}Ar$ plateau age of 514 ± 3 Ma (Series 2, Cambrian; Chew et al. 2010). Dewey and Ryan (1990) interpret this a backstop ophiolite to the Clew Bay Complex accretionary prism, which implies subduction began before 514 Ma (Chew et al. 2010). Garnet mica schists incorporated into the mélange and exhumed at ~482 Ma (Tremadocian) may date the local onset of obduction of the fore-arc marking early stages of the Grampian Orogeny prior to the metamorphic maximum recorded at 475–465 Ma (Floian to Darriwilian) in NW Mayo (Chew et al. 2010).

The Clew Bay and Deer Park Complexes, coupled with the blueschists of the Claggan Formation, provide evidence for a Laurentia-facing subduction accretion complex from mid-Cambrian to mid-Ordovician times.

4.3 Cambro–Ordovician of South Mayo

The Ordovician volcanic and sedimentary rocks of South Mayo (Chapter "The Ordovician Fore-Arc and Arc Complex") rest in the large Mweelrea-Partry Syncline and are of low metamorphic grade. They were deposited in the fore-arc basin (Dewey and Ryan 1990) to the north-facing Lough Nafooey arc. This arc was part of the

Cambrian (Series 2) to mid-Ordovician (c. 510–466 Ma) Laurentia facing intra-oceanic arc that may have stretched from Scotland to Texas (Dalziel and Dewey 2018). The north limb succession, over 8 km in thickness (Dewey and Mange 1999; Graham 2021), comprises Floian to Darriwilian (c. 473–466 Ma) deep water deposits (Locs 6.6–6.12) which pass up conformably into the Darriwilian to Sandbian (464–460 Ma) fluviatile succession of the Mweelrea Formation (Locs 6.4 and 6.5). Derivation was initially from northerly ophiolitic (Loc. 6.9) and Laurentian sources, but by Darriwilian times juvenile metamorphic detritus was supplied from the south and east (Fig. 12, Chap. 6). Interbedded tuffs were sourced from the south. The south limb and fold core contain volcanic complexes whose geochemistry and petrology show a change from intra-oceanic arc deposits in the Tremadoc (c. 480 Ma, possibly older) Lough Nafooey Group (Locs 6.15 and 8.4) and Bohaun Volcanic Formation (Loc. 6.20) to an increasing continental influence on the magma composition in Floian to Darriwilian times (Draut et al. 2004) of the Tourmakeady Volcanic Formation (Loc. 6.17). These complexes are overlain by the deep water Rosroe Formation (Locs 6.15, 6.18 and 6.19), the lutites of the Glenummera Formation and the fluviatile Mweelrea Formation (Loc. 6.4). Proximal sediments of the Rosroe and Mweelrea formations record the stripping of the arc and an orogen (Locs 6.14 and 6.18). The sedimentary evolution of the South Mayo Trough has been linked to collision of the Lough Nafooey arc and fore-arc with the Laurentian margin and subsequent subduction flip (Dewey and Mange 1999).

The South Connemara Group (Locs 7.1 and 7.2) is a poorly dated mid- to late-Ordovician (c. 450 Ma), Iapetus Ocean facing accretionary complex believed to have been formed as a result of this subduction flip and the establishment of a magmatic arc along the Laurentian margin (Ryan and Dewey 2004). The Sandbian ignimbrites of the Mweelrea Formation (Loc. 6.5) may have erupted during the slab detachment which preceded the formation of the South Connemara Group accretionary prism.

The Ordovician of the South Mayo Trough records: the initiation of, or at least the early stages of, subduction within the Iapetus Ocean; the subsequent development of an intra-oceanic fore-arc; arc-continent collision; and consequent subduction flip. The west of Ireland is the only place within the Caledonides of Britain and Ireland where these processes can be studied in such detail.

5 The Grampian Orogeny

The Grampian Orogeny was a short lived, mid-Ordovician (Darriwilian to Sandbian, c. 473–464 Ma) event that affected the Laurentian margin (Dewey 2005). It is attributed to the obduction of the Lough Nafooey arc and fore-arc over the hyperextended Laurentian margin (Chew 2009; Dewey and Shackleton 1984; Ryan and Dewey 1991). The deformation in the NW Mayo inlier (e.g., Loc. 2.8) suggests early top to the north-west or west-facing nappe formation (c. 475–465 Ma, Floian to Darriwilian) that was modified by D2 folding (Locs 2.8, 2.13 and 2.15b) and later (c. 460–450 Ma, Darriwilian to Katian) D3 crenulation and dextral transpression

(Locs 2.12 and 2.15d, e) to produce a facing divergence across the Annagh Gneiss Complex (Chew 2009). The metamorphic maximum was post-D2 and grade generally decreases southwards from the kyanite zone (Loc. 4.5). Much of the region is in the garnet zone (Loc. 2.16). Blueschists are recorded in the Claggan Volcanic and South Carrogarve formations (Locs 2.14 and 2.17) of the Dalradian Supergroup adjacent to the Fair Head–Clew Bay line and are overprinted by low grade, medium P assemblages.

The Slishwood Division (Chapter "The Slishwood Division and Its Relationship with the Dalradian Rocks of the Ox Mountains") was first subjected to eclogite facies conditions, but eclogites decompressed to high pressure granulite facies conditions (Locs 3.1 and 3.9) prior to emplacement of early Grampian tonalites (Locs 3.4 and 3.5; Flowerdew et al. 2005). Late pegmatites (Loc. 3.6) with associated extensional shears may date (455 Ma, Sandbian) D3 collapse of the orogenic pile. Top to the southeast extension tectonically interleaves the Slishwood Division paragneisses with Grampian tonalites, Dalradian schists and metavolcanics and serpentinised ultrabasic rocks (Loc. 3.7). This is interpreted as a suture zone between the Slishwood Division microcontinent, formed during Cryogenian rifting, and the Laurentian margin.

Grampian folding and peak metamorphism in Connemara (Locs 5.1–5.4 and 5.9–5.12) are dated by syn- to post tectonic intrusions and metamorphic minerals. This was a short-lived event within the window ~472 to ~462 Ma (Late Floian to early Darriwilian) by which time the Dalradian was already at garnet grade. Intrusions (Locs 5.s2, 5.8 and 5.13–5.15) comprise a suite of early basic to later acid bodies with the chemical characteristics of arc magmas. They were emplaced in a limited time interval spanning but outlasting pervasive ductile deformation and coincided with peak metamorphic temperatures. The lack of folding of isograds (Fig. 6, Chap. 5) shows the metamorphic peak post-dated major nappe formation. Locally temperatures reached the second sillimanite isograd (Loc. 5.5) with associated generation of migmatitic gneisses (Loc. 5.6); this temperature peak was triggered by the emplacement of the arc magma suite. Shortly after the last magmas were emplaced, Connemara was exhumed over metarhyolites of the Delaney Dome (Loc. 5.18) at c. 457 Ma (Sandbian). These rhyolites were erupted at c. 475 Ma (Floian) and are thus broadly correlatable with volcanics in South Mayo.

The distribution of metamorphic rocks over the study area with blueschists north of the FCL in the NW Mayo inlier, early eclogites immediately to the south of the FCL in the Slishwood Division and migmatites to the south in Connemara argue for a complex collision zone subsequently heavily sliced and modified during orogenic collapse and subsequent tectonism, particularly Acadian (early Devonian) sinistral transpression (Van Staal et al. 1998). However, it should be borne in mind that the metamorphic conditions varied through time even within the Dalradian rocks. Early low grade high-P metamorphism may have been present in many more rocks in NW Mayo, and perhaps even Connemara, while the high-T, low-P metamorphism of Connemara is superimposed on earlier metamorphism similar to that of NW Mayo (Yardley et al. 1987). Nonetheless, there is considerable field, geochemical and isotopic evidence supporting the overall arc-continent collision model involving obduction of an Ordovician arc and fore-arc onto the Laurentian margin (e.g.,

Dewey and Ryan 2016) which has been extended along strike from the Shetlands to Newfoundland (Dalziel and Dewey 2018; Ryan and Dewey 2019).

6 The Silurian

The Silurian of northern Galway and southern Mayo was deposited in three separate, poorly defined, successor basins (Chapter "The Silurian of North Galway and South Mayo"). The first of these, the mid-Telychian to Sheinwoodian (c. 435–430 Ma) Killary Harbour–Joyce Country Succession of north Connemara masks the inferred extensional contact (Locs 8.5 and 8.7) between the Ordovician of South Mayo and the Connemara Dalradian. The succession rests unconformably upon the enigmatic Derryveeny Formation (Loc. 8.1) which in turn rests unconformably on both the Connemara Dalradian and the Ordovician arc rocks. The succession indicates rapid deepening from terrestrial (Loc. 8.2) to turbidite dominated (Locs 8.3 and 8.4) settings and then shallowing (Locs 8.6, 8.9 and 8.10). The second, the Croagh Patrick Succession (Locs 8.12–8.14) was deposited on gently folded Ordovician rocks of the South Mayo Trough. This Llandovery(?) siliciclastic succession of Laurentian provenance was subject to greenschist-facies metamorphism, sinistral transpression and folding during the Acadian Orogeny (Loc. 8.14). The third, the northernmost non-marine Ludfordian or Pridoli (c. 425–420 Ma) Louisburgh—Clare Island Succession (Locs 8.15 and 8.16) rests unconformably on the Clew Bay Complex and is in fault contact with the Croagh Patrick succession.

Andesitic–rhyolitic volcanic rocks (Locs 8.8 and 8.10) and fresh igneous detritus in the Killary Harbour–Joyces Country succession are consistent with deposition in a fore-arc setting within a Iapetus-facing subduction system (Williams et al. 1992). These partially marine deposits rest upon both the hanging wall (Ordovician of South Mayo) and footwall (Connemara Dalradian) of the Grampian Orogen implying that the continental crust had returned to 'normal' thickness by Llandovery times. The preservation of both upper plate (Ordovician) and lower plate (Dalradian) rocks beneath the Silurian provides evidence for late Ordovician orogenic collapse.

7 The Acadian

The Acadian Orogeny in Ireland was an early Devonian event that postdated the complex sinistrally oblique closure of the Iapetus Ocean (Chew and Stillman 2009; Dewey and Strachan 2003). Strain is strongly partitioned into the fault zones separating the major crustal blocks, such as the Fair Head–Clew Bay line (Chew and Stillman 2009). The c. 413 Ma (Lochkovian) Corvock and Ox Mountain Igneous Complex granitoids (Locs 4.1 and 4.2) were emplaced in dilational jogs at depth along this lineament whilst Devonian sediments were deposited in pull-apart basins

(Locs 10.1–10.5) at the surface. The Dalradian succession of the Central Ox Mountains was folded and retrogressed (Loc. 10.1) and high strain sinistral shear zones developed along the Fair Head–Clew Bay line (Locs 4.3 and 4.4). The Ludfordian to Pridoli Clare Island-Louisburgh succession may provide the earliest local evidence for the onset of late Caledonian sinistral transpression related to the docking of Avalonia and the closure of Iapetus (Loc. 8.14) along the Iapetus Suture some 130 km further south (Fig. 1, inset). Ductile folds and a clockwise transecting cleavage, dated by the 413 Ma (Lochkovian; Graham and Riggs 2021) Corvock granite (Locs 6.11 and 8.11), are formed in both the Ordovician (Locs 6.7–6.12) and Silurian (Locs 8.11–8.14) rocks south of Clew Bay.

8 The Fair Head–Clew Bay Line (FCL)

This major magnetic lineament (Max and Riddihough 1975) marks a zone that has a long tectonic history (Chew 2001, 2003; Chew et al. 2003, 2004; McCaffrey 1997) and can be traced into the Highland Boundary Fault of Scotland (Chew 2003). Serpentinite mélanges on Achill (Loc. 2.15f) and mafic volcanic complexes within the Daradian succession (Locs 4.5 and 4.6) attest to this zone being the site of major attenuation of the Laurentian lithosphere during the Ediacaran. Blueschists in the Dalradian Claggan Volcanic and South Carrogarve formations north of the FCL (Loc. 2.14) indicate subduction of the Laurentian margin during the Grampian Orogeny. The Achill Beg fault forms the northern boundary of the Clew Bay and Deer Park accretionary complexes. D2 Grampian nappes within the Dalradian (Loc. 2.24) face downward on the S2 cleavage as the FCL is approached. The FCL experienced late-Grampian dextral transpression until 448 Ma (Katian). The Clare Island–Louisburg succession (c. 424 Ma) provides a minimum age for a switch to late Caledonian sinistral motion along this zone (Loc. 8.14). The Corvock Granite (Locs 6.11 and 8.11) and the Ox Mountains Igneous Complex (Locs 4.1 and 4.2) were emplaced during early Devonian (Acadian) sinistral transpression coeval with ductile folding and sliding within the Dalradian of the Ox Mountains (Locs 4.3 and 4.4). Subsequent late Devonian dextral motion was followed by Carboniferous normal motion on brittle structures. The FCL and structures associated with it mark the northern limit, at the present erosion level, of the obducted oceanic terranes. However, the Dalradian rocks of Connemara (Chapter "The Ordovician Arc Roots of Connemara") and the Tyrone Central Inlier to the north-east (Chew et al. 2008) lie to the south of the FCL. This relationship has been attributed to strike-slip dismemberment of the Grampian Orogen (Hutton 1987) or, more recently, to complexities in the geometry of the rifted margin and the over-riding arc complexes (Chew et al. 2008; Dewey and Ryan 2016). The FCL may have initiated as a major extensional structure during Ediacaran rifting of the Laurentian margin during the opening of the Iapetus Ocean that continued to focus deformation during the closure of this ocean.

9 Siluro-Devonian Granites

The late Silurian to Middle Devonian Galway Granite Complex (GGC) is discussed in chapter "The Late Silurian to Upper Devonian Galway Granite Complex (GGC)". This complex comprises the early (424 Ma, Ludfordian) plutons of Roundstone, Inish, Omey and Letterfrack and the later Galway Batholith (410–380 Ma, Pragian to Frasnian). The Galway batholith is made up of the Carna (410 Ma, Pragian) and Kilkieran (400 Ma, Emsian) plutons (Feely et al. 2010, 2018). The various stages of granite emplacement mark changing tectonic regimes (McCarthy et al. 2015a, b). The early plutons and the Carna pluton were emplaced during regional transpression whilst the Kilkieran and associated plutons were emplaced during regional transtension. The Carna pluton comprises several granodioritic bodies (Locs 9.1–9.6) with a transitional contact (Loc. 9.7) with a K-feldspar porphyritic Errisbeg Townland Granite (Locs 9.11 and 9.12). Late alkali feldspar granites are locally garnetiferous (Loc. 9.7) and are mineralised (Loc. 9.8). The Kilkieran Pluton is the largest in the GGC and is dissected by major faults that expose different levels of the pluton. In the west, a high level megacrystic granite is in intrusive contact with the migmatised Dalradian (Loc. 9.13). Deeper levels in the central region are characterised by mixing and mingling of granitic and dioritic magma (Loc. 9.16). The 380 Ma (Frasnian) Costelloe Murvey alkali feldspar granite (Loc. 9.14) marks the end of granite emplacement in the GGC. The Omey Pluton to the north provides an excellent opportunity to study granite emplacement and contact metamorphism (Locs 9.18–9.24).

The granites are of overwhelmingly of calc-alkaline composition and magma generation is attributed to slab dehydration followed by slab break-off of the north dipping down going slab during and after the final stages of closure of the Iapetus Ocean (Atherton and Ghani 2002). The GGC lies along the projected trace of the Southern Uplands Fault in western Ireland (Leake 1963), the structure which separates the Dalradian of Connemara from the late Ordovician to Silurian south facing Longford—Down—Southern Uplands accretionary prism, which may have provided a conduit for magma ascent. Construction of the GGC was controlled by changes in the regional stress field from sinistral transpression to sinistral transtension between 410 and 400 Ma (Pragian to Emsian). A similar timing for this change in the stress field is observed in the Ox Mountains Igneous Complex (Chapter "The Central Ox Mountains"). The granites of western Ireland document the final stages of Caledonian sinistrally transpressive continent–continent collision and the consequent changes in relative plate motion (Dewey and Strachan 2003).

9.1 Devonian and Carboniferous

The Devonian rocks of the SW Ox Mountains (Chapter "Devonian and Carboniferous") provide a sedimentary record of the sinistrally oblique collision during and immediately after the closure of the Iapetus Ocean. The older Lower Devonian (Pragian to Emsian, c. 410–390 Ma) Islandeady Group (Locs 10.1 and 10.2) has polymict conglomerates at its base (Loc. 10.1). These lie unconformably upon cleaved Ordovician rocks of the Letter Formation a possible correlative of the Upper Derrylea Formation (Mange et al. 2003) suggesting a minimum age for cleavage formation in the Ordovician and Silurian of South Mayo (Loc. 10.1). The younger Middle Devonian (Eifelian, c. 393–388 Ma) Beltra Group (Locs 10.3–10.6) contains conglomerates with sandstone and volcanic clasts. McCaffrey (1997) proposed that these formed as pull-apart basins controlled by faults which abutted a rigid block of cooled Ox Mountains Igneous Complex. It is possible that the switch from Early Devonian sinistral motion to Late Devonian dextral deformation took place after the deposition of the less deformed Beltra Group. These rocks with the Ox Mountains Igneous Complex and the Galway Granites to the south provide evidence for Early to Middle Devonian sinistral deformation associated with ocean closure that ended in the Late Devonian.

The region around Clew Bay remained topographically positive during the early Carboniferous (Locs 10.7–10.10) with subsidence being controlled by major structures within the Caledonian basement (Worthington and Walsh 2011). The marine and non-marine Visean (347–331 Ma) succession of north Mayo (Locs 10.11–10.14) provides evidence that the crust was no longer tectonically thickened. The west of Ireland was now part of the Laurussian Continent (Torsvik et al. 2002). However, the gentle dips and open folds suggest that the lithosphere was sufficiently cooled and thickened to be able to resist far-field stresses from the Variscan Orogeny whose thrust front in Ireland lies 200 km to the south (Fig. 1, inset).

9.2 Outstanding Problems and Future Research

We have been studying the geology of the West of Ireland for nearly two centuries but may topics still inspire debate and invite future research. A few of these are summarised below.

The Laurentian margin is thought to have been hyper-extended and have had outboard microcontinents (Chapters "The Basement and Dalradian Rocks of the North Mayo Inlier"–"The Ordovician Arc Roots of Connemara"). This original architecture has been tectonically modified and much of the outboard portion is buried beneath Carboniferous cover (Fig. 1). Analysis of existing Geological Survey Ireland Tellus geophysical data sets (see section "GSI Tellus Datasets", chapter "Data Sources for use in Pre- and Post-Trip Exercises") along with structural and geochronological studies may help define the distribution of any such partially buried microcontinents and the geometry of their contacts, for example, the unexposed (sub-Carboniferous) boundary between the Annagh Gneiss Complex with the (presumed) basement to the Slishwood Division?

The Ordovician Grampian Orogeny is attributed to a short-lived arc-continent collision. Isotopic dates for individual deformation phases are only available for

Connemara, but even here there are significant uncertainties (Chapter "The Ordovician Arc Roots of Connemara"). More recent techniques utilising spot analysis especially on undated syn-tectonic metagabbros and the Mannin Thrust should resolve these. In North Mayo the present dates (Chapter "The Basement and Dalradian Rocks of the North Mayo Inlier") are only for fabric-forming minerals (e.g. Ar–Ar on micas or hornblende). High-precision U–Pb monazite or Lu–Hf garnet ages are required to allow direct comparison of the deformation and metamorphic history of this region with that of Connemara. If the Grampian Orogen had a similar duration to the ongoing Penglai arc-continent collision in Taiwan (<5 m.y., e.g., Kirstein et al. 2010), accurate and precise geochronology is key.

Many of the major faults discussed in the text have very significant vertical throws (perhaps 20 km or more) in addition to their strike slip components (e.g., Fair Head–Clew Bay Line, Chapter "The Ordovician Fore-Arc and Arc Complex"). In general, their three dimensional orientation in depth is not well known as their surface expression is greatly influenced by later deformational events. Available seismic reflection and refraction studies do not resolve individual structures in the upper crust. There is a need for seismic reflection profiles across major contacts which are believed to be relatively shallowly dipping and separating rocks of high impedance contrast, such as the Mannin Thrust (Chapter "The Ordovician Arc Roots of Connemara"). Gravity surveys may also prove useful, for example, across the Ordovician/Dalradian contact (arc-continent contact) in north Connemara which is buried under Silurian cover (Chapter "The Silurian of North Galway and South Mayo").

Was an early, presumed Grampian, high pressure–low temperature (high P–low T) metamorphism originally experienced widely in the rocks of the NW Mayo and Ox Mountains inliers (Chapters "The Basement and Dalradian Rocks of the North Mayo Inlier" and "The Slishwood Division and Its Relationship with the Dalradian Rocks of the Ox Mountains") and later obliterated in most places by a Barrovian overprint? Or was high P–low T metamorphism only experienced in the rocks closest to the Laurentian margin where it is recorded? Over what time span did this high P event occur? How does the high P history in the north relate to the early metamorphic history of Connemara (Chapter "The Ordovician Arc Roots of Connemara")? These questions may be resolved by a search for high P relics in rocks away from the Laurentian margin, potentially in lithologies such as meta-basites and meta-cherts.

There is considerable scope for further sedimentological and provenance studies (Chapters "The Ordovician Fore-Arc and Arc Complex", "The Silurian of North Galway and South Mayo" and "Devonian and Carboniferous"). For example, The Mweelrea Formation (Chapter "The Ordovician Fore-Arc and Arc Complex") is only known in broad outline as a westward-fast-flowing river system, guided by a Mayoian growing fold that was thought to have been deposited during or just after subduction flip. The abundance of fresh volcanic material, igneous and metamorphic detritus from the emerging Grampian Orogen make this well exposed formation a prime target for understanding the stripping of an orogen.

Western Ireland is thought to contain two plate boundaries, the early- to mid-Ordovician Fair Head–Clew Bay Line in the north and the late-Ordovician to Silurian Skird Rocks—Southern Uplands Fault in the south. There is, however, a paucity of modern quantitative structural techniques over this whole region. The complex history of reactivation of these structures that includes reversal of strike-slip components are as yet poorly understood. An in-depth look at the transpressional strain geometries, kinematics and partitioning in the Ox Mountains (Chapter The Central Ox Mountains), for example, may yield valuable insights.

This is a prime area for the study of granite magmatism (Chapters The Central Ox Mountains and The Late Silurian to Upper Devonian Galway Granite Complex (GGC)). There are still rheological puzzles over the formation of the shears in partially consolidated magma in the Ox Mountains Igneous Complex (OMIC) at Pontoon. Geochronological studies of the intrusive suites within the OMIC and the Galway Granite Batholith (GGB) could show how these igneous complexes were constructed incrementally and how their sources were evolving. Further field and geophysical studies are required to determine the three dimensional shape of the GGB. What is the thickness of each block within the batholith? What is the exact nature of the contact between the Carna and Kilkieran plutons?

9.3 Conclusions

Figure 11.2 links the various tectonic processes described in the literature to the field geology presented in this guide. The west of Ireland probably spent 70% of the period covered (the Palaeoproterozoic to the Lower Carboniferous, c. 1400 m.y.) at or adjacent to a plate boundary. Geological events over this period included: the assembly of the Neoproterozoic supercontinent Rodinia; its subsequent break up leading to the escape of Laurentia and the opening of the Iapetus Ocean by the Early Cambrian; the onset of subduction within the Iapetus Ocean by Late Cambrian leading to the Middle Ordovician Grampian arc-continent collision; subduction flip and the establishment of an active continental margin; the final closure of Iapetus with the collision of Laurentia with Avalonia; subsequent tectonism during the Acadian Orogeny; and final cratonisation. Figure 11.2 is not presented as a detailed tectonic model, but as an illustration of our current understanding of the complex processes that formed the geology of this remarkable region.

References

Atherton M, Ghani A (2002) Slab breakoff: a model for Caledonian, Late Granite syn-collisional magmatism in the orthotectonic (metamorphic) zone of Scotland and Donegal, Ireland. Lithos 62:65–85

Chew D (2003) Structural and stratigraphic relationships across the continuation of the Highland Boundary Fault in western Ireland. Geol Mag 140:73–85

Chew D, Flowerdew M, Page L et al (2008) The tectonothermal evolution and provenance of the Tyrone Central Inlier, Ireland: Grampian imbrication of an outboard Laurentian microcontinent? J Geol Soc Lond 165:675–685

Chew D (2009) Grampian orogeny. The Geology of Ireland, 2nd edn. Dunedin Academic Press, Edinburgh, pp 69–93

Chew D, Stillman C (2009) Late caledonian orogeny and magmatism. In: The geology of Ireland, 2nd edn. Dunedin Academic Press Edinburgh, pp 143–173

Chew D, van Staal C (2014) The ocean–continent transition zones along the Appalachian-Caledonian Margin of Laurentia: Examples of large-scale hyperextension during the opening of the Iapetus Ocean. Geosci Can 41:165–185

Chew DM (2001) Basement protrusion origin of serpentinite in the Dalradian. Irish J Earth Sci 19:23–35

Chew DM, Daly J, Flowerdew M et al (2004) Crenulation-slip development in a Caledonian shear zone in NW Ireland: evidence for a multi-stage movement history. Geol Soc Lond Spec Pub 224:337–352

Chew DM, Daly JS, Magna T et al (2010) Timing of ophiolite obduction in the Grampian orogen. Geol Soc Am Bull 122:1787–1799

Chew DM, Daly JS, Page L, Kennedy M (2003) Grampian orogenesis and the development of blueschist-facies metamorphism in western Ireland. J Geol Soc Lond 160:911–924

Daly JS (1996) Pre-Caledonian history of the Annagh Gneiss Complex North-Western Ireland, and correlation with Laurentia-Baltica. Irish J Earth Sci 15:5–18

Dalziel IW, Dewey JF (2018) The classic Wilson cycle revisited. Geol Soc Lond Spec Pub 470:SP470-1

Dewey JF, Mange M (1999) Petrography of Ordovician and Silurian sediments in the western Irish Caledonides: tracers of a short-lived Ordovician continent-arc collision orogeny and the evolution of the Laurentian Appalachian-Caledonian margin. Geol Soc Lond Spec Pub 164:55–107

Dewey JF, Ryan PD (1990) The Ordovician evolution of the South Mayo Trough, western Ireland. Tectonics 9:887–901

Dewey JF, Ryan PD (2016) Connemara: its position and role in the Grampian Orogeny. Can J Earth Sci 53:1246–1257

Dewey JF, Shackleton RM (1984) A model for the evolution of the Grampian tract in the early Caledonides and Appalachians. Nature 312:115–121

Dewey JF, Strachan R (2003) Changing Silurian-Devonian relative plate motion in the Caledonides: sinistral transpression to sinistral transtension. J Geol Soc Lond 160:219–229

Draut AE, Clift PD, Chew DM et al (2004) Laurentian crustal recycling in the Ordovician Grampian Orogeny: Nd isotopic evidence from western Ireland. Geol Mag 141:195–207

Feely M, Gaynor S, Venugopal N, Hunt J, Coleman DS (2018) New U-Pb zircon ages for the Inish Granite Pluton, Galway Granite Complex, Connemara, Western Ireland. Irish J Earth Sci 36(1):1–7

Feely M, Selby D, Hunt J, Conliffe J (2010) Long-lived granite-related molybdenite mineralization at Connemara, western Irish Caledonides. Geol Mag 147(6):886–894

Flowerdew M, Daly JS, Whitehouse M (2005) 470 Ma granitoid magmatism associated with the Grampian Orogeny in the Slishwood Division, NW Ireland. J Geol Soc Lond 162:563–575

Flowerdew MJ, Daly JS (2005) Sm-Nd mineral ages and PT constraints on the pre-Grampian high grade metamorphism of the Slishwood Division, northwest Ireland. Irish J Earth Sci 23:107–123

Flowerdew MJ, Daly JS, Guise PG, Rex DC (2000) Isotopic dating of overthrusting, collapse and related granitoid intrusion in the Grampian orogenic belt, northwestern Ireland. Geol Mag 137:419–435

García-Navarro E, Fernández C (2004) Final stages of the variscan orogeny at the southern iberian massif: lateral extrusion and rotation of continental blocks. TectonS 23:TC6001, https://doi.org/10.1029/2004TC001646,

Graham J (2021) The geology of south mayo, a field guide. Geological survey of Ireland, Dublin, 158pp

Graham J, Riggs N (2021) A Lower Devonian age for the Corvock Granite and its significance for the structural history of South Mayo and the Laurentian margin of western Ireland. Can J Earth Sci. https://doi.org/10.1139/cjes-2021-0029

Harris AL, Haselock PJ, Kennedy MJ et al (1994) The dalradian supergroup in Scotland, Shetland and Ireland. In: Gibbons W, Harris AL (eds) Revised correlation of Precambrian rocks in the British Isles. Geol Soc Lond Spec Rep 22:33–53

Holland C, Sanders I (eds) (2009) The geology of Ireland, 2nd edn. Dunedin Academic Press, Edinburgh

Hutton DHW (1987) Strike slip terranes and a model for the evolution of the British and Irish Caledonides. Geol Mag 124:405–425

Hynes A, Rivers T (2010) Protracted continental collision—evidence from the Grenville Orogen. Can J Earth Sci 47:591–620

Kirstein LA, Fellin MG, Willett SD, Carter A, Chen YG, Garver JI, Lee DC (2010) Pliocene onset of rapid exhumation in Taiwan during arc-continent collision: new insights from detrital thermochronometry. Basin Res 22:270–285

Leake B (1963) The location of the Southern Uplands Fault in central Ireland. Geol Mag 100:420–423

Mange MA, Dewey JF, Wright DT (2003) Heavy minerals solve structural and stratigraphic problems in Ordovician strata of the western Irish Caledonides. Geol Mag 140:25–30

Max MD, Riddihough RP (1975) The continuation of the Highland Boundary Fault in Ireland. Geology 3:206–210

McCarthy W, Petronis MS, Reavy RJ, Stevenson CT (2015a) Distinguishing diapirs from inflated plutons: an integrated rock magnetic fabric and structural study on the Roundstone Pluton, western Ireland. J Geol Soc London 172:550–565

McCarthy W, Reavy RJ, Stevenson CT, Petronis S (2015b) Late Caledonian transpression and the structural controls on pluton construction; new insights from the Omey Pluton, western Ireland. T Roy Soc Edin-Earth 106:11–28

McCaffrey K (1997) Controls on reactivation of a major fault zone: the Fair Head–Clew Bay line in Ireland. J Geol Soc Lond 154:129–133

Ryan P, Dewey J (1991) A geological and tectonic cross-section of the Caledonides of western Ireland. J Geol Soc Lond 148:173–180

Ryan P, Dewey J (2019) The Ordovician Grampian Orogeny, Western Ireland: obduction versus "Bulldozing" during arc-continent collision. Tectonics 38:3462–3475

Ryan PD, Dewey JF (2004) The South Connemara Group reinterpreted: a subduction-accretion complex in the Caledonides of Galway Bay, western Ireland. J Geodyn 37:513–529

Ryan PD, Sawal VK, Rolands AS (1983) Ophiolitic melange separates ortho-and para-tectonic Caledonides in western Ireland. Nature 301:50–52

Sanders IS (1994) The northeast Ox mountains inlier, Ireland. In Gibbons W, Harris AL (eds) A revised correlation of Precambrian rocks in the British Isles. Geol Soc Lond Spec Rep 22:63–64

Soper N, England R (1995) Vendian and Riphean rifting in NW Scotland. J Geol Soc Lond 152:11–14

Stephenson D, Mendum JR, Fettes DJ, Leslie AG (2013) The Dalradian rocks of Scotland: an introduction. Proc Geologist Assoc 124:3–82

Torsvik TH, Carlos D, Mosar J et al (2002) Global reconstructions and north Atlantic paleogeography 440 ma to recent. BATLAS–Mid Norway plate reconstruction atlas with global and Atlantic perspectives. Geological Survey of Norway, Trondheim

Van Staal C, Dewey J, Mac Niocaill C, McKerrow W (1998) The Cambrian-Silurian tectonic evolution of the northern Appalachians and British Caledonides: history of a complex, west and southwest Pacific-type segment of Iapetus. Geol Soc Lond Spec Pub 143:197–242

Walker JD, Geissman JW, Bowring SA, Babcock LE, compilers (2018) Geologic time scale v. 5.0: Geol Soc Am. https://doi.org/10.1130/2018.CTS005R3C

Wang C-C, Wiest JD, Jacobs J et al (2021) Tracing the Sveconorwegian Orogen into the Caledonides of west Norway: geochronological and isotopic studies on magmatism and migmatization. Precambrian Res 362:106301

Williams DM, O'Connor P, Menuge J (1992) Silurian turbidite provenance and the closure of Iapetus. J Geol Soc Lond 149:349–357

Worthington RP, Walsh JJ (2011) Structure of Lower Carboniferous basins of NW Ireland, and its implications for structural inheritance and Cenozoic faulting. J Struct Geol 33:1285–1299

Yardley BWD, Barber JP, Gray JR (1987) The metamorphism of the Dalradian rocks of western Ireland and its relation to tectonic setting. Philos Trans R Soc A 321:243–270. https://doi.org/10.1098/rsta.1987.0013

Data Sources for use in Pre- and Post-trip Exercises

Paul D. Ryan

Abstract This short chapter lists some of the principal repositories of open source digital geological, geophysical and geochemical data that may be used for further study to compliment the field geology presented in this guide.

1 Geological Maps

The Geological Survey Ireland (GSI) provides bedrock geology maps covering Ireland at scales from 1:2,500,000 to 1:100,000 that can be viewed, queried online using the GSI Map Viewer and accessed as a web service or downloaded for free at: https://www.gsi.ie/en-ie/data-and-maps/Pages/default.aspx. The 1:100,000 map series provides four shapefiles covering bedrock polygons, linework, structural symbols, and outcrop data. In addition, borehole locations with original logs are being added progressively. Quaternary sediments and geomorphology, mineral localities and geological heritage site map data can also be accessed from the same web location.

2 Geophysical Data

2.1 GSI Tellus Datasets

The Tellus Project offers geophysical data sets. These are available at: https://www.gsi.ie/en-ie/data-and-maps/Pages/Geophysics.aspx.

The 2013 'Tellus Border', 2017 'A2' and 2018 'A3' Blocks cover the west of Ireland. The data is in text files in xyz file format. The 'Tellus Border' was a joint

---

P. D. Ryan (✉)
School of Natural Sciences, Earth and Ocean Sciences, National University of Ireland Galway, University Road, Galway, Ireland
e-mail: paul.ryan@nuigalway.ie

© The Author(s), under exclusive license to Springer Nature Switzerland AG 2022
P. D. Ryan (ed.), *A Field Guide to the Geology of Western Ireland*, Springer Geology Field Guides https://doi.org/10.1007/978-3-030-97479-4_12

GSI/GSNI (Geological Survey of Northern Ireland) project (funded by INTERREG IVA). All subsequent blocks are GSI projects. It comprises: airborne magnetic intensity data in Nanotesla; airborne Gamma-Ray data which includes total counts per second, Potassium (%),Thorium (ppm) and Uranium (ppm) equivalents; and airborne frequency domain electromagnetic data (collected at 912, 3005, 11,962 and 24,510 Hz). Magnetic data has been used, in conjunction with other geophysical data, to define the traces of major Caledonian crustal lineaments, partially obscured by cover, across Ireland and into Britain. The Southern Uplands Fault (Fig. 1 in chapter "The Geology of Western Ireland:A Record of the 'Birth' and 'Death'of the Iapetus Ocean") runs through Galway Bay (Leake 1963) and the Fair Head—Clew Bay line (Fig. 1 in chapter "The Geology of Western Ireland:A Record of the 'Birth' and 'Death'of the Iapetus Ocean"), a continuation of the Highland Border Fault zone, runs through Clew Bay (Max and Riddihough 1975).

2.2 DIAS Gravity Data

The Geophysics Section of the Dublin Institute for Advanced Studies (DIAS) has, for over 70 years, collected terrestrial gravity data In Ireland. Data coverage averages ~ 4.2 stations per 10 km square and measurements, with necessary terrane corrections, are often adjacent to Ordnance Survey Ireland bench marks or spot heights. The locations of the gravity stations and details concerning free access to this data for academic purposes are found at https://www.dias.ie/cosmicphysics/geophysics/geo-gravity-data/.

This important data set has been used to define major crustal lineaments (e.g., O'Reilly et al. 1996; Readman et al 1997) and the deep structure of the Irish granites (e.g., Feely 1982).

2.3 BIRPS Deep Seismic Reflection Profiles

The British Institutions Reflection Profiling Syndicate (BIRPS) shot deep, multi-channel seismic reflection north–south profiles, Wire 1 and Wire 3, offshore the west of Ireland in 1987 (Klemperer et al. 1991). The recording depth was to ~ 95km, 18 s two way travel time (TWTT). Raw and processed digital data have been archived by the British Geological Survey (BGS) and are available subject to cost of reproduction and distribution (updated 1st September 2021). Contact enquiries@bgs.ac.uk for further information. The seismic Moho is clearly defined at about 10 s TWTT (~30 km) and north dipping reflectors in the middle and lower crust project up into the surface trace of the Iapetus Suture (Fig. 1 in chapter "The Geology of Western Ireland:A Record of the 'Birth' and 'Death'of the Iapetus Ocean") suggesting that the main architecture of the Irish crust was defined by the end-Silurian closure of the Iapetus Ocean (Klemperer et al. 1991).

3 Geochemical Data

The Tellus geochemical data Block G3 'Tellus West' and Block G1 'Tellus Border' cover the region studied in this guide and are accessible at:
https://www.gsi.ie/en-ie/data-and-maps/Pages/Geochemistry.aspx.

They contain multi-element datasets for stream sediment, stream water, shallow topsoil and deeper topsoil. Data can be downloaded as Excel files. These data have been used to fingerprint regional geology and mineralisation systems elsewhere in Ireland (e.g., Steiner 2018; Moles and Chapman 2019) and the influence of factors, including bedrock composition, on Irish river geochemistry (Lyons et al. 2021).

References

Feely M (1982) Geological, geochemical and geophysical studies on the Galway Granite in the Costelloe/ Inveran sector, western Ireland. Unpublished Ph.D. thesis, University College Galway, National

Klemperer S, Ryan PD, Snyder DB (1991) A deep seismic reflection transect across the Irish Caledonides. J Geol Soc Lond 148:149–164

Leake BE (1963) The location of the S. Uplands fault in central Ireland. Geol Mag 100:420–423

Lyons WB, Carey AE, Gardner CB, Welch SA, Smith DF, Szynkiewicz A, Diaz MA, Croot P, Henry T, Flynn R (2021) The geochemistry of Irish rivers. J Hydrol 37:100881

Max MD, Riddihough RP (1975) The continuation of the Highland Boundary Fault in Ireland. Geology 3:206–210

Moles N, Chapman R (2019) Integration of detrital gold microchemistry, heavy mineral distribution, and sediment geochemistry to clarify regional metallogeny in glaciated terrains: application in the Caledonides of Southeast Ireland. Econ Geol 114:207–232

O'Reilly B, Readman P, Murphy T (1996) The gravity signature of Caledonian and Variscan tectonics in Ireland. Phys Chem Earth 21:299–304

Readman P, O'Reilly B, Murphy T (1997) Gravity gradients and upper-crustal tectonic fabrics, Ireland. J Geol Soc Lond 154:817–828

Steiner B (2018) Using Tellus stream sediment geochemistry to fingerprint regional geology and mineralisation systems in southeast Ireland. Irish J Earth Sci 36:45–61

GPSR Compliance
The European Union's (EU) General Product Safety Regulation (GPSR) is a set of rules that requires consumer products to be safe and our obligations to ensure this.

If you have any concerns about our products, you can contact us on

ProductSafety@springernature.com

In case Publisher is established outside the EU, the EU authorized representative is:

Springer Nature Customer Service Center GmbH
Europaplatz 3
69115 Heidelberg, Germany

www.ingramcontent.com/pod-product-compliance
Ingram Content Group UK Ltd.
Pitfield, Milton Keynes, MK11 3LW, UK
UKHW021255180426

11947UKWH00010B/789